# AGRICULTURE GEOGRAPHY OF JHARKHAND

*Raj Kamal Tiwari*

**RAJESH PUBLICATIONS**
NEW DELHI - 110 002

# AGRICULTURE GEOGRAPHY OF JHARKHAND

*Published by :*
**Sanjay Gupta**
**RAJESH PUBLICATIONS**
1, Ansari Road, Darya Ganj, New Delhi - 110002
Ph.: 011-23274550, 43084725
Website : www.rajeshpublications.com
Email : rajeshpublications@gmail.com



First Published in India 2021

ISBN : 978-93-91139-22-3

Price : Rs. 900.00

*Printed in India by :*
R.K. Offset Process
Delhi - 110095

*DEDICATED TO THE PLOUGHMEN OF*

*JHARKHAND WHO*

*FEED US*

*&*

*ALL STUDENTS OF AGRICULTURE*

# Acknowledgement

To Department of Economic Affairs, Govt of India, New Delhi for the publication of Economic Survey; Planning cum Finance Department, Govt. of Jharkhand for the publication of Jharkhand Economic Survey; Indian Council of Agricultural Research, New Delhi for the publication of Jharkhand Agriculture Development Vision (Draft); Ministry of Fisheries, Animal Husbandry and Dairying, Department of Fisheries, Krishi Bhawan, New Delhi, Govt. of India for publication of Handbook on Fisheries Statistics; Directorate of Economics & Statistics, Ministry of Agriculture, Govt. of India, New Delhi for the publication of Agriculture Informatic Division and all the Authors and publishers whose name has been mentioned in either reference or Bibliography of this work without which this work could probably never have been completed.

To Dr. Lal G.S.N. Shahdeo, Supervisor of this work, presently Head, University Department of Geography, Ranchi University Ranchi and present CCDC of Ranchi University for their keen interest in supervising this work and writing of research papers for journals.

To Dr. Ram Kumar Tiwari, Former Head, Department of Geography, Ranchi University, Ranchi and Editor of 'Geographical Outlook' which is Department Research Journal, who is my father and teacher who stimulated my interest for Agriculture Geography.

To Dr. J. Shukla and Dr. G.K. Singh, Assistant Professor, University Department of Geography, Ranchi University Ranchi for their constructive suggestions and help in different academic work.

To my graceful and kind mother Sujata Tiwari for her all-time support and encouragement.

To the great sacrifice and support of my amenable wife Nitu Tiwari who assisted me by shouldering most of the family responsibilities while I remained engaged with my studies. At the same time, I must acknowledge my sweet, radiant, adorable daughter Ishita Tiwari whose joyous gestures kept on motivating me all the time.

**Raj Kamal Tiwari**

# Contents

# List of Maps

# List of Tables

# Introduction

- Selection of Topic
- Objective of Study
- Hypothesis
- Methodology
- Previous Literature
- Study Area
- Research Design
- Delimiting the Subject

## Introduction

India is one of the leading agricultural countries of the world, and one of the biggest sources of its wealth is its produce from land. There is hardly any crop of the tropical, sub-tropical or temperate zone which is not grown in this country. (Singh, 1989, P.21). Agriculture forms the backbone of Indian economy and more than 70 percent of the population of the country depend on this sector for their livelihood. Even after planning and giving utmost importance to programs of industrialization, approximately 69 percent of working population was engaged in agriculture in 1971 and this sector has contributed more than 40 percent of the national income (Chand and Puri, 1993, P.273). Thus, we can say that agricultural development is the sine qua non of economic development and without it all talk of economic progress is meaningless and futile. And when we talk of agricultural development, we imply the development of the whole rural sector in an integrated manner and not merely increase in agricultural production and productivity. Programs in this direction were initiated at some places in our country in the pre-independence period though planning on a national scale was initiated in the post-independence period and most specifically in April 1951 when the first Five Year Plan was launched. Notable experiments in the pre-independence period included the Martandham project initiated in 1921, the rural development scheme initiated in 1932 in Baroda state, the experiments conducted by F.L. Brayne in Gurgaon during 1920-28, by R.N. Tagore in Shanti Niketan, and by Mahatma Gandhi in 1920 at Sevagram and then in 1938 at Wardha. In 1935-36 the Govt. of India initiated a rural development program with a grant of Rs. 1 crore for rural construction through an integrated development of village industries, agricultural improvement, village communications, rural sanitation etc. Most of the above-mentioned project were comprehensive in nature. A real begining in these directions were made in post-independence period when on 2nd October 1952 the program of community development was launched. (Chand & Puri, 1993, P. 273-274).

That time Jharkhand was a portion of Bihar. Bihar had only 17 districts from which only 5 districts were under Chotanagpur and whole Santhal Pargana was one district under the area of Jharkhand State (now 24 districts). Capital of Bihar was at Patna from where all development program were being prepared for this remote southern area. One can easily imagine the pace of development of this area through the agricultural data of that period. Cropping area, crop production, irrigation system all were in critical situation. All aspects of agriculture were in back foot. The pace of development in agriculture has accelerated after the formation of Jharkhand. The figures/data of both periods distinguish it well.

Present work is the geographical study of agriculture in Jharkhand (Changing Pattern of Agriculture in Jharkhand: A Geographical Analysis). According to Johnston (1985) Agricultural Geography has been defined as the study of spatial variations in agricultural activities, involving both the description of such variations and attempts to explain them. Symon (1970) has defined agricultural geography as men's husbandry of land. Agriculture Geography has also been defined as the science which deals with the regional or spatial variations in the distribution of agricultural entities and to explain the causes of such variations (Husain, 1996, P.19).

## Selection of Topic

- At first the reason behind selecting this topic for study is a fact embedded in the depth of my mind since the period of doing Master's Degree in Geography. All my text books of agricultural geography were based on the agricultural geography of either Haryana or UP. Since then, I had a thought in my mind that I shall pursue a comprehensive work on the agriculture of Jharkhand if the opportunity presents.
- Secondly, no geographical work has been carried out on the entire state related to agriculture. Whatever work is done is in a form of a research paper or focused on the agriculture of a district. As mentioned in the previous literature in the coming paragraphs, my work will fulfil a gap in the field of study of agriculture as well as in the field of geography. Though we are too delimited in the sphere and range of the subject.
- Thirdly, the academic allotment of agricultural study with respect to Jharkhand.

## Objective of Study

Almost all major text books of Agriculture Geography like Singh & Dhillon, Husain (1996), Goutam (2011) have established these six objectives. We shall also unravel all these in context of Agriculture of Jharkhand

i. To examine the spatial distribution of crops, livestock and fisheries of Jharkhand.
ii. To ascertain the spatial concentration of agricultural phenomena of Jharkhand.
iii. To evaluate the crop association and livestock combination of Jharkhand and their change in space and time.
iv. To analyze the inter district variation of the performance of various crops in Jharkhand.

v. To diagnose the causes of existing agricultural backwardness and to suggest suitable strategies to enhance productivity.
vi. To evaluate the situation of Jharkhand whether agriculture has achieved the status of agribusiness or not.

## Hypothesis

Hypothesis is a proposed explanation for a phenomenon. It is based on previous observation. We have framed three hypotheses for this work:

1. Farmers of Jharkhand are using chemical fertilizers on the farms for all food crops regularly. Null hypothesis of this work is that the use of fertilizer over Rice and Maize (which are larger crops) is same according to their ratio of hectare of area and production.
2. State government tries to improve the quality of seeds and instruments of agriculture frequently. Null hypothesis of this work is that the changing ratio of improved seeds and instruments of agriculture is same.
3. Finally, the null hypothesis is that cropping intensity of the nation and state is same. It is assumed that highest intensity region lies within the triangle of DHANRAJ (Dhanbad, Ranchi, Jamshedpur)

## Methodology

Methodology is a system of methods used in a particular field of research. It is a way to systematically solve the research problem, hypothesis and analysis of different data used in the research. For the study of this work we have applied different techniques which have been described by the geographers in the different books, thesis and papers published by them. The main objective of this study is the investigation of the spatial arrangements and distribution of agricultural phenomena, the degree of correspondence between them and the delimitation of weaker areas. It will also take into account the regional imbalances in levels of irrigational facilities, resource use, land use efficiency and agricultural production.

Apart from the methods of data collection, data processing and analyzing, questionnaires preparation, we have also used some statistical methods to interpret the data. Such as population density, agricultural density, computation of growth rate, irrigation potential, assessment of farm power availability at district level, cropping intensity, irrigation intensity, correlation, crop diversification, crop concentration, crop combination, cropping pattern, livestock combination etc. This data has been utilized to prepare maps and diagrams for better representation and description.

## Previous Literature

The research studies related to agriculture by the Geography Dept. of Ranchi University are the best means of present study. Hence, a brief review has been made by us. No doubt this review will enable us easily to formulate the research problems. First work on agriculture was done by Awdhesh Singh (1986) in Ranchi University Geography Dept. on the topic, 'Dry land Agriculture. Bahura Ekka (1990) was the second scholar with the topic, 'Transformation and development of agriculture in Oraon dominated areas of Upper South Koel basin'. J.K.Mahto (1991) has added 'Land use and planning with agriculture in Jumar basin'. Jayashree Shahdeo (1992) studied 'Agro potential of land in Bero Block'. Work of Pramod Kumar Sinha (1992) was, on 'Palamu District for Agricultural Development'. With a similar topic R.K Sinha (1992) has done a work on 'Khunti sub-division'. Again Archana Agarwal (1992) did her work on Ramgarh Block. S.P. Singh (1993) has done his work on 'Torpa Block for Agricultural potential'. Awadhesh Mishra (1993) takes up the work on 'Dry land Agriculture: A case study of Mandu Block'. This is the second work on Dry land agriculture after Awdhesh Singh. Third work on dry land farming was completed by Basudeo Prasad (1993) on the topic, 'Prospects and problems of Dry land farming in Hazaribagh District' Similarly, Rani Jha (1993), Simanti Kumari (1993), Madhurima Pratihar (1993) and Pratibha Shrivastava (1993) completed their work on the topic. Agriculture in Lohardaga District'; Simdega sub-division'; Land use and agricultural planning in Singhbhum District' and 'Pattern Trends and problems of agricultural land use- Chash' respectively. Jaikishore Nath Shahdeo (1994) worked on North-Koel Basin. Sanjay Kumar Singh (1994) took a new topic on 'Cultivation of Lac in Chhotanagpur'. Yashodha Kumari (1994) also took a new problem of Agriculture related to tribal community. Study of Marya Grace Toppo (1995) was 'Agricultural Innovation and problems of Diffusion'. Sunita Rani Sinha (1998), Chandi Charan Mahto (1999), Balandina Lakra (2000), Rajesh Kumar Diwan (2001) also studied on Agricultural land use of Bokaro District, Latehar and Sisai Block. Similarly, Alok Kumar (2002), Praveen Zahra (2003), Md. Abu Amir (2006), Shashi Kumar Sahu (2006), Rajiv Ranjan Shrivastava (2006), Sr. Jyoti Kispotta (2007), Mariam Aind (2008), Shashi Kanta Toppo (2010), Ajita Gulab Minj (2011), Arun Nath Deogharia (2011), Asha Bhagat (2012), Madhu Beck (2013), Naresh Kumar Pandey (2015), Arpita Roy (2015) have studied Agricultural problems of a part of Jharkhand. All above studies were associated with the block level area, Basin area or District area of the state. Thus, we can say that many accounts of the agriculture of Jharkhand are available, but they all suffer from lack of substantive research and inadequate methodology. This work will fill up the all above stated serious gap to get a wholesome study of agriculture in Jharkhand.

## Study Area

So far, the description of the study area is concerned the Jharkhand extends between 21°58'10" to 25°18' Northern latitude and 83°19'50" to 87°57' East longitude. The state is surrounded by Bengal in East, U.P. and Chhattisgarh in west, Bihar in the north and Orissa in the south. Jharkhand extends on an area of 79714 square kilometers. In Jharkhand a series of plateau of different heights are there. The western plateau is known as 'Pat Region' which is the highest part of state. Second plateau is 'Ranchi plateau' and 'Hazaribagh plateau'. Third plateau is spread over Ramgarh, Hazaribagh, Giridih, Bokaro & Dhanbad Districts. Fourth plateau area is outer plateau which extends in three sides of the States i.e., north,

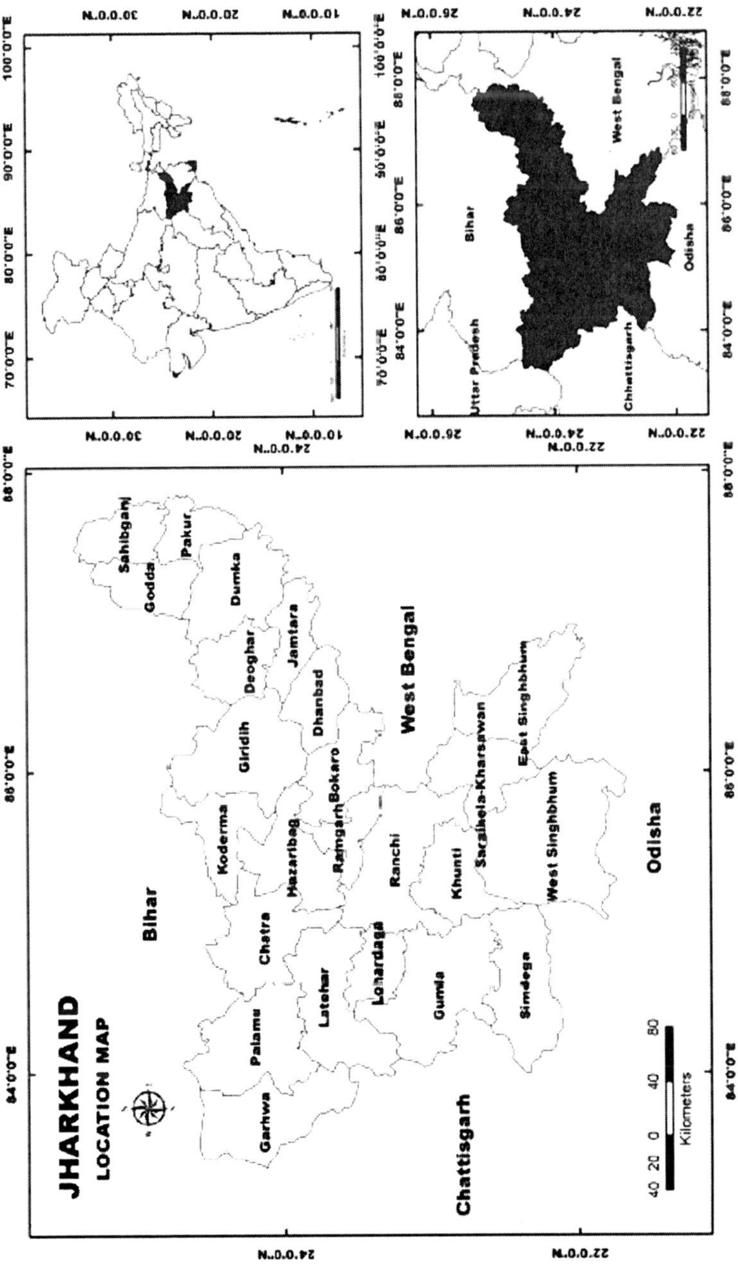

JHARKHAND
LOCATION MAP
Bihar
West Bengal
Odisha
Chattisgarh
Garhwa
Palamu
Latehar
Chatra
Koderma
Hazaribag
Ramgarh
Bokaro
Giridih
Dhanbad
Deoghar
Jamtara
Dumka
Godda
Sahibganj
Pakur
Lohardaga
Gumla
Simdega
Ranchi
Khunti
Saraikela-Kharsawan
East Singhbhum
West Singhbhum
Kilometers
Uttar Pradesh
Chhattisgarh

east and south. Each of these series of plateau descends to the next lower one not by gradual but by rather narrow and steep slopes known as scarps.

The whole surface of the plateau region is thrown up into long undulation, the depressions between the ridges of which have been cut up for purposes of cultivation into innumerable terraces rising one above the other like steps. The surface of the plateau is further diversified in appearance by expansion of arid, gravelly land, exposures of naked rock, range of wooded hills and here and there gapping ravines and gullies and the remnants of stately Sal forests.

Here I would also like to address a common question of this context that when agricultural activities were started in Jharkhand? We know that the whole Jharkhand was a no man land area till the end of Paleocene or Pre- Historic period (before 10 Lakh years ago). During Pleistocene period or beginning of Paleolithic age people began to live in caves. This was the period of hunting and gathering activities of cave men. History says that before the advent of tribal people in Jharkhand some civilized community were present here. This was upper Paleolithic age (10,000 to 6000 B.C.). They were practicing agriculture in Jharkhand. During Mesolithic age (6000- 4000 B.C.) there were several small settled villages near the water body or high land area. After 4000 B.C. or during the Neolithic period, the copper age of Jharkhand, Asur entered into the area of Jharkhand. Several references of copper age are available in Jharkhand that Asur, Birjhia, and Birhors were making copper during 4000-3000 B.C. in Singhbhum area. 2000-1000 B.C. was the Iron Age of Jharkhand and the same tribal people were related to this age. Simultaneously after their advent they were involve in agricultural activities as well as fishing, hunting and gathering. Thus, we conclude that two types of agricultural practice were prevailing in Jharkhand-

**First agricultural practices** were initiated by those so-called civilized people or Aryan people who were settled before advent of tribal people during 10000-6000 B.C. or Mesolithic age.

**Second agricultural practices** were initiated by the tribal people after their advent in Jharkhand during Neolithic age to Post-Vedic period (4000 B.C.to 800 B.C), (Tiwari, 2004). The agricultural data of the above period is not available but situation can be compared between two long periods.

The net area sown in Jharkhand was 23.5 per cent in 1955-56 but has gradually decreased to 22.68% in 2002 which is negligible area and it is almost constant during these 50 years. Total cropped area was 30.6% in 1955 and it is about 26% in 2002. 4.6 percent area has decreased within 50 year. Fallow land area has increased from 5 per cent to 7 per cent. Fallow land was 8.7 per cent in Jharkhand during 1955 and it grew to 16.5 per cent in 2002. All these agricultural data show the change within a span of five decades, sown area is same, total cropped area has decreased and fallow land has increased. Reasons behind this change have been discussed neither by district gazetteer nor by agricultural reports and economic survey reports. There are several books on the geography of the state for example Bihar of Dr. Ahmad (1965), Rural settlement of Chotanagpur of A. Prasad (1973), Jharkhand Ka Bhoogol by Ram Kumar Tiwari (2001) and Chotanagpur Survey by F. Ivern (1969). All books have an exhaustive chapter on agriculture but all narrate the general situation of agriculture and answer of above questions are still unsolved that why there is changing pattern in the area of fallow land and total cropped area where as net sown area is same. Present work will answer all these questions with an adequate methodology. Through this work we will present a full picture of agricultural geography of Jharkhand.

## Research Design

Research design is the conceptual structure within which research is conducted. In fact, research design means the sampling design, observational design, statistical design and operational design etc. (Kothari, 2008, P. 31-32). But all designs together form the thesis in different chapters. Hence research design is known in other words as research organization or chapterisation. Organization of this work has been completed in eight chapters having an introduction in the beginning and conclusion at the end.

**Chapter - One (Geographical background.of agricultural relevance)** describes the general features of Jharkhand which are closely associated with agriculture; divided into two parts - Physical aspects and Cultural aspects. These are the essentials for the agriculture in state.

**Chapter- Two (Land tenure and size of holdings)** describes the tenure system of Jharkhand or narrate the story of the actual tillers of the land. It determines the rights and responsibility of holdings. Right of holding or relationship has a magical power. A proverb by Arthur Young is worth mentioning here, "The magic of private ownership turns dust into gold. Give a person a secured right to a large mountain and he will turn it into a garden. However, if the garden is leased to him for nine years, he will turn it into a desert."

In this chapter different size of holding have been studied in different groups i.e., marginal holding, small holding, medium holding and big holding.

**Chapter - Three (Farm Technology)** discusses about the use of technology in the farms for the purpose of agriculture. Technology is the knowledge being applied by human being to improve production or marketing process. It is the knowledge used to operate equipment to produce goods and services. We have studied the progress of agriculture in Jharkhand towards new technology.

**Chapter-Four (Intensity of Cropping)** has studied the relationship between the net area sown and the total cropped area of the state. This relation is known as cropping intensity. It has been studied in district level for the purpose of micro study.

**Chapter- Five (Cropping Pattern and Diversification)** analyzes the different crops of the state in relation to their area and production. At the same time diversity index has been calculated as well.

**Chapter - Six (Agricultural Regions)** has demarcated the

whole state of Jharkhand into different regions through a specific method given by different geographers. Here crop concentration, crop combination and cropping pattern has been studied in detail.

**Chapter - Seven (Animal Husbandry)** examines the number of livestock in the state and their produces. Fisheries, Poultry, Dairy have been studied in this chapter. At the end of the chapter livestock combinations has been prepared by us.

**Chapter - Eight (Problems and Prospects)** is devoted to highlight the problems and prospects of agriculture and animal husbandry with fisheries and dairy.

## Delimiting the Subject

It is well known to geographers that Agriculture is now a days declared as an industry (Though government has not yet accepted the fact) where two sectors are studied - one is plant science and another is animal science.

- Under plant science agriculture studies - 1. Agronomy, 2. Horticulture, 3. Silviculture, 4. Plantation crops, 5. Condiments and Spices, 6. Ponology & Fruit preservation, 7. Olericulture, 8. Floriculture and 9. Arboriculture.
- Under animal science agriculture studies - 1. Livestock (Cattle, Buffalo, Horse, Sheep, Goat, Pig, Poultry etc.), 2. Fisheries or Pisciculture, 3. Apiculture, 4. Sericulture, 5. Lac Culture etc.

However, in spite of the vast practice on all of the above sectors in Jharkhand, we have limited our study to crop husbandry, animal husbandry, fisheries and dairy.

## References

Chand, Mahesh (1993); Regional Planning in India, Allied Publishers Pvt ltd, New Delhi.

Goutam, Alka (2011); Krishi Bhugol, Sharda Pustak Bhawan, Allahabad.

Husain, Majid (1996); Systematic Agricultural Geography, Rawat Publications, Jaipur.

Kothari, C.R. (2008); Research Methodology, New Age International Publishers Pvt. Ltd, New Delhi.

Singh, R.L. (Ed) (1989); India: A Regional Geography, National Geographical Society of India, Varanasi.

Symon, L (1970); Agriculture Geography, Boulder, Colorado, Westview Press.

Tiwari, R.C. (2016); Krishi Bhugol, Prawalika Publications, Allahabad.

# 1

# Geographical Background with Agricultural Relevance

- Physical Environments of Agriculture
  - Geology
  - Terrain
  - Physiographic Units
  - Climate
  - Soil
  - Drainage
- Cultural Environments of Agriculture
  - Population
    - Distribution of Population
    - Density of Population
    - Agricultural Density 2001 & 2011
    - Tribal Population
    - Tribal Growth
    - Working Force in Jharkhand
      - Cultivator
      - Agricultural Laborers
  - Land Utilization
  - Transportation
  - Irrigation

Geography of Agriculture is highly different in structure and spatial processes and patterns than that of Industrial and service sector. Agriculture is land based, spatially wide flung and spread in different types of physical environments. (Singh,2012, P-270). Variations in terms of climate, soils, water resources, relief and physiography or terrain etc. At the same time the cultural environment is also a more forceful determinant. Cultural determinants are population, land utilization, transportation, irrigation and markets etc. Here we shall discuss all these factors gradually in brief as all are very important for the agriculture of the area.

## Physical Environments of Agriculture

### Geology

In the study of cultural landscape, like agriculture, geological background of the region may not appear quite necessary but it becomes desirable for a better and fuller appreciation of the various cultural features upon which the geology of the area has distinct but definite bearing. It is specially so in Jharkhand which is a geomorphic unit and a geological province (Prasad, 1973, p.3). The relationship between geology and physiography, agricultural field-pattern, distribution and density of settlements, alignment of routes etc. have been referred to at suitable places in the following pages.

The basement of this region is made of gneissic granite known as Bengal Gneiss or Chotanagpur Gneiss. With these granitic rocks are associated sedimentary, metamorphic, igneous intrusive and extrusive rocks. Most of the rocks in this area belong to three important geological formation of India viz. Archean, Gondwana and Jurassic-Cretaceous (Prasad,1973, P. 3).

The Archeans are by far the most important formation which cover more than 90 percent of the plateau surface. The Archeans are represented by two facies: (i) Sedimentary and (ii) Granitic and basic intrusives. The sedimentary rocks have been grouped under the name of Dharwarian System.

The Dharwarian rocks are found in the extreme southern and northern portions of the plateau. They are best developed in Singhbhum Districts. The component rocks are quartzite, sandstone, conglomerate, phyllite, schists and limestone. In Singhbhum district the quartzite is interbedded with thick masses of iron ore and is called banded hematite quartzite. It is the most important source of iron ore in India and for this region the Dharwarian formation in this part has been named Iron-ore Series. With these rocks are associated basic lavas and intrusives which at places, have been altered to epidiorite and various types of schists. Dharwarian rocks present a broad spectrum of metamorphism and have variously responded to erosion in outlining the surface (Prasad,1973, P3). Granitic rocks in Jharkhand do not represent the basement complex that constituted the primordial crust. They appear to have been re-melted, interbedded and highly metamorphosed at a later date, perhaps later than Iron-Ore Series. The most outstanding characteristics of the Chotanagpur granite is its gneissic appearance. But at places, it assumes schistose character and it frequently contaminated by the inclusion and absorption of the schistose materials which have led to the development of gneiss like banding in the granite rocks. (Prasad,1973, P4).

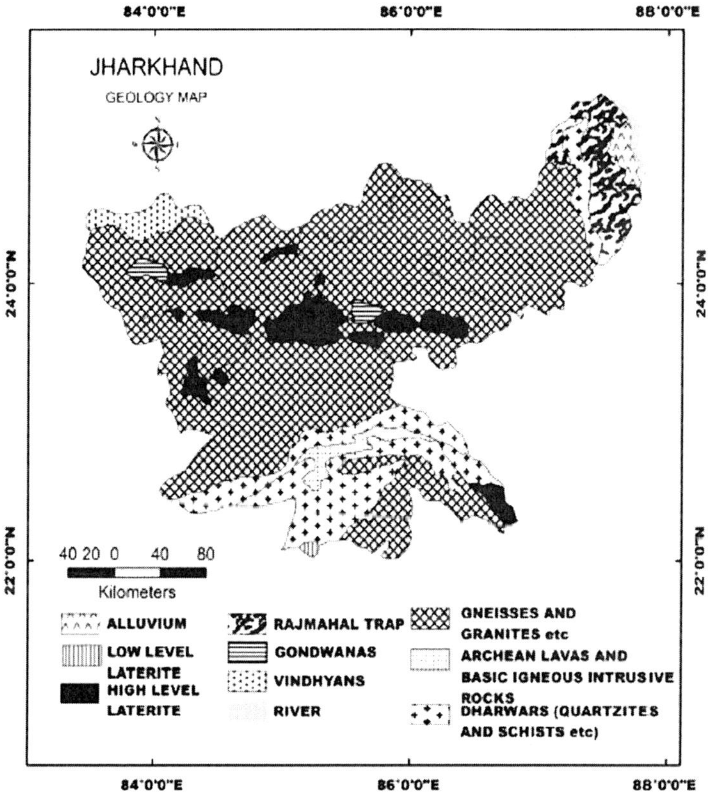


**Map 1.1: Geological Structure of Jharkhand**

The third important formation of the plateau is known by the name of Gondwana System. This system of sedimentary rocks was originally deposited in a trough-like sinking basin of tectonic origin. The principal basin covers an elongated area conforming approximately to the Auranga-Damodar Valley. Within the principal basin there are six 'sub-basins' of varying sizes viz. Hutar, Auranga, Karanpura, Ramgarh, Bokaro and Jharia. Outside the Gondwana Trough there are other basins of Gondwana Formation which are important coalfields. The principal component rocks of the system are sandstone, shale and basal conglomerate. Sandstone is the dominant rock which is interbedded with the coal seams. The alternation of sandstone with shale has led to the evolution of ridge-in valley configuration that characterizes the coalfields.

The basic extrusive rocks are found in three separate tracts. They are the Western part of Ranchi District, The Rajmahal Hills and the bordering areas of the districts of Singhbhum and Ranchi. The volcanic rocks of the western Ranchi District are actually the outliers of the Deccan trap. The Rajmahal Lava that belongs to the Jurassic period overlies the Gondwana rocks. The Singhbhum lava, poured out in the Archean period, has been thoroughly metamorphosed, folded and uplifted. Under hot-humid climate of this area, the volcanic rocks specially in their upper layers, have been altered to laterite bauxite.

## Terrain

Arthur L. Bloom (1978) has mentioned in his book 'Geomorphology': A Systematic Analysis of late Cenozoic landforms (page-5) the clear concept of 'Terrain' and 'Terrane'. He has mentioned that a useful distinction is made between the **terrane**, an area of certain structure or rock type, as in the expressions "granitic terrane" or "Limestone terrane" and the **terrain,** a landscape or "the lay of land". A military tactician must appreciate the terrain; he need not be concerned with the terrane. But for a student of Agriculture both 'Terrane' and 'Terrain' are useful because terrane decides the quality of soil and terrain determine the shape and size of farms.

Here we have already discussed about the terrane under the heading of Geology. Now terrain will be discussed as structure, physiography and relief of the area.

The structural frame work of Jharkhand is in general, linked up with the peninsular upland. Jharkhand, like most of its parent body, the peninsular India, was not submerged beneath the sea since the Cambrian period, nor could it succumb to the great orogenic movements of the Mesozoic and Tertiary eras. To these movements it reacted epirogenetically and still preserves some very significant marks of its reactions which outline the geomorphology of the area and are responsible for the distinctive physical features, bold reliefs and present elevation of plateau (Dunn & Ahmad 1941 &1958, P 8-14 and 16-22).

Jharkhand does not consist of one plateau, but it is a series of plateaus. Each plateau occurs at a height different from that of adjoining plateau. There are four plateaus or erosion surfaces in Jharkhand. The highest surface is formed by western Ranchi plateau or 'Pat' region, 2500 to 3600 ft or 762 to 1097 meters. It covers the Latehar, Lohardaga and Gumla districts. Except the Parasnath hill it is the highest region in Jharkhand. It is believed to be composed of Deccan lava, but as a result of weathering the lava has been converted into laterite and bauxite. The pat region consists of flat-topped dissected hills rising mostly above 1600 ft or 488 meters above the lower plateau to the east, known as Ranchi plateau. Netarhat is located in this pat area.

The second or the next lower plateau is known as the Ranchi Plateau and it covers the whole Ranchi and Khunti districts and partly Lohardaga and Gumla districts except the area covered by 'pat' region. This plateau, composed mostly of gneisses and granites, is about 2000 ft or 610 meters above the sea level. The topography is roughly undulating with terraced slopes ending on one hand on the banks of a local stream and on the other in a local upland which may or may not have low hill; or which may or may not consist of solid rocks exposed bare of soil. With the minor variations the Ranchi plateau characteristically lies at an elevation of 2000 ft or 610 meters above sea level. The Ranchi plateau is separated from another surface of the same elevation by Damodar trough. It is known as Hazaribagh plateau and is probably a continuation of the Ranchi plateau.

The third plateau is laying at an elevation of 1000 ft or 305 meters above the sea level known as the outer plateau or lower chotanagpur plateau. It consists mostly of gneisses and granites, but partly of schists and other Dharwad rocks. The plateau covers a wide area on the outer parts of chotanagpur. In Hazaribagh it has been called the lower Hazaribagh Plateau, but it girdles the 2000 ft or 610 mts on most sides from palamu via Hazaribagh into Singhbhum districts.The fourth or next lower erosional surface formed by the river valleys, plains and lower parts of the outer lower plateau laying between 500 ft and 1000 ft or 150 mts or 305 mts above sea level. They again consist of gneisses and granite and partly of schists etc. The basaltic Rajmahal hills belong to this erosion level.

Each of these series of plateau i.e., 3600, 2000, and 1000-feet plateaus descend to the next lower plateau not by gradual slopes but by rather narrow and steep slopes known as scrapes (Ahmad, 1973, P 23-24).

It is worthwhile to mention here that physical origin of Jharkhand is the result of three upliftment over the peneplain. It is well known that the whole chotanagpur was a great peneplain before the tertiary period where only three major tectonic features were present i.e., Damodar trough of Paleozoic era (carboniferous period 35 crore years ago); Rajmahal volcanic eruption of Mesozoic era (Jurassic period 18 crore years ago) and Lava flow in Western Ranchi or Pat region during cretaceous period (13.5 crore years ago) which is the north east end of Deccan Trap (Tiwari, 2019, P 33). On such peneplain or low undulating plain rivers would flow slowly through wide meandering or winding channels. There can be no scarps or waterfalls or prominent hills on such a surface. Then the question arises as to what has happened to chotanagpur since early tertiary times that the area instead of being a low peneplain now consists of a series of plateaus separated by steep scarps on which rivers descend through waterfalls. These rivers instead of flowing through broad and shallow channels occupy relatively deep cut meandering channels. Reason behind this was that when a peneplain uplifted, the meandering broad river channels become deep-cut channels and along the lines of uplift scraps may appear on which the rivers descend through waterfalls. Thus, the present topographic features of Jharkhand are the clue to the past, and geographers and geologists think that before the Himalayan movement started in Tertiary period, Jharkhand was a low peneplain. As a result of the violent Himalayan movement, parts of peninsular upland in general and chotanagpur peneplain in particular began to be successively uplifted. The Himalayan movement occurred three times during early and late tertiary and Pleistocene times and chotanagpur peneplain probably was also concurrently subjected to three successive uplifts. (Ahmad, 1973, P 24).

The highest pat region was uplifted above pre-existing peneplain in early tertiary times. Then towards late tertiary times Ranchi plateau and upper Hazaribagh plateau was uplifted with higher Pat region leaving the rest peneplain. The line of this block uplift is marked by a steep scarpes that surrounds the Ranchi and upper Hazaribagh plateau where well-known waterfalls e.g., Hundru, Hirni and Jonha developed. Subsequently the entire hilly region on Jharkhand including Rajmahal hills uplifted by about 1000 ft or more, bringing the outer plateau to their present elevation and as a result of this upliftment pat region and Ranchi plateau got an additional uplift of 1000 ft by the final uplift in the Pleistocene period.

At the end of this discussion, it is also worthwhile to mention the view of the scholars that the physiography of Jharkhand thus shows four erosional surfaces where the upper three surfaces are quite distinct and well- marked by the presence of steep escarpment, while the fourth one, the lowest, is recognized by the alignment of the nick-points in the long profiles of the rivers. Great controversies exist as to the evolution of the physiography of Jharkhand and it is outside the scope of the present work to discuss and examine whether the four surfaces are uplifted peneplain (Ahmad, 1951, & Dunn 1941) or have partially resulted from 'differential erosion' (Chatterjee, 1949, P-7)

## Physiographic Units

A long list of geographers analyzing the influence of terrain upon agriculture is on record, although some present ones are much less inclined to record terrain a weighed role in Land use variations. The debate notwithstanding, the form of terrain throws powerful illumination on land use and on the growth and distribution of crops in areas where relief

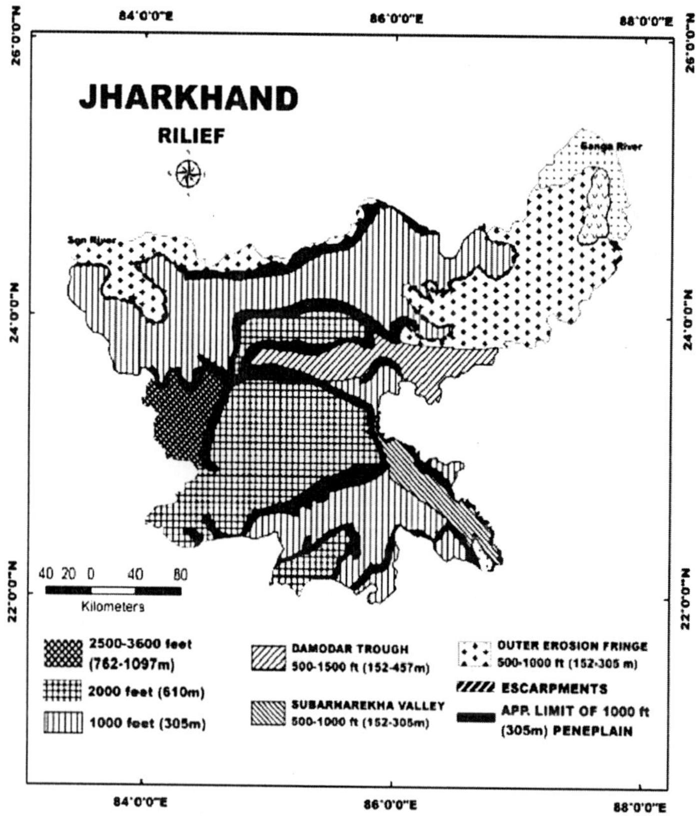


Map 1.2: Relief

encompasses complex and diverse lithology and landforms. In such case the geography of agriculture being with the morphology of an area. Relief of the land exercises a direct influence on land use, particularly through elevation, ruggedness and slope, while on lower ground it effects flooding. Relief also influences farming by modifying the climate and affecting the ease of cultivation and the degree of accessibility (Singh, 1974, P 1). Any geographical study of landscape must include a rational division of the land surface upon which other distributions have been fashioned, agriculture in particular (Singh, 1976, P 34).

Here Jharkhand has been classified into 24 physiographic units on the basis of contour forms, varied elevation and geomorphological features as under:

1. The North Koel Basin
2. The Pats
3. The Ranchi Plateau
4. The Hazaribagh Plateau
5. The Kodarma Plateau
6. The Rajmahal Highlands
7. The Damodar Basin
8. The Panch Pargana Plains
9. The Porahat Highland
10. The Dhalbhum Plain
11. The Dhanjori Range
12. The Kolhan Upland
13. The Singhbhum Plain
14. The Simdega Plateau
15. The Dalma Range
16. The Chaibasa Range
17. The Saranda Hilly Region
18. The South Koel Basin
19. The Sankh Basin
20. The Ajay Valley
21. The Dissected Deoghar Region
22. The Chatra Plateau
23. The Chattarpur Plain
24. The Northern Scarp

For Convenience of study the above physiographic units can be grouped into six categories:

1. Central Plateau
2. Lower Plateau
3. South Koel - Sankh Basin.
4. Dharwarian Country
5. Damodar Basin
6. Rajmahal Uplands.

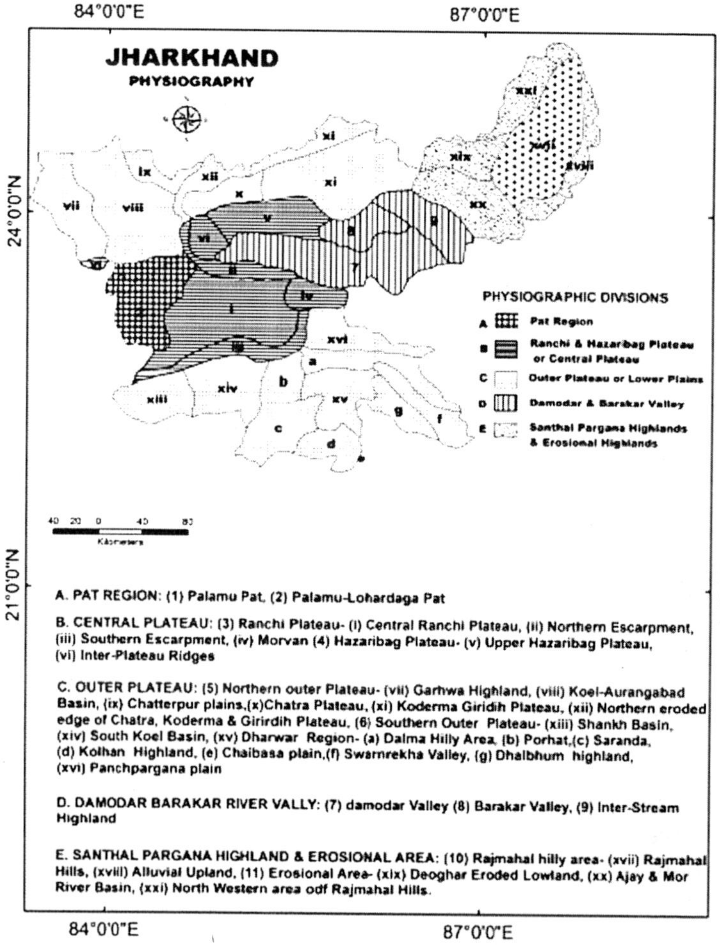


**Map 1.3: Physiography of Jharkhand**

1. **Central Plateau:** Central Plateau physiographic unit covers the other small physiographic units like pat region, Ranchi plateau, Hazaribagh plateau. It consists of the areas above 2000 ft.

2. **Lower Plateau:** Lower plateau physiographic unit contains the sub units like North koel Basin, Koderma plateau, The Ajay Valley, Chatra Plateau, Chhatarpur plains, The northern Scrap. It comprises the third higher surface of Jharkhand laying between 1500 ft to 1000 ft. It covers a very wide area that girdles the central plateau (2000 ft) on all sides from Palamu to Santhal Pargana and Singhbhum.

3. **South Koel-Sankh Basin:** South from latitude 22°50" extending across Khunti District is a dissected region which is really a prolongation of the chotanagpur plateau (Dunn, 1941, P 11). This dissected tract is drained by two large rivers - the south koel and the Sankh, which beyond the borders of Jharkhand join to form the mighty Brahmani river. Of all the rivers that drain Jharkhand, The Brahmani river is the largest.

4. **Dharwarian Country:** This region comprises the whole of Singhbhum districts ie the East and the West Singhbhum with Saraikela. This physiographic unit contain the small physiographic units like porahat highland, Dhalbhum Plains, Dhanjori range,.Kolhan Uplands, Singhbhum plain, Simdega plateau, Dalma range, Chaibasa plain and Saranda hilly region. The entire region is made of the Dharwarian rocks which have been variously eroded and dissected to form a hilly upland tract containing hills alternating with valleys, steep forest-clad mountains and in the river basins, some stretches of comparatively low level of undulating country (D.G.Singhbhum, 1911, P- 2).

5. **Damodar Basin:** This basin comprises the areas lying between the plateaus of Ranchi and Hazaribagh and the valleys of Barakar and the Jamunia rivers. Besides, the two broad spurs projecting from Hazaribagh and Koderma plateaus are also included. From the Damodar-Auranga divide in west to the state boundary in the east, the level falls from 1750 ft to 450 ft. Residual hills and spurs further add to

the variety of the topographical features and bring about strong contrasts in different sections that justify a subdivision of the basin. (Prasad, 1973)

6. **Rajmahal Uplands:** The Rajmahal hills are the only considerable mass of hills which approach the Ganga in its entire course from the Himalayas to the sea. (Dayal, 1953, P-12). From Sahibganj in the north to the state boundary in the south these hills form a central block of hilly country which covers an area of 2000 square miles or 5180 sq kms.

Bedded basaltic structure can be seen in the whole hill mass which is made of post Gondwana origin and attain a thickness of 2000 ft with 100 ft of intertrappean beds. Dark dolerite rock can be seen in the trap while the intertrappeans consist of hard white and gray shales, carbonaceous shales, white and gray sandstone and hard quartzose grit (District Gazetteers, Santhal Pargana, 1938, P-5). The volcanic rocks have been at places weathered into laterites, which cap the higher flat-topped hills.

The hills are divided into two elongated chains by the longitudinal valleys of Burhait and Gumani rivers. Burhait valley is made up of two narrow valleys of the Meral and Gumani river that after joining at Burhait debouch into the Ganga. This axial valley of the hills extends over 24 miles from north to south and constitutes a five miles wide ribbon of cultivated fields and settlements between the two parallel hilly chains which are continuous from Sahibganj to Dumka for about 80 miles and are crossed only by the five narrow passes that connect the piedmont plains with Deoghar erosional depression and the Ganga plain. The eastern face of the hills is comparatively steep and rocky. It overlooks the alluvial surface. Among all these Santhals and Paharias have their villages situated on the brow of steep hills with cultivated fields and grasslands stretching beyond them (Prasad, 1973, P-29).

## Climate

In the climatic map of India, Jharkhand is shown to form the part of the region which is characterized by "heavy rainfall and considerable humidity" (Kendrew, 1961, P-185)

When we examine the climatic classification by Koppen and Thorntwaite we get two and three types of climate respectively in the Jharkhand area. Koppen's classification shows that Jharkhand has only two types of climate - (i) Cwg = Humid mesothermal warm dry winter climate and (ii) Aw = Tropical rainy savanna climate. Northern portion of Jharkhand has Cwg type of climate while southern portion has Aw type of climate. In the western portion of the state if a line crosses in between the boundary of Gumla and Latehar districts and further it touches the boundary of Bokaro and Hazaribagh district and at the eastern portion it crosses in between the Jamtara and Dumka then this line will divide

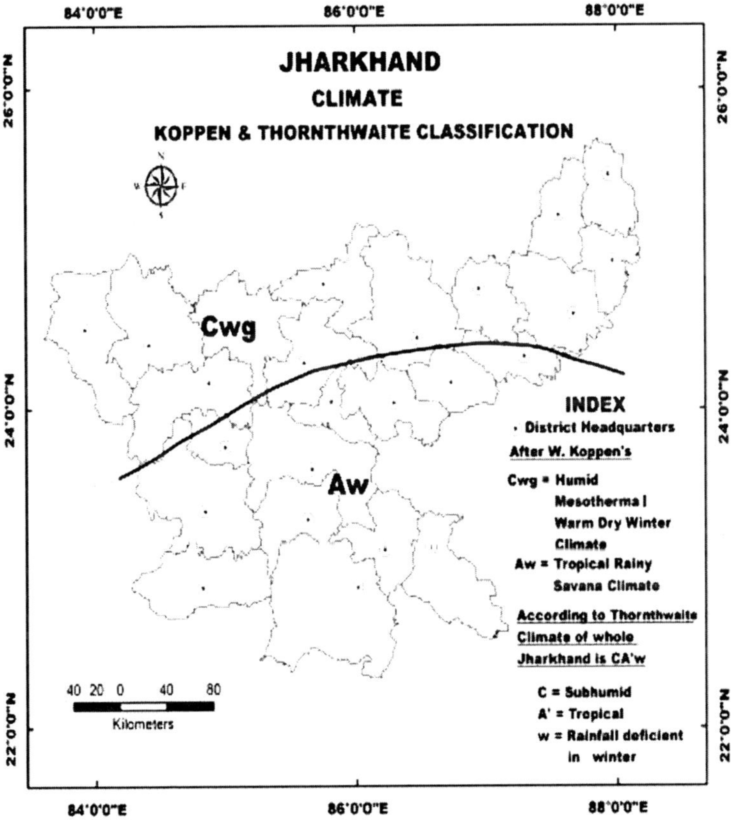


Map 1.4: Climate-Koppen & Thornthwaite Classification

Jharkhand into two divisions according to the Koppens climatic classification.

According to the Thornthwaite classification the whole Jharkhand experiences the climate like CA'w type. Here C= Subhumid, A' = Tropical and w= Rainfall deficient in winter. Thus, it is humid meso thermal warm dry winter climate or Monsoon type with dry winter type of climate according to Koppen and According to Thronthwait it is sub-humid Tropical Rainfall Deficient in winter type of climate.

In general, the State of Jharkhand faces four types of seasons every year on the rational basis of the reversal of winds. There are two main seasons: of general rains due to South-West monsoon (mid-June to mid- September) and that of continental winds or cold weather season (November to February) In between these two principal seasons there are the transitional seasons of the hot weather (march to mid-June) and of retreating monsoon (mid-September to October) (Ahmad, 1973, P 41-43)

**South-West Monsoon:** The total rainfall of this period varies from over 6" in East-Singhbhum and Eastern Santhal pargana to less than 2" in Palamu, Garhwa and Latehar districts. Ranchi gets a rainfall of 4.18" while Hazaribagh 2.45" and Netarhat 3.24". By middle of June the monsoon sets in. One of the immediate effects of the 'burst of monsoon' is the fall in temperature. Grasses sprout up from their fried, weathered roots and parched surface becomes uniformly green turf. From June onwards, with the increase in precipitation, temperature continues to fall. In may the mean temperature at Ranchi is 86.7° F. It falls to 83.1° F in June, 79.0° F in July, 75.3° F in August and 74.2° F in September: In all the alluvial upland of Santhal Pargana and Eastern Singhbhum, this fall in temperature is much less marked, perhaps, due to high humidity. Dumka for example has a mean temperature of 88.2° F in May, 86.3° F in June, F in July and 83.1° F in Aug which amounts to total fall of 5.7° F only. With the cessation of rain by middle of October the sky becomes clear and days are sunny.

Jharkhand is one of the rainiest tracts. The rainfall is however not uniformly distributed all over the plateau. It varies from less than 50" in the North Koel Valley and northern edge of the state to 80" on the higher pats. Most of the plateau receives a rainfall over 50" and a good part over 60". A small area comprising the district of Dhanbad and Ajay basin looks like a dry island receiving an annual rainfall of less than 50 ". Similarly, the districts of Garhwa and Palamu gets relatively lesser rainfall than Chaibasa. Pat region receives the highest amount of rainfall.

**The Winter Season:** The high pressure formed in the north western India allows the wind to blow from the north-west to south-east. The temperature remains between 60° F and 70° F. The winter of this region has charming, clear and fine weather with pure air and blue sky. The days remain slightly warm but nights are quite cool or even cold during the months of December and January. Under the effect of cyclone, the region receives smaller amount of rainfall. The rain is, however of immense importance for Rabi crops. In January, some parts of Jharkhand get cold winds and frost. The wind speed in this season is very low, blowing from north-west and west.

**The Summer Season:** In April temperature goes on rising, the relative humidity decreases further and the air becomes hot and dry. Barometer continues to fall and the wind gains in velocity. April and May become typically hot. May is the hottest month at all stations, except Dumka. Even in May the mean temperature, except Chaibasa plain and the North Koel valley, remain fairly bellow 100° F. The plateau of Ranchi and Hazaribagh with a mean temperature of 87.7° F and 88.9° F respectively remain comparatively cool throughout this period. But the coolest part of the plateau is the pat region. From the pat region towards all directions, temperature rises as altitude decreases. It is 87.7° F at Ranchi, 88.9° F at Hazaribagh, 91° F at Chaibasa, 91.5° F at Jamshedpur and 91.7° F at Daltonganj and 88.2° F at Dumka.

**Retreating Monsoon:** Retreating Monsoon period in Jharkhand falls during min-September to October. The retreat is complete by the end of the month of October which is followed by cold weather season. On the whole, the winter is a rainless period. The little amount of rainfall that occurs in this period is associated with western disturbances. These disturbances enter the plateau from west and north-west and usually die out before they cross the eastern boundary of state.

**Table 1.1**
**District wise Rainfall pattern of Jharkhand**

| District | Annual Rainfall (mm) | Seasonal Rainfall (mm) | Date of Start | Date of Termination | Number of Rainy Days |
|---|---|---|---|---|---|
| Deoghar | 1202 | 1087 | 10 June | 7 Oct | 41 |
| Dumka | 1412 | 1377 | 15 June | 7 Nov | 15 |
| Jamtara | 1376 | 1200 | 20 June | 11 Oct | 15 |
| Godda | 1095 | 954 | 10 June | 19 Oct | 45 |
| Pakur | 1725 | 1550 | 15 June | 10 Oct | 47 |
| Sahebganj | 1468 | 1077 | 15 June | 10 Oct | 30-35 |
| Bokaro | 1250 | 1063 | 15 June | 10 Oct | 46 |
| Chatra | 1309 | 1188 | 24 June | 10 Oct | 32 |
| Dhanbad | 273 | 128 | 5-15 June | 10 Oct | 20 |
| Giridih | 1329 | 1187 | 13 June | 18 Oct | 41 |
| Hazaribagh +Ramgarh | 716 | 575 | 22 June | 25 Sep | 45 |
| Koderma | 301 | 218 | 15 June | 7 Oct | 23 |
| E. Singhbhum | 1230 | 1102 | 9 June | 23 Sep | 38 |
| Gumla | 1386 | 1141 | 15 June | 30 Oct | 52 |
| Simdega | 1047 | 985 | 15-25 June | 30 Oct | 20-30 |
| Lohardaga | 1250 | 1084 | 15 June | 18 Oct | 65 |
| Ranchi +Khunti | 1480 | 1232 | 15 June | 5 Oct | 35 |
| W. Singhbhum | 1250 | 1109 | 15 June | 15 Oct | 40 |
| Saraikela Kharsawan | 194 | 98 | 10 June | 19 Oct | 23 |
| Palamu | 189 | 106 | 15 June | 15 Oct | 15 |
| Latehar | 189 | 106 | 15 June | 15 Oct | 15 |
| Garhwa | 237 | 110 | 25 June | 30 Sept | 09 |

**Source:** District Agricultural offices (Jharkhand) (Socioeconomic Dynamics of Rice Production in Jharkhand, R.K. Singh, 2003, P-26)

## Soil

At the beginning of his work on political Geography, Ratzel made a far- reaching statement that 'every nation is a bit of soil and humanity' (Quoted by Klages, 1958 P-44, 111 and 189) Hence, no one forgets for an instant the fundamental importance of soil. It is the source of practically all man's food, clothing and an ever-increasing list of other needs. So much so, that man gets nearly all of his food from the soil, less than one percent of what he eats being fish (Pearson and Harper, 1945, P-5). In any agricultural operation soil is of the utmost importance as it is the cradle for all the crops and plants. The top soil or upper layer having an average thickness of about 20 cm is the principal feeding zone of the crops, providing food for human ingestion or livestock feeding (Husain, 1996, P- 103). The standard of living of the people depending on agriculture is often determined by the fertility and productivity of soils. The soils of any region are developed on the underlaying parent rock. Jharkhand consists of its parent rocks like granite, gneiss, sandstone, shale, limestone, schists, laterite and trap. These rocks are eroded by the different soil forming factors like climate i.e., temperature and rainfall, running water, wind, vegetation etc. In Jharkhand there are four types of soil -

1. Red and Yellow soil
2. Loose sandy soil
3. Black clayey soil
4. Laterite soil

All above mentioned soils have different chemicals, physical and biological properties. The colors of soils indicate their organic content such as clay, sand particles and silt etc. Contents like organic matter and chemicals determine the color of the soil. High content of ferric oxide and bauxite makes the soil look like red in color. Soils vary in color from laterite of plateaus to yellow grey loam and black soil in valley beds. All these because of constituents like lime content, compounds of silicon, aluminum and iron. Soil differs in properties based on dissolved substances due to percolation of water and intensity of human interference. So, the main soil forming factors become the parent materials, climate with its alternation, humidity and desiccation, natural vegetation, fauna, land terrain and human interference. In the formation of soil, the role of different rocks is no less important. The rock like granite, gneiss, schist, laterite, quartzite, sandstone, limestone, slate etc. help in formation of the soil. Similarly, the vegetative cover- the plant fossils- influences the soil formation. Hence the areas having considerable forest cover have soil containing more organic content. The rate of decomposition determined by climatic conditions like moisture and temperature also matters. Influenced by all these features, Jharkhand has displayed various types of soils. Four main types of soil as mentioned above are distributed in the entire state.

1. **Red and Yellow Soil:** This soil is found all over the state barring Rajmahal highland and a narrow belt of Gondwana Sandstones and Shale in Damodar valley. This soil gets its red color from decomposition of rocks like granite, gneiss etc. The higher areas have soils with light red color but the lower areas have dark red color. The proportion and size of sand differ from region to region. It has less Nitrogen, Phosphorous, acid content and humus, but more potash and lime content. The red soil derived from the acid granite and gneiss are poor in general fertility. They usually constitute the third and second class gora-lands all over Jharkhand. "On the plateau country the granitic soil frequently contains a certain amount of Iron-Hydroxide which cause the soil in dry season to set almost as hard as cement at the immediate surface" (Dunn, 1941, P-31). Such soil is completely useless for agricultural purposes and forms the barren tanr-land of the region. The red soil derived from the acid rocks are in general poor in lime, magnesia, phosphates, nitrogen and humus but fairly rich in potash derived from muscovite and orthoclase. (Ahmad, 1973). At places where mica is replaced by hornblende in the granite composition, the lime content is sufficient.

2. **Loose Sandy Soil:** In Damodar valley the soils are laid down by Gondwana rocks. Damodar valley area is covered

with the rocks like sand-stone and shales. These rocks are interbedded with coal seams and ironstone shales. When weathering occurs on these rocks it rises in a comparatively immature soils of less variety and fertility than the surrounding area (Wadia, 1953, P-360). The predominance of massive and coarse sandstones of Barakar, Panchet and Mahadev series have led to the formation of loose soils with a very high proportion of coarse sands. The soil that results from them are rich in Iron content. There are few beds of limestone and shale that have brought about local variations in soil. On the whole the soil derived from the Gondwana rocks are pale, thin and sandy. They are deficient in humus and other fertilizing constituents and at places, have given rise to bed lands (Prasad, 1973, P-50). Here the carbonaceous soil becomes remarkably loose in summer and regularly feeds the dust rises in the summer days. The shale outcrops, generally forming elongated lowlands between sandstone-ridges, have given rise to good agricultural soil with a very high clay-factor. The areas of clayey soils are given to paddy and are much valued by the farmers (Prasad, 1973, P-50).

3. **Black Clayey Soil:** Black soil is comparable to regur or black cotton soil of western India. Black soil in Jharkhand is a type of residual soil. This soil has developed on the Rajmahal trap and the basic gneisses in Santhal Pargana. The area of black soil is by no means continuous with the boundary of the Rajmahal trap. It is only the western flank of the hills which is capped with black soil. On the eastern flank of the Rajmahal hills the development of this soil has been obstructed by steeper slope and heavier rainfall causing regular removal of the top soil at faster rate. Black soil is highly argillaceous with a very large clay-factor. This soil has exceedingly high capacity to retain moisture. The characteristics are related to high clay-proportion which means larger aggregate of particle surface than in a less clayey or coarser soil (Ahmad, 1973, P-21). Black soil consists of a variable proportion of potash, phosphate, nitrogen, humus, calcium, magnesium, carbonates, Iron oxide, alumina etc. Here the soil consists of higher proportion of clay than

the normal, looks deep, dark and becomes exceptionally fertile because of 'secondary enrichment' with the accumulation of finely assorted sediments brought down by rain washing from above. The area with thick black soil constitutes some of the best paddy lands of the region. On the hill sides and tops, the soil thins out and becomes almost infertile. "A very considerable extent (of the hilly tract) is occupied by mere rock, totally incapable of cultivation" (Buchanan, 1922, P-7).

4. **Laterite Soil:** Laterite soil is a peculiar soil in India and some other tropical countries too. In Jharkhand it is found in three distinct tracts. They are - Pat Region, The Rajmahal hills and the south-eastern corner of East Singhbhum district. It is found on different altitudes "from the cappings on 3000 feet plateau in western Lohardaga, Gumla and Latehar districts down to the level of Laterites at 150 feet along the eastern side of the Rajmahal trap." (Dunn, 1941, P-20). Laterite itself, something of the nature of soil cap, is not a soil. "It is a compact vesicular rock composed essentially of a mixture of the hydrated oxide of aluminum and iron with small amount of manganese oxides, titania etc." (Wadia, 1935, P-373). About the origin of laterite, it is said that under the monsoon conditions of alternating dry and wet seasons, laterite capping can develop on any rock surface that contains some alumina and iron. It is however, formed far more readily over rocks like traps and ferruginous shales which contain a very high proportion of alumina and iron. But further modification resulting particularly from tillage and actions of biologic and other soil forming agencies, laterite is converted into a red colored soil. (Prasad, 1973, P-49). Laterite in Singhbhum has been altered from tertiary sediments. In Santhal Pargana it is drift soil formed by the deposition of lateritic sediments from the hill tops. Laterite of Pat area is very hard and is typically pisolitic in structure (Dunn, 1941, P-21). Laterite soil are poor in lime and magnesia and deficient in nitrogen and potassium. Phosphorous and organic contents sometimes become high and for this reason the soil attain agricultural importance. The area of such soil like Santhal Pargana are cultivated and given to paddy. (Prasad, 1973, P-50).

## Drainage

In geography drainage refers to the manner in which precipitation falling within an area or brought from outside is drained off. Such water generally takes two forms. It may flow through rivers or accumulate in lakes or natural ponds etc. All these surface features of drainage are called surface water forms or water bodies. The water that goes underground forms the underground reservoir of water. It is also generally drained from higher to lower areas, but its flow may terminate at the edge of hill or the bed of a river. This drainage is not so simple as the surface drainage. It is known as underground drainage. (Ahmad, 1973, P-27).

Thus we can say that the drainage of this region is rainfed basically and there are hardly any other sources of water giving rise to surface flow of water. At the same time this region doesn't get water through rivers from outside except little area of Son valley in the north-west. The flow of water is more influenced by the variation in amount of rainfall as well as diversity of terrain. The diversified areas have drainage of different nature than that of relatively homogeneous regions. Underground drainage mainly depends upon the composition of rocks and alignment of beds which determine the percolation as well as underground storage of water. It may be noted that this region comprises of varied types of rocks whose percolation capacity differ. This region slopes in all directions except in the west because the land has uplifted more in the west. It is therefore the flow of river is in all directions except in the west. Local slopes have also determined the pattern of drainage in some parts of the region. Since the region has considerable slope it leads to quick runoff of rain water and hardly allows water stagnation in the form of water bodies. Certainly, artificial reservoirs and water bodies have been made to store rain water or river water for different uses. This stagnation of water also helps in the formation of underground drainage. Areas having joints, faults, cracks in rocks etc. In the region have generally allowed the percolation of water. The vegetal cover also influences the surface drainage of water.

In addition to the aforementioned facts the drainage of this region has also been considerably influenced by the scarps formed during different upliftment. As a result, most rivers have given rise to a number of waterfalls. The varied hardness of rocks found in the river courses also determines the vertical and horizontal process of erosion and flow of water (Sinha, 2003, P-23).

The rivers of this region have some special characteristics. Except a few rivers most of them remain dry except during rainy season. Certainly, these rivers become turbulent during rainy season particularly after heavy rainfall. Areas having high slopes provide quick run-off water and for few hours the rivers become furious. The rivers have been flowing in wider courses and these rivers form drainage patterns of different types. Most of the rivers of the state feed the major rivers discharging into the Bay of Bengal. To the north, the main drainage axis runs from west to east before turning to north-east in Rajmahal highlands, forming a divide between north and south-east flowing streams (Singh, 1970, P-15). The center of the drainage axis lies on an elevated tract of the plateau south of piska- nagri near Ranchi. Three drainage axes radiate from this elevated tract of the state which divides the main river basins. The east ward drainage axis forms watershed between the Damodar and Subarnarekha Basins. The eastward drainage axis forms a semi-circular water divide till it takes a south eastern loop to turn west and reaches eastern limit of the state.

**Drainage System:**

As the Jharkhand region is a plateau region, here several river basins have developed their basins/valleys in different directions mainly in south, east and north. These river basins are mainly Damodar, Barakar, Subarnrekha, North-Koel and South-Koel etc. Here we shall learn in brief about their extent, potential and services for the agriculture of the area.

1. **Damodar System:** The Damodar River rises in the Palamu hills of Chotanagpur at an elevation of about 609.75 m. It flows in a south-easterly direction entering the deltaic

plains below Raniganj. Near Burdwan, the river abruptly changes its course to a southerly direction and joins Hooghly about 48.27 km below Calcutta. Its slope during the first 241.35 km is about 1.89 m/km, during the next 160.9 km about 0.568 m/km and during the last 144.8 km about 0.189 m/km. The total length of the river is nearly 547 km and a total drainage area is 22,005 sq. km. The principal tributary Barkar joins the Damodar after traveling for about 241 km. The catchment area of the river above the confluence is fan shaped and as such is susceptible to the concentration of flood flows but the catchment below the confluence is narrow and has an average width of 16.09 km. The upper catchment is rough hilly areas denuded of forest and vegetal cover and is subject to erosion while the lower catchment is silt covered and fertile. There are no irrigation facilities in the upper catchment and the cultivation is solely dependent on monsoon rains. However, the lower portion had irrigation from the Anderson Weir situated at Rhondia on the Damodar nearly 19 km below the Durgapur Barrage.

Damodar basin lies in the states of Jharkhand and West Bengal. Based on the 2001 census, the population of the basin is close to 14.25 million. For this population, domestic demand has been estimated at 338.46 MCM/year. This demand is likely to increase to 507 MCM/year by the year 2021 (Roy et al. 2004). For the industrial sector, the present water demand is 663 MCM/year which is about 15.36% of the utilizable water potential of the basin. This demand is likely to increase to about 884 MCM/year by the year 2021 due to new thermal power plants, brick fields, rice mills, and cold storages etc.

Two irrigation canals, the Left Bank Main Canal and the Right Bank Main Canal take off from the Durgapur barrage. The cultivable command area in the basin is 3,940 sq. km. Water demand for agricultural uses in the Damodar basin for the year 2001 was nearly 652.41 MCM/year. This demand is projected to sharply rise to 1,948 MCM/year by the year 2021. At present, the infrastructure for irrigation is highly inadequate in the basin.

**Tributaries of Damodar:** The tributaries and sub tributaries of the Damodar include Konar, Barakar, Haharo, Bokaro, Ghari, Jamunia, Khadia, Guaia and Bhera. The biggest tributary of the Damodar River is the Barakar. The source of Barakar is located in the vicinity of Padma in Hazaribagh district. Subsequently, the river runs through Jharkhand prior to joining the Damodar close to Dishergarh in West Bengal. The Barakar and the Damodar get the Chotanagpur flat terrain divided into three parts. The rivers flow through mountainous regions with enormous might, overwhelming anything that comes in their way.

**Climate:** The climate of the area consists of very mild winters and hot-wet summers. During the months of December to March the general flow of surface air is northeasterly. It is of continental origin and is less humid. During June to September, the flow of wind is southwesterly, from sea to land, and this period has high humidity. Transition seasons of the hot weather months of April and May and the retreating monsoon months of October and November lie between the above two principal seasons. In general, the months of June to October are known as monsoon season and nearly 90% of the mean annual rainfall of about 1,250 mm occurs in this period. The months of November to May are known as non-monsoon or dry period when rainfall is small. For the five sub-catchments, namely, Tilaiya, Konar, Maithon, Panchet, and Durgapur, the average annual precipitation is 1,117.7 mm, 1,320.8 mm, 1,141.7 mm, 1,142.0 mm, and 1,320.8 mm respectively.

Heavy rainfall occurs generally in the Damodar Valley due to:

The monsoon depression from head bay moving in a north west direction over the catchment or west-north-west direction passing near the catchment area., and the land low developing over Gangetic West Bengal or Bihar Plateau and the remaining stationary or moving slowly across or near the Damodar Valley.

The average yearly precipitation on the Chotanagpur

terrain is approximately 1400 mm, and nearly the whole amount of it takes place in the monsoon season, between June and August. The enormous quantity of water that runs through the Damodar River and its tributaries at the time of the monsoons is mighty and the flow is quite vehement in the higher basins of the plateau. However, in the lower basins, the river floods its banks and inundated other big localities nearby.

In the past, the Damodar River used to cause frequent and immense flood damages in Bihar and West Bengal, so much so that the river came to be associated with sorrow and sufferings. To overcome these, the then government decided to build a series of dams on the pattern of developments in the Tennessee Valley in the USA. Accordingly, five dams were constructed in the Damodar Valley, namely, Maithon, Panchet, Konar, Tilaiya and Tenughat and a barrage was constructed at Durgapur. Further, the Damodar Valley Corporation (DVC) was established in 1948 for management of the Damodar Valley water resources. The dams in the Damodar system serve for different purposes, viz., domestic and industrial water supply, flood control, irrigation, and hydropower generation.

**Present Scenario**

At present, the Damodar is the most contaminated river in India, due to the various industries that have mushroomed on its riverbanks, which are good resources of minerals. There are a number of coal-oriented industries that are scattered over the Damodar basin. Majority of them are government-owned coke oven plants; coal washeries, important iron and steel plants in India; glass, zinc, and cement plants; and thermal power plants. The contamination was a result of excessive excavation, oil, fly ash, poisonous metals, as well as coal dust.

**2. The North-Eastern Drainage System:**

**1. Ajay River:** The Ajay river originates from Batpar from Chakai block of Jamui district in Bihar. It then enters Jharkhand near Devipur (a proposed industrial area of

Deoghar) and flows through Jharkhand and enters West Bengal at Simjuri, near Chittaranjan. It first forms the border between Paschim Bardhhaman district and Jharkhand and then between Paschim Bardhhaman district and Birbhum district, and finally it enters Katwa subdivision of Purba Bardhhaman district at Nareng village in Ketugram police station. It then joins the Bhagirathi River at Katwa Town. Total length of the Ajay is 288 kilometers (179 mi), out of which 152 kilometers (94 mi) are in West Bengal. The catchment area of Ajay River is 6,000 square kilometers (2,300 sq mi).

The important tributaries of the Ajay are Pathro and Jayanti in Jharkhand, and Tumuni and Kunur in Bardhhaman district of West Bengal. The upper reaches of the Ajay pass through hilly regions with laterite soil. It is only from Ausgram in Bardhhaman district that the Ajay flows through alluvial plains. The Ajay valley was densely forested with sal, piyasal and palash trees till recent times when mining and other activities led to the clearing of forests. Recently, Government of India (Ministry of Shipping) declared Ajay River as National Waterway 7, under The National Waterways Act, 2016.

2. **Mayurakhshi River:** Mayurakshi River is a major river in West Bengal, India, with a long history of devastating floods. It has its source on Trikut hill, about 16 kilometers (10 mi) from Deoghar in Jharkhand state. It flows through Jharkhand and then through the districts of Birbhum and Murshidabad in West Bengal before flowing into the Hooghly River. The river is about 250 kilometers (160 mi) long.

Mayurakshi though named after its crystal-clear water of the dry seasons, floods its valley during the monsoons. Even after the construction of the Massanjore dam, it wreaks havoc with its floods, washing away embankments.

It is a rain fed river. Annual rainfall over the basin varies between 765 and 1,607 millimeters (30 and 63 in) with an average of 1,200 millimeters (47 in) of which 80 percent occurs during the monsoon season from June to September.

**Massanjore Dam:** Massanjore dam (also called Canada Dam), across the Mayurakshi, was commissioned in 1955. Unfortunately, the Massanjore dam located near Dumka in the state of Jharkhand (erstwhile Bihar) was not allowed to have any flood reserve. Simultaneously with construction of dams the state government in 1956, selectively took over flood control embankments till then maintained by the landlords or local bodies. Massanjore dam is about 38 kilometers (24 mi) upstream from Siuri in West Bengal. It is 47 meters (155 ft) high from its base and is 660 meters (2,170 ft) long. The reservoir has an area of 67.4 square kilometers (16,650 acres) when full and has a storage capacity of 620,000,000 cubic meters (500,000 acre. ft). It had cost Rs. 16.10 crore.

**Tilpara Barrage:** Apart from the Massanjore Dam there is a barrage, some 32 kilometers (20 mi) downstream, at Tilpara, near Siuri. The barrage is 309 meters (1,013 ft) long and had cost Rs. 1.11 crore. Since 1960 In the four decades between 1960 and 2000 only five years could be identified as flood-free years, when only less than 500 square kilometers (190 sq mi) of area were inundated. After major floods in 1978, West Bengal suffered consecutively in 1998, 1999 and 2000. In 1978, seventy-two hours of continuous and concentrated rainfall over the western river basin areas of the Bhagirathi viz. from the Pagla-Bansloi to the Ajay, generated so huge flood volume that all embankments on the eastern side of the Bhagirathi were almost washed away and the whole of Nadia district, a larger part of Murshidabad district and northern areas of North 24 Parganas district were flooded and remained underwater for a long time. Construction of embankments is the only structural measure available for the provision of relief to the people. Major embankments line long stretches of such rivers as Mayurakshi, Dwarka, Brahmani and Ajay.

**Irrigation and Power:** Massanjore dam has ensured irrigation of some 2,400 square kilometers (600,000 acres) of land with an estimated resultant increased yield of approximately 400,000 tons of food annually and generation of 2,000 kW of electric power.

**Tributaries:** The Mayurakshi is fed by tributaries Brahmani, Dwaraka, Bakreshwar and Kopai.

3. **Brahmani River:** The Brahmani is a major seasonal river in the Odisha state of Eastern India. The Brahmani is formed by the confluence of the Sankh and South Koel rivers, and flows through the districts of Sundargarh, Deogarh, Angul, Dhenkanal, Cuttack, Jajapur and Kendrapara. Together with the river Baitarani, it forms a large delta before emptying into the Bay of Bengal at Dhamra.

**Sources:** The Brahmani is formed by the confluence of the rivers South Koel and Sankh near the major industrial town of Rourkela at 22 15'N and 84 47' E. The Sankh has its origins near the Jharkhand-Chhattisgarh border, not far from the Netarhat Plateau. The South Koel too arises in Jharkhand, near Lohardaga, on the other side of a watershed that also gives rise to the Damodar River. Both of these sources are in the Chotanagpur Plateau. The site of the Brahmani's origin is mythologically reputed to be the place where Sage Parashara fell in love with the fisherman's daughter, Satyavati who later gave birth to Ved Vyasa, the compiler of the Mahabharata. The place is thus called Ved Vyasa.

South Koel river near Jaraikela, Orissa. One of the two tributaries of Brahmani River, after assuming the name of Brahmani, the river crosses the Tamra and Jharbera forests, skirting along National Highway 23. It then passes the town of Bonaigarh in Sundargarh district before being dammed at Rengali in Anugul district. A large reservoir of the same name is created as a consequence. It then flows through the towns of Talcher and Dhenkanal before splitting up into two streams. The main stream flows by the town of Jajpur Road beyond which it is crossed by National Highway 16 and the Kolkata-Chennai mainline of East Coast Railway. The branch stream called Kimiria receives the waters of the Birupa (a distributary of the Mahanadi, Kelua and Genguti before re-joining the main stream at Indupur. It then flows crisscrossing through Pattamundai. The river then receives the Kharsuan, on its left bank before merging with the Baitarani, a major river, to form the Dhamra estuary. A distributary called

Maipara branches off here to join the Bay of Bengal a short distance away while the main stream proceeds northward for a few km more before ultimately meeting the sea near Chandbali at Palmyras Point. The Brahmani delta is the site of the Bhitarkanika wildlife sanctuary, famous for its estuarine crocodiles. Patrapur Bridge on Brahmani River in Pattamundai along SH-9A

**Length and catchment:** At about 480 kilometers (300 mi) long, the Brahmani is the second longest river in Odisha after the Mahanadi. However, if its constituent rivers are included its length extends to about 799 kilometers (496 mi), of which 541 kilometers (336 mi) are in Odisha. It has a catchment area of about 39,033 square kilometers (15,071 sq mi) in Odisha alone.

**Irrigation projects:** Apart from the Rengali dam mentioned above, there also exists a barrage shortly before Talcher called Samal Barrage. The Jokadia and the Jenapur anicuts are the other irrigation projects on the river. Near its mouth, the river is also crossed by the Odisha Coast Canal that takes water from it.

4. **Gumani River:** The Gumani River Basin, Jharkhand (24°-25015' N, 87°29' -87°45' E), lies within the Rajmahal Volcanic Province, a Continental Flood Basalt Province (CFBP) of the early Cretaceous age (McDougall and Mc.Elhinny, 1970; Duncan, 1992:610). The Gumani rises in the northwestern slopes of Singarsi range and flows northwestward up to Simlong. Thereafter it flows northeastward up to Berhait and finally turns east from south of Borna Pahar and debouches into the alluvial plains of the Ganga (Fig. 1). The basin can be approached by Sahibganj Loop Line of Eastern Railway from Kolkata, located at a distance of about 300 km to its east-southeast and also from Bhagalpur lying to its northwest. The principal towns and railway stations of the area are Sahibganj in the north-central and Barharwa in the east-central margins of the Rajmahal hills.

5. **Bansloi River:** The Bansloi River originates on Bans Hill in Sahebganj district of Jharkhand, flows through Pakur

district of Jharkhand and Birbhum and Murshidabad districts of West Bengal before flowing into the Bhagirathi north of Jangipur.

The western half of Murshidabad district slopes eastwards toward the Bhagirathi; a number of the hill streams do not find their way directly into that river; they are intercepted by beels or marshes. The large beels act as reservoirs during a flood absorbing some of the excess water carried down by these streams, and also drain the excess water through the streams, emanating out of them. The Bansloi enters the northern part of the district.

The combined catchment area of the Pagla-Bansloi river system is 2,200 square kilometers (850 sq mi).

**Floods:** A number of rivers that originate on the Chota Nagpur Plateau or in Jharkhand, including the Bansloi, flow into Birbhum district. These rivers then enter the Murshidabad and Bardhhaman districts before joining the Bhagirathi. They generate flood because of high rainfall in their basins and limited carrying capacity of the river Bhagirathi between the towns of Jangipur and Kalna. In this reach the Bhagirathi has discharge carrying capacity of maximum 3,100 cubic meters per second (110,000 cu ft/s). If all the rivers receive rainfall simultaneously in their catchment areas, they can generate run-off volume of any amount up to 17,000 cubic meters per second (600,000 cu ft/s) at their outfall at the Bhagirathi. Thus, floods are caused regularly.

### 3. The Subarnarekha System:

Translated literally, Subarnarekha means 'streak of gold'. With a drainage area of 1.93 million ha this smallest of India's major inter-state river basins is a mute host to effluents from various uranium mining and processing units. While most rivers in the country are classified -- depending on the pollution load -- on a 'best designated use basis, the Subarnarekha defies any classification, as the existing parameters do not include radioactivity.

The rain-fed Subarnarekha originates 15 kms south of Ranchi on the Chhotanagpur plateau draining the states of

Jharkhand, Orissa and West Bengal before entering the Bay of Bengal. The total length of the river is 450 kms and its important tributaries include the Raru, Kanchi, Karkari, Kharkai, Garra and Sankh rivers.

As per tradition, gold was mined near the origin of the river at a village named Piska near Ranchi. This is why it was named Subarnarekha, meaning "streak of gold". Legend has it that traces of gold were found in the river bed. Even now, people look for traces of gold particles in its sandy beds. The name is a combination of two words meaning gold and line/ streak in Indian languages.

**Course:** After originating near Piska-Nagri, near Ranchi, the capital of Jharkhand, the Subarnarekha traverses a long distance through Ranchi, Saraikela- Kharsawan and East Singhbhum districts in the state. Thereafter, it flows for shorter distances through Paschim Medinipur district in West Bengal for 83 kilometers (52 mi) and Balasore district of Odisha. There, it flows for 79 kilometers (49 mi) and joins the Bay of Bengal near Talsari. The total length of the river is 395 kilometers (245 mi).

The basin of the Subarnarekha is smaller than most multi-state river basins in India. The rain-fed river covers a drainage area of 18,951 square kilometers (7,317 sq mi)

**Tributaries:** The prominent tributaries of the Subarnarekha are Kharkai, Roro, Kanchi, Harmu Nadi, Damra, Karru, Chinguru, Karakari, Gurma, Garra, Singaduba, Kodia, Dulunga and Khaijori. The Kharkai meets the Subarnarekha at Sonari (Domuhani), a neighborhood of Jamshedpur.

**Hundru Falls:** Hundru Falls is created on the course of the Subarnarekha, where it falls from a height of 98 meters (322 ft). The spectacular sight of water falling from such a great height is a sight to behold. The different rock formations due to erosion by the constantly falling water have added to the beauty of the place.

**Pollution:** The Subarnarekha passes through areas with extensive mining of copper and uranium ores. As a result of the unplanned mining activities, the river is polluted. The

Subarnarekha has been the lifeline of tribal communities inhabiting the Chhotanagpur region and water pollution affects their livelihood.

**Flood:** Several areas in the lower reaches of the Subarnarekha, particularly the coastal areas of Odisha and West Bengal, are within the flood hazard zone. The Subarnarekha in Odisha had crossed its previous Highest Flood Level (HFL) of 12.2 meters (40 ft) in 2007, surpassing the earlier record of 1997. In 2009, the Subarnarekha witnessed flash floods following heavy rainfall in the upper catchment areas of the river. During the flood, large areas of Jaleswar, Bhograi and Baliapal blocks and a small pocket of Basta block in Balasore district of Odisha were affected. Certain areas of Medinipur in West Bengal are also affected by floods

**4. South-Koel River System:**

South Koel River runs across Jharkhand and Odisha states in India. It originates on the Ranchi plateau a few miles east of Ranchi, and conjoins the Belsiangar and Singbhum Rivers. The Koel is fed by three streams in Jharkhand, namely the North Karo, South Karo and Koina. The South Koel enters Odisha and joins with Sankh River at Vedavyas near Rourkela from where it is named as Brahmani.

**West Singhbhum drainage scenario:** Singhbhum is drained by three river systems - Subarnarekha, Baitarani and Brahmani. The watersheds of these three systems originate near Gamharia in the Kolhan and radiate north-west, south-west and east respectively from their common center. These watersheds divide the Subarnarekha and its feeders from the Baitarani and its tributaries, and the latter again from the South Karo and Deo rivers, which feed the Brahmani through the South Koel.

The tunnel on the South Eastern Railway divides the Subarnarekha and Brahmani systems, and at this point the watershed leaves the Kolhan, continuing in a northerly direction through Porahat and finally merging in the Ranchi plateau between the Bicha and Tatkora hills. Of these three great rivers the Subarnarekha alone flows through the district. The Baitarani forms for about 12 kilometers (7 mi) the boundary between the Kolhan area and Keonjhar (in Odisha) while the Brahmani drains the west of the district through its tributary, the South Koel, and its feeders, the North Karo and the South Karo, and the latter of which in its turn is fed by the Deo river.

**Koel-Karo project:** The Koel-Karo project is located in the Ranchi and West Singbhum districts. The 710 MW power project involves the construction of two earth dams-one 44 meters (144 ft) high, across the South Koel river near Basia, and the other 55 meters (180 ft) high, across the North Karo river near Lohajima. The two dams will be linked by a trans-basin channel, with six units of 115 MW each in the underground powerhouse at Lumpu- ngkhel and one unit of 20 MW at Raitoli. About 120 villages are to be affected, displacing over one lakh (100,000) people. About 22,000 hectares (54,000 acres) of land will be submerged. Of the total land acquisition, 12,000 hectares (30,000 acres) are reported to be agricultural land, while 10,000 hectares (25,000 acres) comprise forests. Dispute arose over the compensation package for the displaced people and protests started in 1974 and have been continuing since then. The project was finally shelved in 2003 bowing to protests.

5. **North-Koel River System:**

The North Koel rises on the Chotanagpur plateau and enters Latehar district, below Netarhat near Rud. After flowing nearly due west for about 32 kilometres (20 mi), it turns north at an almost complete right angle through a gorge at Kutku, and flows through the center of the district until it falls into the Son a few miles north-west of Haidarnagar.

From its source to its junction with the Son its length is about 260 kilometers (160 mi), and since it drains a catchment area of at least 9,100 square kilometers (3,500 sq mi), it naturally contributes a large supply of water to the Son during the rains; at other times the stream is not deep enough to enable cargo boats of even small dimensions to make their way up to Daltonganj. In many places the reaches of this river present scene of great beauty and sometimes even of grandeur, such as the rocky bed and rapids north of Hutar and the gorge at Kutku.

Mention may be made of the metamorphic rocks which form the watershed between the North Koel and the Damodar to the west of the Chandwa-Balumath road.The North Koel, along with its tributaries, meanders through the northern part of Betla National Park.

**Tributaries:** The principal tributaries are the Auranga and the Amanat, both of which join it from the east, the former at Kechki, 16 kilometers (10 mi) south and the latter 8 kilometers (5 mi) north of Daltonganj. Another tributary is the Burha, which joins the North Koel above Kutku at Bagechampa.

**North Koel Project:** The North Koel project is being implemented at Mandal near Kutku. The project has a 64.82-metre (212.7 ft) high dam, from which water will be released for picking up by the Mohammadganj Barrage and Indrapuri Barrage. This Dam will be known as Kutku Dam. Hydro-electric power generation would be 2 x 12 MW. The project would flood 56 villages, of which 8 are within the core area of the Palamu Tiger Reserve.

In 1974, this project was started promising to irrigate 120,000 hectares (300,000 acres) of farmlands of Palamu, Latehar, Garhwa of Jharkhand as well as of Gaya, Aurangabad, Jehanabad and Arawal of neighboring Bihar.

The recent push has come following the efforts of MPs from Chatra Sunil Singh, Palamu's VD Ram and of Aurangabad, Gaya and Jehanabad. The concerted effort has brought Bihar in agreement on reducing height of the dam by four meters. "Bihar Government has agreed to it and this would reduce submerged area by 860 hectares by bringing it to 51136 hectares," said an official of Water Resources Department, adding that catchment area would go up to 128 km.

The dam over North Koel River is said to have had completed 90% of its work in 1993 when it stopped. The reason was the amendment of the Forest (Conservation) Rules 1981 the previous year, which tightened the conditions for getting forest clearances. Till that time, there was no need for detailed forest clearances. The new requirements however mandated that any proposal for converting forest land to non-forest land had to be evaluated by a Committee who would assess whether the proposed land formed a part of a nature reserve or national park. Since the nearby Palamu Tiger Reserve was under threat of flooding because of the project, the government at the time chose to abandon the project.

The unfinished dam has a height of 67.86 meters and length 278 meters. Nine gates had to be stalled and most of the turbines were brought to the place. Following the abandonment of the project, the entire setup and resources

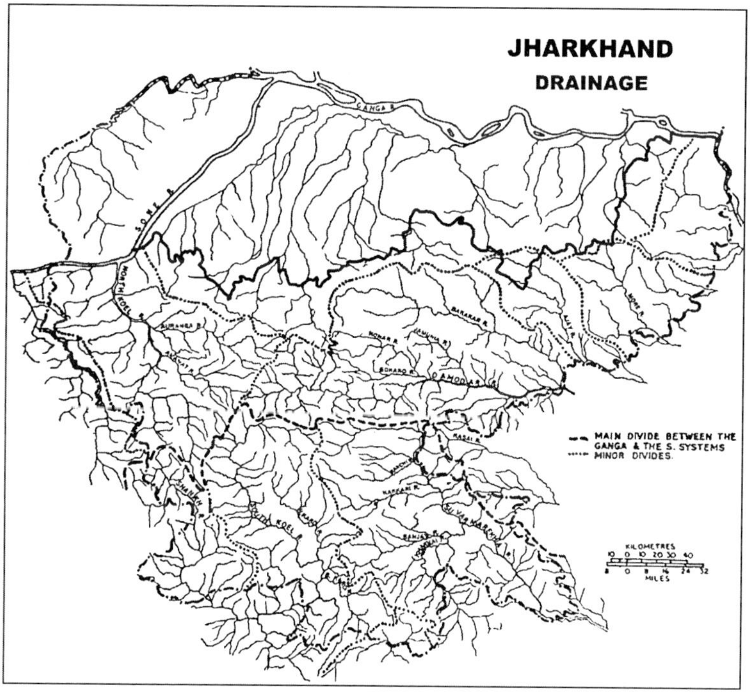


Map 1.5: Drainage

have vanished now, leaving the locals in peace to continue their traditional way of life.

Apparently, out of 634 families of 17 submerged villages, a majority of 555 had been paid compensation. However, locals remaining in the villages today claim that the compensation was received by their forefathers. It is important that they also receive adequate money and land to compensate for the hardships they would face as a result of flooding of their homes. They are not principally opposed to the project, only because there are promises that the project will benefit them by irrigating dried land and generating 24 MW hydro power.

Therefore, it is up to the government to ensure not only that the proposed benefits of the project actually reach the local people, but also that they are adequately compensated and rehabilitated before being displaced. For this it is important to take consensus and engage with the locals.

**Ground Water Level:** In Jharkhand state ground water levels of 333 Ground Water Monitoring Wells (GWMW) were monitored four times in the year 2015 - 2016 as a part of regime monitoring of phreatic aquifer in different hydrogeological and agro-climatic zones. The water level monitoring was carried out in the months of May'15, August'15, November'15, and January'16, and ground water samples were collected in pre-monsoon period (May 2015) for chemical analysis. In the state the phreatic aquifer consists of weathered mantle, saprolite zone, and fractures in hard rocks underneath. Over 78% area of the state is underlain by rocks of Chotanagpur Gneissic Complex (CGGC) suit. Hence, most of the GWMW represented water level in weathered CGGC. A few GWMW represented water level of phreatic aquifer of Gondwana Super Group and Tertiary Formation.

The observed water level data had been grouped into four categories viz. 0 -2m, 2-5m, 5-10m and >10m. Thematic maps depicting ground water levels measured in different periods have been prepared. The water levels have been further analyzed to study its change with respect to measurement of pre-monsoon period of the same year, previous year water level data of corresponding period, and decadal mean water level data of the corresponding period. The fluctuations have been shown under rise and fall categories. In each category there are three groups viz. 0-2 m, 2-4 m and >4 m. Thematic maps had been prepared for each category.

During 2015-16 the water level in the State ranges between 0.1 to 18.73 mbgl. The minimum and the maximum depth to water levels during pre- monsoon have been recorded as 0.5 m bgl at Dhanbad district and 18.73 m bgl at E.Singhbhum and in general the water level throughout the State varies in the range of 5 - 10 m bgl. During post-monsoon the minimum and the maximum depth to water levels have been recorded as 0.39 m bgl at Amrapara, Pakur district and 12.28 m bgl at Kajri, Palamu district and in general the water level throughout the State varies in the range of 2 - 5 m bgl. The annual fluctuation of water level between May 2015 and November 2015, the major part of the state shows general rise in water level in the range between 0-2 m bgl (47.6%) and fall (31.2%). The fluctuation of water level of May 2015 with respect to decadal mean water level of May indicate the fall (51.7%) as well as rise (32%) in water level in the range of 0 - 2 m. The fluctuation of water level of November 2015 with respect to decadal mean water level of November indicate the fall (44.3%) as well as rise (45%) in water level in the range of 0 - 2 m. However, overall regional fluctuation of water level in the entire state is mainly restricted within 2 m only which is normal phenomenon and no abnormal rise or fall in water level is observed except in few localized well. The pH of ground water ranged between 6.79 and 8.96. The water was mildly alkaline in nature in most of the wells. The Electrical Conductivity (EC) varies between 8.7 microS/cm (Gandey, Giridih) and 3705(Daru, Hazaribagh dist.) micro-S/cm. Spatially in major part of the state EC rested in the range of 61-500 micro-S/cm. In majority of the samples the concentration of chloride in ground water is within the desirable limit for drinking water of 250 mg/l.

## Scenario of Depth of water level (2015-16)

### May 2015

Water levels during May 2015 were monitored from 333 wells (out of 417 existing wells). The minimum and the maximum depth to water levels have been recorded as 1.20 m bgl and 16.50 m bgl in Dhanbad and Garhwa district respectively. In general, the water level throughout the State varies in the range of 5 - 10 m bgl and has been observed in the 175 wells (67.3%) out of 20 analyzed wells. Secondly, water level >10 m bgl has been observed in the 31 wells (12%). The water level in the range of 2- 5 m bgl has been observed in the 18.5% of the wells. The water level below 2 m has been observed only in 6 wells, out of which 3 wells located in E Singhbhum and rest 3 well in Ranchi, Gumla and Dhanbad district.

The entire state shows water level varying between 5 and 10 m bgl except few patches where water level is more than 10 m bgl. Including few patches in the State, an area covering the parts of Gumla, Simdega and W Singhbhum has shown water level less than 2 m bgl.

### August, 2015

Water levels during August, 2015 were monitored from 307 Dug wells. The minimum and the maximum depth to water levels have been recorded as 0.1 m bgl in Ranchi district and 21.18 m bgl in Gumla district. In general, the water level throughout the State varies in the range of 2 - 5 (58%) m bgl from 178 analyzed wells. Secondly, the water level in the range of 5 - 10 m bgl has been observed in the 18 % of the wells. Water level >10 m bgl has been observed only in the 4 wells (1.3%) one each in Chatra, Deoghar Bokaro, and E Singhbhum district. Major part of the State shows water level varying 2 - 5 mbgl. Water level above 5 mbgl is observed mainly in northern and north western part of the state whereas the water level less than 2 m bg has been observed in eastern and western part.

### November, 2015

Minimum and the maximum depth to water levels have

been recorded as 0.50 m bgl and 12.65 m bgl in Dumka and Bokaro district respectively. In about 50% of GWMW, water level rests in range of 5-10 m bgl which covers almost entire Jharkhand state. The water level in the range of 2-5 m bgl has been observed in the 123 wells (44%) which occur in localized areas of south-eastern parts of the state. Ground water level of 0 - 2 m bgl depth range has been observed only in 9 wells at different locations. Only 7 wells have shown water level more than 10 m bgl.

**January, 2015**

To study the water levels of recession period data were collected during

January, 2016 from 292 wells. The minimum and the maximum depth to water levels in the State have been recorded as 0.97 m bgl and 19.25 m bgl in E Singhbhum. The water level in general varies between 2 and 5 mbgl as the water level in the range of 5 - 10 m bgl has been observed in the 71% (206 wells) of the wells analyzed and covered almost entire State. Few patches of water level range from 10 to 20 m bgl has been observed. About 12% of the wells analyzed has shown water level in the range of 2- 5 m bgl occurred in the 6 localized areas. The water level below 2 m has been observed in 3 wells located in E Singhbhum district.

**Scenario of Annual Fluctuation in water level in Jharkhand during 2015-16**

The annual fluctuation in water levels for the periods of (1) May 2014 and May 2015, (2) August 2014 and August 2015, (3) November 2014 and November 2015 and (4) January 2015 and January 2016 have been analyzed to study the net status of ground water conditions during the previous and current year.

**May 2014 and May 2015**

The annual fluctuation in water level between May '2014 and May '2015 indicates the net status of ground water condition during the previous year and current summer measurement. The major part of the state shows general fall in water level in the range between 0-2 m bgl (54.5%) and rise (27%). Total 126 wells out of 231 analyzed wells, comes under 0- 2 m falling zone category, on the other hand 62 wells show rise within 2 m, which may indicate that the regional fluctuation of the state (82%) is mainly restricted within 2 m. The next higher magnitude of fluctuation is of 2 -4 m bgl fall in water level in the state (7.4%) is observed in some part of the state.

Overall, the entire State is covers under falling zone category (126 wells out of 231 analyzed well), which may indicate the slightly less rainfall (2014) and ground water level decline during the analyzed period.

**August 2014 and August 2015**

The annual fluctuation in water level between Aug 2014 and Aug 2015 indicates the net status of ground water condition during the previous year and current monsoon measurement. A general fall in water level (44%) has been found in major part of the State whereas rise in water level recorded in 34% wells within 2 m. About 8% wells ranges between 2 - 4 m rise and 7.3 % wells have shown fall out of 205 analyzed well. Out of 205 analyzed wells fall in water level is recorded in 137 wells within 2 mbgl.

Overall, the entire State is covered under the category of within 2 m fluctuation (44% fall and 34% rise) which may be due to less rainfall in respect to during previous year.

**November 2014 and November 2015**

The Annual fluctuation in water level between November 2014 and November 2015 indicates the net status of ground water conditions during the previous and current post-monsoon year.

The comparison of fluctuation in water level between November 2014 and November 2015 shows fall in 52.4% GWMW as well as rise in 23% GWMW of the total 252 analyzed wells during the period. The major part of the state shows a general fall in water level within 2.00 m. Out of 177 wells showing fall, 2 - 4 m and > 4 m fall has been observed in the 39 wells and 6 wells respectively. Only 29% wells analyzed has shown rise in water level. of the total analyzed wells, out of which 14 wells have shown fall within 2-4 m and only one well have shown higher magnitude of annual fluctuation > 4m during the period.

**January 2015 and January 2016**

The annual fluctuation in water level between January 2015 and January 2016 indicates the net status of ground water condition during the previous year and current measurement.

The major part of the state shows general fall (52.6%) in water level in the range between 0-2 m bgl as well as rise (18.5%). Total 162 wells out of 216 of the analyzed well comes under 0-2 m falling zone category, on the other hand 40 wells show fall within 2 m, which may indicate that the regional fluctuation (83%) of the state is mainly restricted within 2 m. The rise in water level above 2 m bgl has been observed in isolated patches of the State.

## Cultural Factors

Under the heading of Geographical Background of Agricultural Relevance now we shall study the cultural factors. Cultural factors deal with population, Land Utilization, Irrigation, Industries, transportation and market etc.

### Population

The people of Jharkhand, Chotanagpur and Santhal Pargana share a common geography and a common history. Here people are of two major groups, one is Sadan i.e., Non-Tribal and other is Tribal. While analyzing the early account of people in the state of Jharkhand, it becomes difficult to portray early accounts due to lack of data. Census enumeration initiated only in the year 1872 and since then limited population data is available for analyzing population characteristics. Prior to this census, the long history of human habitation has no documented accounts and other authentic sources of information. Certainly, on the basis of historical record, archeological investigation, carbon dating of the remains of the fossils and some description in the religious books and epics, scholars have given sketchy accounts of population. Moreover, the state of Jharkhand, being considered as a land of forest, relatively less suitable for ideal human habitation and untapped resources, remained unmolested by the incoming migrants who inhabited the vast plains and brought significant transformation in economic pursuits and cultural developments. Jharkhand remained occupied for long mainly by the aboriginals and only in later historic periods, some migrants from the neighboring plain are inhabited the forest clad region may be for safety or some other reasons and settled there.

Historical records, archeological evidences, cultural attributes substantiate human habitation in this region in the paleolithic stone age (200000 - 20000 BC). Relics of human habitation have been discovered in the Jharia coal fields and other areas of the state. The discovery of quartzite axes and spearheads in the Jharia coalfields, Chert flanks and Knives along the banks of rivers at Chaibasa and Chakradharpur and at Buradih in Singhbhum and eastern part of Tamar pargana in Ranchi attest the paleolithic habitation in this region. The habitation of Neolithic age is evidenced by and earthing of a pot full of Indo-Scythian copper coins at Gulka, south of Chaibasa and traces of Neolithic culture at Chakradharpur, Parasnath, Ranchi and Hazaribagh. At the time of Aryanization of the Ganga Plain the northern dissected fringe of Chotanagpur plateau was only influenced when several kingdoms developed in the Ganga Plain. The pre-history of human habitation in this region is also shrouded in mystery and it is not so easy to provide definite account by a few hazy legends of the aboriginals which give some clues about the Asurs and indicate the triumph of that race over the original inhabitants of the country, possibly the ancestors of small iron smelting tribe of Iron age of that name which still speaks a language akin to Mundari. Certainly, clear cut evidences are lacking to fix the date at which the Munda, Oraons, Santhals, Hos etc. inhabited this region. It seems probable that they were the first to enter the region from west and north-west. It is opined that Munda belonged to the Austro-Asiatic race. They are found in considerable numbers in the south-eastern portion. The aboriginal races such as Khawars, Oraons, Ceros, Hos etc. particularly ruled over this tract. At the time of Aryan and Non-Aryan rivalries in the plains non-aryans encroached in this region especially in accessible areas. Afterwards, some aryan habitation also developed in the adjoining plains of the Rajmahal and the northern portion of the koel basin.

Contrary to this thought that tribal are the first settlers to this tract, Dr Ram Kumar Tiwari (Assoc Prof University Dept of Geography, Ranchi University, Ranchi) has explained his views during their lectures in the class room for the Jharkhand paper that "In Geography place names are one of the tools to identify the time period on any place or its antiquity. In Jharkhand name of rivers are Damodar, Subarnarekha, Shankh etc. the name of a mountain is Parasnath, all these names are similar to Sanskrit. Which reveals that speakers of such language have settled in this tract before the tribal.

During the Buddhist and Mahavir period this region did not get much impact of the habitation of Buddhist followers except the areas around the Parasnath hills, part of the Koel basin. This is why there are villages having nomenclature related to Buddhism and Jainism in part of Palamu, Chatra, Hazaribagh and Giridih districts.

With the downfall of Buddhism and Jainism the region again underwent unrest and distributional pattern got jolted. Stability in habitation was restored only after Muslim conquest over major area of the country. Towards the end of 10th century AD this region got habitation of some Rajput immigrants.

By the development of transport network distributional pattern was influenced the demographic pattern of the region. We shall discuss here the distribution, Density, Growth and Composition of population of Jharkhand as the agriculture is related to all these phenomena directly or indirectly.

**Distribution of Population:**

While analyzing the distributional pattern of population in this region, a few areas have emerged as focal points of human concentration but the remaining areas portrayed sparse distribution. The agglomeration of rural population is visualized by the crowding of dots showing population and urban concentration is reflected by the increasing size of circles and concentration of urban agglomeration in limited areas. Unevenness in population distribution becomes more obvious when the relief and population distribution maps are overlapped over each other. This also becomes obvious by the analysis of man-land ratio and density of population of both rural and urban.

The table of population distribution 2011 (Table 1.2) reveals the share of area as well as population of the district in the state. The comparison of the share of the area and share of population reveals the unevenness in the distribution of population. Districts consisting of more population in comparison with are included Dhanbad, Bokaro, Ranchi, E. Singhbhum, Hazaribagh, Giridih, Deoghar, Godda, Jamtara, Pakur, Palamu, Ramgarh, Sahebgunj. The district of Dhanbad consisted of 2.56 percent are of the state but has 8.14 percent population share of the state. Similarly, Bokaro has 3.62 percent, E. Singhbhum 4.47 percent and Ranchi 6.39 percent area but the population share is 6.25 percent, 6.95 percent and 8.83 percent respectively.

The districts having sparse population because of more land than population share included W.Singhbhum, Gumla, Latehar, Khunti, Chatra, Dumka, Garhwa, Koderma, Lohardaga, Saraikela-Kharsawan. The W. Singhbhum district has 9.06 percent area but 4.55 percent of population of the state because of considerable forest cover. Similarly, Gumla has percent, Latehar 5.38 percent, Khunti 3.18 percent area but the population share is 3.11 percent, 2.20 percent and 1.61 percent respectively.

## Table 1.2
## Distribution of Population: Jharkhand (2011)

| District | Area of the State (%) | Population of the State (%) Total Population (%) | Rural Population (%) | Urban Population (%) |
|---|---|---|---|---|
| Bokaro | 3.62 | 6.25 | 4.31 | 12.40 |
| Chatra | 4.66 | 3.16 | 3.91 | 0.79 |
| Deoghar | 3.11 | 4.52 | 4.92 | 3.26 |
| Dhanbad | 2.56 | 8.14 | 4.49 | 19.67 |
| Dumka | 4.72 | 4.01 | 4.91 | 1.14 |
| Garhwa | 5.13 | 4.01 | 5.00 | 0.88 |
| Giridih | 6.22 | 7.41 | 8.93 | 2.62 |
| Godda | 2.84 | 3.98 | 4.99 | 0.81 |
| Gumla | 6.72 | 3.11 | 3.83 | 0.82 |
| Hazaribagh | 4.41 | 5.26 | 5.82 | 3.47 |
| Jamtara | 2.27 | 2.40 | 2.85 | 0.95 |
| Koderma | 3.19 | 2.17 | 2.29 | 1.78 |
| Khunti | 3.18 | 1.61 | 1.94 | 0.57 |
| Latehar | 5.38 | 2.20 | 2.69 | 0.65 |
| Lohardaga | 1.88 | 1.40 | 1.61 | 1.94 |
| Pakur | 2.27 | 2.73 | 3.32 | 0.85 |
| Palamu | 5.51 | 5.88 | 6.84 | 2.85 |
| W. Singhbhum | 9.06 | 4.55 | 5.13 | 2.75 |
| E. Singhbhum | 4.47 | 6.95 | 4.07 | 16.07 |
| Ramgarh | 1.68 | 2.88 | 2.12 | 5.28 |
| Ranchi | 6.39 | 8.83 | 6.61 | 15.85 |
| Sahibgunj | 2.59 | 3.49 | 3.95 | 2.01 |
| Saraikela-Kharsawan | 3.33 | 3.23 | 3.22 | 3.26 |
| Simdega | 4.73 | 1.82 | 2.22 | 0.54 |
| **Jharkhand** | **100% 79714 sq. Km.** | **100% 32988134 persons** | **100% 25055073 persons** | **100% 7933061 persons** |

When we examine the distribution pattern of rural urban population in the state the under mentioned table reveals that the district of Dhanbad has got only 4.49 percent rural population of the state but has 19.67 percent urban population of the state. Similarly, the district of E. Singhbhum has 4.07 percent rural and 16.07 percent urban population of the state.

The district of Ranchi has 6.61 percent rural population of the state but 15.85 percent urban population. The district having high concentration of rural population included Giridih (8.93 percent) Palamu (6.84 percent), Ranchi (6.64 percent), Hazaribagh (5.82 percent) etc. The districts having high concentration of urban population included Dhanbad (19.67 percent), E.Singhbhum (16.07 percent), Ranchi (15.85 percent), Bokaro (12.40 percent). The controlling factors for larger concentration of urban population in Dhanbad, E. Singhbhum and Ranchi are industrial developments as well as the development of other economic and cultural characteristics. The percentage of rural population is large in Giridih, Palamu, Hazaribagh etc. which are comparatively large in area and less developed.

**Table1.3**

**Rural-Urban Population in Jharkhand (2011)**

| District | Total Population | Rural Population | Urban Population | Rural Population (%) | Urban Population (%) |
|---|---|---|---|---|---|
| Bokaro | 2062330 | 1078686 | 983644 | 52.30 | 47.70 |
| Chatra | 1042886 | 979932 | 62954 | 93.96 | 06.04 |
| Deoghar | 1492073 | 1233712 | 258361 | 82.68 | 17.32 |
| Dhanbad | 2684487 | 1124093 | 1560394 | 41.87 | 58.13 |
| Dumka | 1321442 | 1231264 | 90178 | 93.18 | 06.82 |
| Garhwa | 1322784 | 1253114 | 69670 | 94.73 | 05.27 |
| Giridih | 2445474 | 2237450 | 208024 | 91.49 | 08.51 |
| Godda | 1313551 | 1249132 | 64419 | 95.10 | 4.90 |
| Gumla | 1025213 | 960132 | 65081 | 93.65 | 06.35 |
| Hazaribagh | 1734495 | 1459188 | 275307 | 84.13 | 15.87 |
| Jamtara | 791042 | 715296 | 75746 | 90.42 | 09.58 |
| Koderma | 716259 | 575013 | 141246 | 80.28 | 19.28 |
| Khunti | 531885 | 486903 | 44982 | 91.54 | 08.46 |
| Latehar | 726978 | 645120 | 51858 | 92.87 | 07.13 |
| Lohardaga | 461790 | 404379 | 57411 | 87.57 | 12.43 |
| Pakur | 900422 | 432910 | 67512 | 92.50 | 07.50 |
| Palamu | 1939869 | 1713866 | 226003 | 88.35 | 11.65 |
| W.Singhbhum | 1502338 | 1284304 | 218034 | 65.49 | 14.51 |
| E. Singhbhum | 2293919 | 1019328 | 1274591 | 44.44 | 55.56 |
| Ramgarh | 949443 | 530488 | 418955 | 55.87 | 44.13 |
| Ranchi | 2914253 | 1656918 | 1257335 | 56.86 | 43.14 |
| Sahebganj | 1150567 | 990901 | 1599666 | 86.12 | 13.88 |

| | | | | | |
|---|---|---|---|---|---|
| Saraikela-Kharsawan | 1065056 | 806310 | 258746 | 75.71 | 24.29 |
| Simdega | 599578 | 556634 | 42944 | 92.84 | 07.16 |
| Jharkhand | 32988134 | 25055073 | 7933061 | 75.95 | 24.05 |

**Source:** Primary Census Abstract, Total Table for Jharkhand, 2011

## Density of Population:

The term 'Density of Population' refers to the number of persons per square kilometer. In other words, density of population indicates the man-land ratio. This is calculated by dividing the number of persons of a country or region by the total land area. In case the land area is small for a given population, the density will be high; but if the land area is large, then the density will be low. Density depends on many natural and human factors, such as soil, rainfall, climate, economic resources, the stage of economic growth and so on. Since those factors differ in many places, density will also differ. Density measures the degree of population concentration in a particular area. There are many types of density of population. They are as follows:

1. Arithmetic Density
2. Economic Density
3. Agricultural Density
4. Nutritional Density and
5. Critical Density etc. (Ghosh, 1987, P. 28-30)

Here it is worthy to mention the views of Dr. Ram Kumar Tiwari regarding the word Distribution and Density which has been mentioned in his book Jansankhya Bhoogol (2015, P.56). He has explained that both the words have different meanings and have different methods of analysis. Distribution denotes only the placing of population in numbers whereas Density is the ratio of population to the area of Land. Distribution can be measured through the percentage method, concentration index method, location quotient method and Lorenz curve method etc., but density can be measured by different methods apart from the following i.e., Arithmetic, Physiological, Agricultural, Economic and Nutritional density etc.

At the same time Dr. Tiwari has also suggested regarding the type of density which is somehow different from the general. He has mentioned in this same book (page 57) that when we divide any thematic items to the area then it is a sub-division of Arithmetic density or Mathematical density or crude density such as Ethnic Density, Sex Density, Literacy Density, Rural-Urban Density, Industrial Density and density of poverty etc. Different authors have mentioned all these as a type of density but Dr. Tiwari's classification brings more clarity and removes any confusions. Despite these many types we shall discuss only two types of density i.e., Arithmetic Density for general understanding and Agricultural Density as it is directly related to our work here.

**Arithmetic Density:** Crude density, also known as arithmetic density, is the most commonly used measure of population density. It is expressed as the number of people divided by the total area. It has been found that higher density of population is recorded near cities, industries and mining areas of Dhanbad, Bokaro, suburbs of Jamshedpur and Ranchi. The district of Dhanbad has registered the highest density of population i.e., 1316 persons per sq. km. (2011), but there are some development blocks where per sq. km. density has been very high. Dhanbad-cum-Kenduadih- Jagta development block had the density of over 3000 persons per sq. km in 2011. It may be noted that development block of Dhanbad has very high density of population because of urban and industrial influence. In the C.D. Block of Golmuri-Jugsalai in East Singhbhum district again shows density of 1441 persons per sq. km because of the influence of Tatanagar industrial centers. Similarly, there are some areas of high density near Ranchi, Hazaribagh and Bokaro. In the C.D. Block of Kanke the density of population is 703 persons per sq. km. On the basis of density figures of 2011, the density of the state can be categorized into three groups:

- High Density Areas (Above 500 persons per sq. km)
- Moderate Density Areas (499-300 per sq. km.) and
- Low density areas (below 300 persons per sq. km)

1. **High Density Areas:** In Jharkhand, the areas having high

density are Dhanbad (1316 persons per sq km) followed by Bokaro (715 persons per sq km), Ramgarh (708 persons per sq km), E. Singhbhum (644 persons per sq km), Deoghar (602 persons per sq km), Godda(580 persons per sq km), Ranchi (572 persons per sq km) and Sahebganj (558 persons per sq km). In these districts the density is very high as all those areas are either industrial or mining dominated region. Deoghar, Godda and Sahebganj district have high density due to suitability of the land for agricultural purposes and other forest products.

2. **Moderate Density Areas:** Districts having moderate density ranging from 500 to 300 persons per sq km include Pakur (497), Hazaribagh (488), Palamu (442), Jamtara (437), Saraikela-Kharsawan (401), Dumka (351) and Garhwa (323) respectively. All these districts are either mining based, industrial based or agro-based with moderate resources.

3. **Low Density Area:** Low density areas having less than 300 persons per sq. km. include those districts of the state where considerable areas have rough terrain, negligible mineral resources, vast forest stretch and fewer economic and industrial activities. These districts include Koderma (282), Chatra (281), Khunti (210), W. Singhbhum (208), Gumla (191), Latehar (169) and Simdega (159) respectively. In these districts neither agriculture resource- based population have concentrated nor non-agriculture-based people are found in considerable proportion.

**Table 1.4**
**Population Density: Jharkhand**

| District | Area (in sq. km) | Density (2001) Person/Sq. Km. | Density (2011) Person/Sq. Km. | Net Change |
|---|---|---|---|---|
| Bokaro | 2833 | 621 | 715 | 94 |
| Chatra | 3718 | 213 | 281 | 68 |
| Deoghar | 2477 | 468 | 602 | 134 |
| Dhanbad | 2040 | 1167 | 1316 | 149 |
| Dumka | 3761 | 290 | 351 | 61 |
| Garhwa | 4093 | 256 | 323 | 67 |
| Godda | 2266 | 496 | 580 | 84 |

| | | | | |
|---|---|---|---|---|
| Gumla | 5360 | 148 | 191 | 43 |
| Hazaribagh | 3555 | 370 | 488 | 118 |
| Jamtara | 1811 | 364 | 437 | 73 |
| Koderma | 2540 | 380 | 282 | 98 |
| Khunti | 2535 | - | 210 | - |
| Latehar | 4291 | 154 | 169 | 15 |
| Lohardaga | 1502 | 244 | 307 | 63 |
| Pakur | 1811 | 388 | 497 | 109 |
| Palamu | 4393 | 296 | 442 | 146 |
| E.Singhbhum | 3562 | 560 | 644 | 84 |
| W.Singhbhum | 7244 | 240 | 208 | 32 |
| Ramgarh | 1341 | - | 708 | - |
| Ranchi | 5097 | 362 | 572 | 210 |
| Sahebganj | 2063 | 580 | 558 | 22 |
| Saraikela-Kharsawan | 2657 | 308 | 401 | 93 |
| Simdega | 3774 | 138 | 159 | 21 |
| Jharkhand | 79716 | 338 | 414 | 7.6 |

**Source:** Census of India 2011

## Agricultural Density:

Agricultural Density means the number of agricultural people per unit of cultivable land. Agricultural density is, thus, the ratio between the number of people earning their living or subsistence from working of the land and the total amount of farmland.

Here the term agricultural population denotes the number of cultivators and agricultural laborer together. Similarly, the term total cropped are means the area sown and area sown more than once together. The Agricultural population and total cropped area, both are intimately and directly related to each other. In other words, agricultural density shows the pressure of agricultural population on each unit of food producing resource i.e., cultivated land. Thus, the above-mentioned Agri. Density are the indicators of the agricultural status i.e., commercial or subsistence of a region and sometimes it also helps to estimate the food surplus in the region.

Now we shall discuss the situation of agricultural density for the period of 2001 census and 2011 census gradually and compare their changing pattern.

## Agricultural Density: 2001

The agricultural density of census 2001 shows that the state has an average density of 3.67 agricultural population per hectare of total cropped area. During the census of 2001 there were only 18 districts in the state. Among the 18 districts, four have below-average density of the state. They are Hazaribagh (3.47 persons per hectare), E. Singhbhum (3.25 persons per hect.), Lohardaga (2.59 persons per hect.) and Ranchi (2.59 persons per hect.). It has been noted from the different books of economics and Land-Use that economically advanced area has very low agricultural density as compared to the less advanced area. Hence, we can say that in Jharkhand Hazaribagh, E. Singhbhum, Lohardaga and Ranchi districts are very advanced or developed as indicated by the agricultural density.

The agricultural density having very high concentration included Bokaro (6.06 persons per hect.), Dumka (5.97 persons per hect.), Palamu (5.70 persons per hect.), Giridih (5.71 persons per hect.) and Koderma (5.30 persons per hect.) all these districts are less advanced due to the highest level of agricultural density have been depicted here. Rest nine districts of the state where agricultural density is seen moderate are Chatra, Deoghar, Garhwa, Godda, Gumla, Pakur, W. Singhbhum, Dhanbad and Jamtara. In these districts the pressure of agricultural population on the cropped area per hectare are neither very high or low; hence these districts may be treated as neither advanced nor less developed.

**Table 1.5**
**Agricultural Density: Jharkhand - 2001**

| Districts | Agricultural Population/ (Cultivator+ Agri Labourers) | Total Cropped Area/ (Net area sown+Area sown more than once | Agricultural Density (No. of people per hectare) |
|---|---|---|---|
| Bokaro | 226882 | 37432 | 6.06 |
| Chatra | 243663 | 51711 | 4.71 |
| Deoghar | 277288 | 70564 | 3.93 |

| | | | |
|---|---|---|---|
| Dhanbad | 158779 | 40521 | 3.92 |
| Dumka | 636567 | 106668 | 5.97 |
| E. Singhbhum | 319088 | 98057 | 3.25 |
| Garhwa | 337971 | 79883 | 4.23 |
| Giridih | 478019 | 83724 | 5.71 |
| Godda | 340807 | 76006 | 4.48 |
| Gumla | 570724 | 140529 | 4.06 |
| Hazaribagh +Ramgarh | 487162 | 140334 | 3.47 |
| Jamtara | - | 46807 | - |
| Koderma | 109887 | 20745 | 5.30 |
| Latehar | - | 62836 | - |
| Lohardaga | 123798 | 47837 | 2.59 |
| Pakur | 207976 | 54539 | 3.81 |
| Palamu | 632331 | 110957 | 5.70 |
| Ranchi+Khunti | 678278 | 261549 | 2.59 |
| Sahebganj | 245946 | 40660 | 6.05 |
| Saraikela +Kharsawan | - | 67086 | - |
| Simdega | - | 91633 | - |
| W. Singhbhum | 665637 | 142098 | 4.68 |
| Jharkhand | 6740803 | 1836176 | 3.67 |

**Source:** Primary Census Abstract. Series-21, 2001 and calculation by author.

## Table 1.6
## Agricultural Density: Jharkhand - 2011

| Districts | Agricultural Population/ (Cultivator+ Agri Labourers) | Total Cropped Area/ (Net area sown+Area sown more than once | Agricultural Density (No. of people per hectare) |
|---|---|---|---|
| Bokaro | 263927 | 39027 | 6.76 |
| Chatra | 307810 | 38925 | 7.91 |
| Deoghar | 338349 | 47523 | 7.12 |
| Dhanbad | 200606 | 13543 | 14.81 |
| Dumka | 496486 | 87521 | 5.67 |
| E. Singhbhum | 321861 | 82121 | 3.91 |
| Garhwa | 442234 | 63262 | 6.99 |
| Giridih | 732523 | 58902 | 12.43 |
| Godda | 452370 | 73954 | 6.11 |
| Gumla | 419190 | 124802 | 3.65 |
| Hazaribagh | 414845 | 56396 | 7.35 |
| Jamtara | 241675 | 42964 | 5.62 |

| | | | |
|---|---|---|---|
| Kodarma | 145433 | 11541 | 12.60 |
| Khunti | 215738 | 77781 | 2.77 |
| Latehar | 246422 | 54134 | 4.55 |
| Lohardaga | 180620 | 50194 | 3.59 |
| Pakur | 255362 | 39957 | 6.39 |
| Palamu | 514561 | 111972 | 4.59 |
| Ranchi | 584847 | 121304 | 4.82 |
| Ramgarh | 137585 | 15702 | 8.76 |
| Sahebganj | 316877 | 48177 | 6.57 |
| Saraikela-Kharsawan | 259697 | 73706 | 3.52 |
| Simdega | 248251 | 90467 | 2.74 |
| W. Singhbhum | 511647 | 129726 | 3.94 |
| Jharkhand | 6618761 | 1553691 | 4.26 |

**Source:** Primary Census Abstract, Jharkhand-Series-21, 2011 and Agricultural Stats from Agri Directorate

## Agricultural Density: 2011

A general scenario of Agricultural Density can be seen in comparison to 2001 which has been changed during 2011 in Jharkhand. The number of agricultural population and total cropped area both has grown positively in almost all the districts except six districts i.e., Dumka, Gumla, Hazaribagh, Palamu, Ranchi and W. Singhbhum. The cause of negative growth in these districts are administrative and not demographic. It is well known that during 2001 census there were only 18 districts in the state whereas it became 24 in number in the census of 2011. District Simdega was separated from Gumla, Ramgarh from Hazaribagh, Latehar from Palamu, Khunti from Ranchi, Saraikela from W. Singhbhum, and Jamtara from Dumka district. Hence the negative growth has no meaning in this context.

During 2011 Agricultural Density ranges from 2.74 in Simdega to 14.81 in Dhanbad in the state. We divide it into three groups i.e., High (more than 10), Moderate (10-5) and Low (below 5). District Dhanbad had recorded the highest pressure of 14.81 agricultural persons per hectare of cropped area followed by Giridih (12.43 persons/hect) and Kodarma (12.60 persons/hect). The highest agricultural density has been noted in these three districts because of concentration of a good number of urban centers which have developed on account of the development of the coal resources and related industries.

The moderate agricultural density of population has been noted in the district of Ramgarh (8.76 Agri population/hect), Chatra (7.91 Agri population/hect), Deoghar (7.12 Agri population/hect), Hazaribagh (7.35 Agri population/hect), Garhwa (6.99 population/hect), Pakur (6.39 Agri population/hect), Godda (6.11 Agri population/hect), Godda (6.11 Agri population/hect), Dumka (5.67 Agri population/hect) and Jamtara (5.62 Agri population/hect). These districts have comparatively less industrial development hence the pressure of agricultural population over the cropped area is moderate.

They districts having agricultural density of population less than the average of the state (4.26 Agri Population per hect of cropped area) consists of Ranchi (4.82), Palamu (4.59), Latehar (4.55), E. Singhbhum (3.91), Lohardaga (3.59), Saraikela-Kharsawan (3.52), Gumla (3.35), Khunti (2.77), and Simdega (2.74). These districts have very low agricultural density which reflects their better. Industrial development while agricultural condition very poor. Hence, we suggest to develop agricultural position in the state of Jharkhand.

**Tribal Population:**

No doubt, agriculture is most dominating activity among the population of Jharkhand, at the same time about one third tribal population (26.21 percent) actively participate in agricultural activities in Jharkhand and therefore the study of this population is relevant in the study of Agriculture.

The state of Jharkhand came into being because of the fact that a large number of tribes live within these boundaries. Presently 32 tribal groups live in the state out of which 8 are PTGs (Primitive Tribal Groups). All these tribes of the state have been divided in different racial groups i.e.

- Protoaustroloid - Asur, Birhor, Khariya, Kora, Sabar, Lohra, Bhumij and Khond.
- Austroloid (Pre Dravidian - Black) - Munda, Souriya paharia, Mal Pahariya, Korwa, Ho, Santhal, Pahariya, Gond, Bathudi, Karmali, Kawar, Koi, Kisan

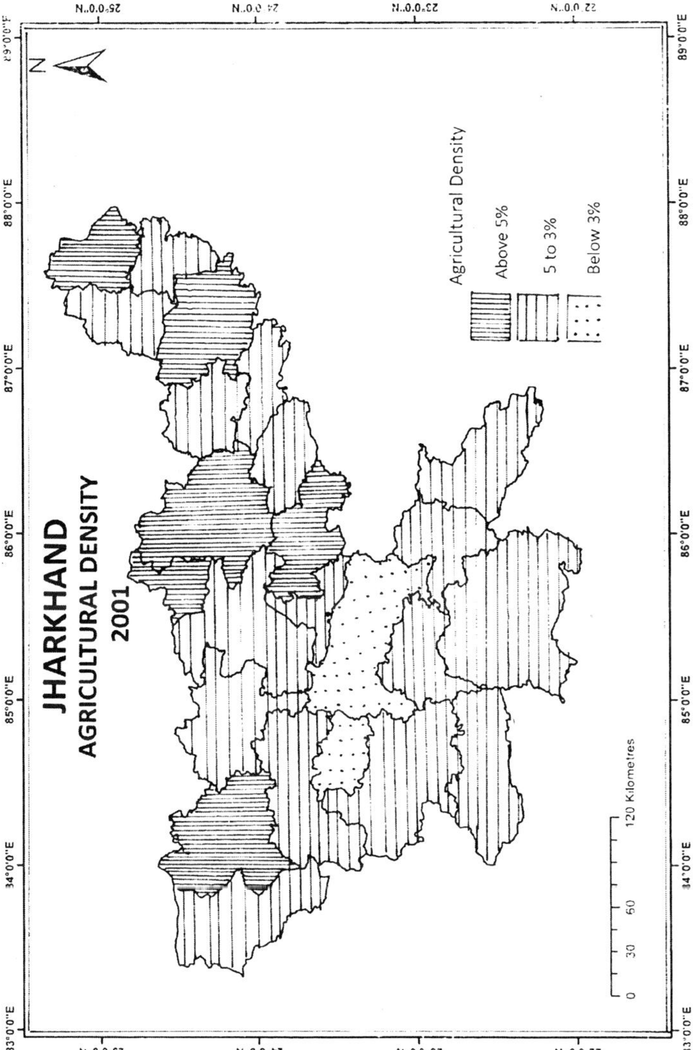


Map 1.6: Agricultural Density (2001)

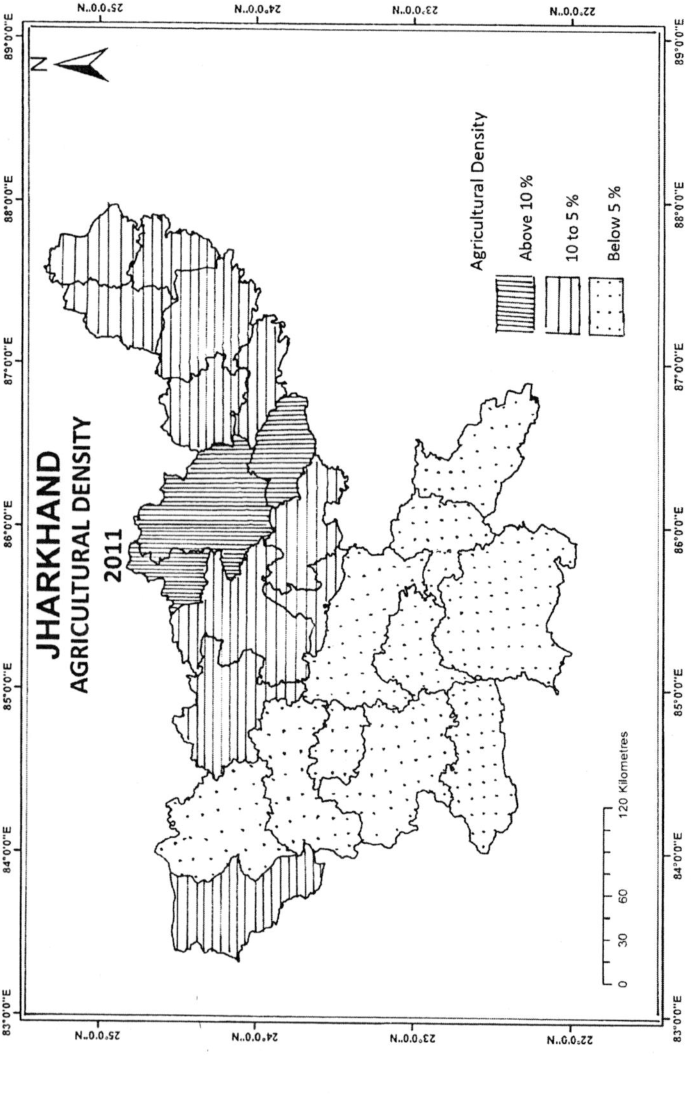


**Map 1.7: Agricultural Density (2011)**

- Dravid (Caucosoid - White) - Oraon, Chero, Kharwar, Kisan, Binjhia, Bedia, Chik Baraik, Gorait, Mahli, Baiga, Birjhiya.
- Primitive Tribes - Asur, Parhiya, Sabar, Birhor, Birjhiya, Korwa, Malpahariya, Souriya Paharia.

Most of the tribal population are involved either in agricultural activities as cultivator and agricultural laborers or in-service sector where they are working as skilled or unskilled workers. Several populations among different groups have migrated from Jharkhand to different parts of the country or even world in search of jobs or any work as most of them are landless or have very small uneconomic land holdings. Those tribal or even non-tribal people have their agricultural land but no sufficient income which leads them to migrate seasonally from Jharkhand to neighboring area or states.

**Table 1.7**
**Total Population and Tribal Population in Jharkhand 1901-2011**

| Decade | Total Popula- (lakhs) tion | Tribal Popula- (lakhs) tion | Percentage of Tribal Popula- tion (%) | Decadal difference in Total Popula- tion | Decadal difference in Tribal Popula- tion |
|---|---|---|---|---|---|
| 1901 | 60.68 | 18.29 | 30.14 | - | - |
| 1911 | 67.47 | 20.88 | 30.95 | +11.18 | +14.16 |
| 1921 | 67.67 | 18.92 | 27.96 | +0.31 | -9.38 |
| 1931 | 79.08 | 30.07 | 38.02 | +16.85 | +58.93 |
| 1941 | 88.68 | 32.49 | 36.64 | +12.13 | +8.05 |
| 1951 | 96.97 | 34.92 | 36.01 | +9.35 | +7.48 |
| 1961 | 116.06 | 39.38 | 33.93 | +19.69 | +12.77 |
| 1971 | 142.06 | 45.69 | 32.16 | +22.40 | +16.04 |
| 1981 | 176.12 | 53.29 | 30.25 | +23.97 | +16.61 |
| 1991 | 218.43 | 60.44 | 27.66 | +24.02 | +13.41 |
| 2001 | 269.45 | 70.87 | 26.30 | +25.35 | +17.25 |
| 2011 | 329.88 | 88.45 | 26.21 | +22.42 | +21.98 |

During first century tribes were king of the area later they became landlords and zamindars and now a days very poor due to their extravagant nature. Now agriculture is the main means of their stay in Jharkhand. Each family owns homestead land, agricultural land, cattle shed and agricultural equipment like Kudal, Jham, Khurpi, Plough, Juath, Henga, Bullock cart etc. They do not hire labor at the time of agriculture. They do it by exchange of labor with other neighboring families.

They have agricultural land of two types - Don and Tanr. In Don land cultivation of Jorahan Dhan is done which is harvested in the minth of Aghan. In tanr land, cultivation of Maize, Tisi, Marua, Kurthi, Til, Arhar, Tisi, Mustard etc. is done. They categorize crops as Bhadai, Aghani, Rabi and Garma as it is in common practice in Jharkhand. Thus, agriculture provides tribal people engagement for 6-8 months in a year. For rest 6-4 months they depend upon wage earning. They work as casual labors in digging soil, road construction, brick kiln, field and forest. They do not get work regularly.

**Tribal Growth:**

As depicted in the above table there were only 18.29 lakh population in Jharkhand which was 30.14 percent of the total population of the state. Now during census 2011 they are 86.45 lakh in number which is only 26.21 percent of the total population. But the decadal growth shows that during the decade of 1901-1911 their growth was only 14.16 percent but now during the decade of 2001-2011 their growth has been recorded as 21.98 percent. When we compare to the decadal growth of total population of the state it was only 11.18 percent during 1901-1911 but it has been depicted 22.42 percent during 2001-2011 census. No doubt the growth of tribal population has become similar to the growth of total population.

**Working Force in Jharkhand:**

The state of Jharkhand exhibits variation in the percentage of workers in different occupations based on some diversities in mining, industrial, trade and commerce and other non-agricultural activities. Virtually agriculture related involvement of workers is dominant in all districts especially in rural areas but the mineralized and industrialized area have involved workers engaged in mines and industries respectively. In the census of 2011 marginal workers and non-workers have further been categorized into cultivator, Agricultural Laborer, Household workers and other workers.

In the census records workers are considered those persons who are participating in any economically productive activity. Such participation may be physical or mental in nature. Work involves not only actual work but also effective supervision or direction of work. Main workers are those who have worked for the major part of the year preceding the date of enumeration. Major part of the year means six months i.e., 180 days or more during the reference year. Marginal workers are those workers who have not worked for the major part of the year i.e., those who have worked for less than six months. Non-workers are those who have not worked any day during the concerned year. The above-mentioned categories of workers are also broadly grouped into primary, secondary and tertiary workers. In case of primary workers human activities concern with natural products in which no transformation of natural gift is done. In other words, we can say that harvesting of nature is primary activity. But in case of secondary activity construction and manufacturing activities are included in which natural products are transformed into different articles and tertiary activities include other services and work other than primary and secondary.

While considering the workers involvement in different categories of occupation, different areas have portrayed significant variations. It becomes clear by the analysis of the above table. The percentage of workers is quite similar across the districts of the region. The highest percentage of workers to the total population has been recorded in the districts of Khunti (48.88 percent) and districts having percentage of workers above 40 percent include Simdega (48.14%), Lohardaga (47.93 %), Gumla (47.55%), Dumka (47.28%), W.

Singhbhum (46.25%), Pakur (44.93%), Garhwa (43.20%), Latehar (43.11%), Sahebganj (42.61 %), Giridih (42.38%), Godda (42.04%), Jamtara (41.29%) and Saraikela- Kharsawan (40.38%). Districts having higher percentage of workers are more in agriculture dominated districts. In case of agricultural occupation, even elderly people and teenagers remain involved especially in main agriculture season but in case of industrial and mining dominated areas only certain groups of people of working age group remain involved. In case of rural workers, the percentage of workers in different districts has not recorded much variation.

Marginal workers in respect to the total population are not very significant in this region. The highest percentage has been recorded in Gumla (29.96 %) being followed by Dumka (29.31%), Latehar (26.97%), Jamtara (25.81%), Lohardaga (25.62 %), Giridih (25.49%) etc. The less percentage of marginal worker has been recorded in Bokaro (14.78%), Sahebganj (14.57%), E. Singhbhum (13.41 %), Ranchi (13.27%) and Dhanbad (11.09%). All above mentioned districts where a lesser percentage of marginal workers have been recorded are bigger urban centers where only a few persons prefer to work irregularly but in rural areas a good number of persons work in the main seasons and remain idle during non-agricultural reasons.

When we study the scenario of Main workers and marginal workers together under which Cultivator, Agricultural Laborer, Household workers and other workers have been subdivided. For the agriculture purpose we study the agricultural population (Cultivator + Agri Laborer) separately.

**Cultivator:** This is wise to note the method of calculation for the percentage of Cultivators. At first, we add the number of cultivators from the main workers to the marginal workers and then get the percentage in respect to the number of total workers of the district.

In Jharkhand there are 29.12 percent of cultivators in average. It is 34.82 percent in rural areas and only 2.68 percent

**Table 1.8**

**Percentage distribution of workforce and occupational structure in Jharkhand 2011**

| District | Total Population | Workers (%) | | | Non-Workers | Category of Workers (Main and Marginal) (%) | | | |
|---|---|---|---|---|---|---|---|---|---|
| | | Total Workers | Main Workers | Marginal Workers | | Cultivators | Agri Laborers | Household Workers | Other Workers |
| Bokaro | T-2062330 | 33.23 | 18.44 | 14.79 | 66.77 | 20.07 | 18.44 | 3.16 | 58.33 |
| | R-1078686 | 38.26 | 15.40 | 22.87 | 61.74 | 30.56 | 28.51 | 3.29 | 37.24 |
| | U-983644 | 27.71 | 21.78 | 5.93 | 72.29 | 3.57 | 3.20 | 2.96 | 90.27 |
| Chatra | T-1042886 | 38.13 | 18.75 | 19.39 | 61.87 | 30.00 | 47.40 | 3.57 | 19.03 |
| | R-979932 | 38.78 | 18.48 | 20.30 | 61.22 | 31.34 | 49.39 | 3.48 | 15.92 |
| | U-62954 | 28.04 | 22.93 | 5.11 | 71.96 | 3.85 | 4.65 | 5.62 | 85.87 |
| Deoghar | T-1492073 | 36.96 | 19.91 | 17.04 | 63.04 | 23.33 | 38.02 | 7.14 | 31.51 |
| | R-1233712 | 38.75 | 18.79 | 19.96 | 61.25 | 26.46 | 43.62 | 7.46 | 23.69 |
| | U-258361 | 28.39 | 25.27 | 3.12 | 71.61 | 2.93 | 4.49 | 5.05 | 87.53 |
| Dhanbad | T-2684487 | 31.46 | 20.37 | 11.09 | 68.54 | 11.50 | 12.25 | 3.07 | 73.18 |
| | R-1124093 | 36.58 | 17.36 | 19.22 | 63.42 | 21.87 | 22.80 | 3.30 | 27.71 |
| | U-1560394 | 27.77 | 22.53 | 5.24 | 72.23 | 5.39 | 7.20 | 9.25 | 78.06 |
| Dumka | T-1321442 | 47.28 | 17.97 | 29.31 | 52.72 | 30.96 | 48.51 | 3.70 | 16.84 |
| | R-1231264 | 48.52 | 17.56 | 30.96 | 51.48 | 32.17 | 50.35 | 3.64 | 13.84 |
| | U-90178 | 30.39 | 23.59 | 6.80 | 69.61 | 4.47 | 8.48 | 4.87 | 82.17 |
| Garhwa | T-1322784 | 43.20 | 13.24 | 29.96 | 56.80 | 19.87 | 57.52 | 2.54 | 20.07 |
| | R-1253114 | 49.93 | 12.95 | 30.98 | 50.07 | 20.42 | 58.70 | 2.51 | 10.84 |
| | U-69670 | 30.07 | 18.39 | 11.68 | 69.93 | 5.50 | 26.41 | 3.12 | 64.97 |

| District | Total Popula-tion | Workers (%) | | | Non-Workers | Category of Workers (Main and Marginal) (%) | | | |
|---|---|---|---|---|---|---|---|---|---|
| | | Total Workers | Main Workers | Marginal Workers | | Cultivators | Agri Laborers | Household Workers | Other Worker |
| Giridih | T-2445474 | 42.38 | 16.88 | 25.49 | 57.62 | 35.70 | 34.99 | 3.14 | 26.17 |
| | R-2237450 | 43.65 | 16.31 | 27.33 | 56.35 | 37.58 | 36.79 | 3.07 | 22.55 |
| | U-208024 | 25.72 | 23.02 | 5.70 | 71.28 | 4.83 | 5.60 | 4.28 | 85.30 |
| Godda | T-1313551 | 42.04 | 17.95 | 24.09 | 57.96 | 26.19 | 55.74 | 3.39 | 14.69 |
| | R-1249132 | 42.79 | 17.68 | 25.10 | 57.21 | 26.77 | 57.23 | 3.37 | 12.63 |
| | U-64419 | 27.07 | 23.07 | 4.47 | 72.93 | 8.70 | 10.83 | 4.08 | 78.11 |
| Gumla | T-1025213 | 47.55 | 27.21 | 20.34 | 52.45 | 65.90 | 20.09 | 2.62 | 11.39 |
| | R-960132 | 48.92 | 27.52 | 21.39 | 51.08 | 68.14 | 20.62 | 2.45 | 5.79 |
| | U-65081 | 27.41 | 22.53 | 4.88 | 72.59 | 6.90 | 5.95 | 7.12 | 80.03 |
| Hazaribagh | T-1734495 | 37.74 | 18.87 | 18.87 | 62.26 | 37.04 | 26.33 | 2.47 | 34.16 |
| | R-1459188 | 39.65 | 18.04 | 21.61 | 60.35 | 41.45 | 29.38 | 2.45 | 26.72 |
| | U-275307 | 27.61 | 23.26 | 4.35 | 72.39 | 3.43 | 3.15 | 2.60 | 90.82 |
| Jamtara | T-791042 | 41.29 | 15.48 | 25.81 | 58.71 | 32.45 | 41.54 | 3.54 | 22.47 |
| | R-715296 | 42.42 | 14.69 | 27.73 | 57.58 | 34.69 | 44.27 | 3.58 | 17.46 |
| | U-75746 | 30.66 | 22.93 | 7.73 | 69.34 | 3.30 | 5.84 | 2.94 | 87.95 |
| Koderma | T-716259 | 35.94 | 20.01 | 15.93 | 64.06 | 32.29 | 24.97 | 3.08 | 39.66 |
| | R-575013 | 37.97 | 19.57 | 18.40 | 62.03 | 37.29 | 28.37 | 2.87 | 31.46 |
| | U-141246 | 27.68 | 21.80 | 26.97 | 72.32 | 4.37 | 5.96 | 4.25 | 85.41 |
| Khunti | T-531885 | 48.88 | 28.78 | 20.10 | 51.12 | 63.96 | 19.02 | 3.56 | 13.46 |
| | R-575013 | 50.52 | 29.28 | 21.24 | 49.48 | 26.12 | 19.80 | 3.48 | 09.00 |
| | U-44982 | 31.13 | 23.34 | 7.79 | 68.87 | 14.49 | 5.38 | 4.82 | 75.31 |

| District | Total Population | Workers (%) | | | Non-Workers | Category of Workers (Main and Marginal) (%) | | | |
|---|---|---|---|---|---|---|---|---|---|
| | | Total Workers | Main Workers | Marginal Workers | | Cultivators | Agri Laborers | Household Workers | Other Worker |
| Latehar | T-726978 | 43.11 | 16.14 | 26.97 | 56.89 | 33.09 | 45.55 | 2.94 | 18.43 |
| | R-675120 | 14.47 | 15.76 | 64.27 | 85.53 | 34.57 | 47.24 | 2.89 | 15.30 |
| | U-51858 | 30.21 | 21.08 | 9.13 | 69.79 | 4.93 | 13.30 | 3.86 | 77.90 |
| Lohardaga | T-461790 | 47.93 | 22.30 | 25.62 | 52.07 | 45.08 | 36.52 | 2.01 | 16.38 |
| | R-404379 | 50.52 | 22.12 | 28.41 | 49.48 | 48.40 | 39.06 | 1.89 | 10.65 |
| | U-57411 | 29.67 | 23.64 | 6.03 | 70.33 | 5.25 | 6.09 | 3.44 | 85.22 |
| Pakur | T-900422 | 44.93 | 27.99 | 16.94 | 55.07 | 26.04 | 37.08 | 4.51 | 32.07 |
| | R-832910 | 45.71 | 27.67 | 18.04 | 54.29 | 27.56 | 39.12 | 4.52 | 44.56 |
| | U-67512 | 35.32 | 31.91 | 3.41 | 64.68 | 1.73 | 4.53 | 9.48 | 84.26 |
| Palamu | T-1939869 | 37.76 | 14.62 | 22.14 | 63.24 | 18.73 | 53.42 | 2.79 | 25.06 |
| | R-1713866 | 37.85 | 14.01 | 23.84 | 62.15 | 20.27 | 57.26 | 2.66 | 19.80 |
| | U-226003 | 28.51 | 19.28 | 9.23 | 71.49 | 3.19 | 14.78 | 4.10 | 77.94 |
| W. Singhbhum | T-1502338 | 46.25 | 22.83 | 23.42 | 53.75 | 35.14 | 38.49 | 3.57 | 22.79 |
| | R-1284304 | 48.78 | 22.22 | 26.55 | 51.22 | 38.77 | 42.14 | 3.60 | 05.49 |
| | U-216034 | 31.38 | 26.42 | 4.96 | 68.62 | 1.90 | 5.11 | 3.28 | 89.71 |
| E. Singhbhum | T-2293919 | 36.50 | 23.09 | 13.41 | 63.50 | 13.39 | 25.06 | 2.56 | 58.99 |
| | R-1019328 | 43.84 | 16.92 | 26.92 | 56.16 | 43.39 | 21.63 | 2.83 | 32.15 |
| | U-1274591 | 30.62 | 28.02 | 2.61 | 69.38 | 0.54 | 1.66 | 2.09 | 95.49 |
| Ramgarh | T-949443 | 32.87 | 23.29 | 9.59 | 67.13 | 29.08 | 15.00 | 2.67 | 53.25 |
| | R-530488 | 36.99 | 23.27 | 13.72 | 63.01 | 43.39 | 21.63 | 2.83 | 32.15 |
| | U-418955 | 27.67 | 23.31 | 4.35 | 72.33 | 4.86 | 3.77 | 2.40 | 88.97 |

| District | Total Population | Workers (%) Total Workers | Main Workers | Marginal Workers | Non-Workers | Category of Workers (Main and Marginal) (%) Cultivators | Agri Laborers | Household Workers | Other Worker |
|---|---|---|---|---|---|---|---|---|---|
| Ranchi | T-2914253 | 39.22 | 25.95 | 13.27 | 60.78 | 27.78 | 23.39 | 3.36 | 45.46 |
| | R-1656918 | 45.34 | 25.78 | 19.55 | 34.66 | 41.07 | 33.53 | 3.22 | 22.19 |
| | U-1257335 | 31.15 | 26.16 | 4.99 | 68.85 | 2.29 | 3.95 | 5.67 | 90.11 |
| Sahebganj | T-1150567 | 42.61 | 28.04 | 14.57 | 57.39 | 22.43 | 42.21 | 8.81 | 26.55 |
| | R-990901 | 44.53 | 28.53 | 16.00 | 55.47 | 24.35 | 45.57 | 8.97 | 21.11 |
| | U-159666 | 30.65 | 24.96 | 5.69 | 69.35 | 5.13 | 11.92 | 7.39 | 75.56 |
| Saraikela-Kharsawan | T-1065056 | 40.38 | 21.46 | 18.91 | 59.62 | 27.14 | 33.24 | 2.38 | 37.23 |
| | R-806310 | 43.58 | 19.56 | 24.02 | 56.42 | 32.95 | 40.19 | 2.45 | 24.41 |
| | U-258746 | 30.40 | 27.39 | 3.01 | 69.60 | 1.19 | 2.24 | 7.39 | 94.47 |
| Simdega | T-599578 | 48.14 | 25.94 | 22.20 | 51.86 | 54.75 | 31.26 | 2.38 | 11.22 |
| | R-556634 | 49.45 | 26.05 | 23.40 | 50.55 | 56.73 | 32.02 | 2.45 | 8.45 |
| | U-42994 | 31.18 | 24.41 | 6.78 | 68.82 | 13.89 | 15.59 | 2.57 | 67.95 |
| Jharkhand | T-39988134 | 39.71 | 20.67 | 19.04 | 60.29 | 29.12 | 33.87 | 3.47 | 33.53 |
| | R-25055073 | 43.01 | 19.50 | 23.51 | 56.99 | 34.82 | 40.26 | 3.52 | 21.40 |
| | U-7933061 | 29.26 | 24.35 | 4.91 | 70.74 | 2.68 | 4.17 | 3.27 | 89.89 |

in urban areas. Study of districts depicted that highest concentration of cultivator recorded by Gumla District (65.90 %) followed by Khunti (63.93%), Simdega (54.75%) as first second and the third. The lowest concentration has been recorded by Dhanbad (11.50%), E. Singhbhum (13.39%) and Palamu (18.73%) as first second and third. This shows that cultivators are more in number in area where industrial area has less effect but in the industrial districts their percentage has been reducing gradually.

**Agricultural Laborers:** In Jharkhand average share of agricultural laborers is 33.87% which is greater than the cultivators. Agricultural laborers are those who works on another person's land for wages in money, Kind or share. He has no risk in cultivation, but merely works on another person's land for wages. An agricultural worker has no right of lease or contract on land on which they work. Share of agricultural laborers in rural area is 40.26 percent and 4.17 percent in urban areas. The highest percentage of agricultural laborers has been recorded in Garhwa (57.52 %) followed by Palamu (53.42 %) and Dumka (48.51%) as first second and third. Similarly, lowest number of agricultural laborers has been recorded in the district of Dhanbad (12.25%), Ramgarh (15.00%) and Bokaro (18.44 %) and Khunti (19.02 %) etc.

It is very much clear that only those workers are agricultural laborers who have either small land holdings which are not economically sufficient income bearing or who are landless. Some other workers who are not engaged in any household industry or others are working as agricultural laborers.

**Table 1.9**
**Comparison of workforce in Jharkhand (%)**

| Census Year | Total Workers | Main Workers | Marginal Workers | Non-Workers |
|---|---|---|---|---|
| 1991 | 36.81 | 31.24 | 5.57 | 63.19 |
| 2001 | 37.64 | 24.02 | 13.62 | 62.36 |
| 2011 | 39.70 | 52.05 | 47.94 | 60.29 |

**Table 1.10**
**Comparison of Main Working Forces in Jharkhand (%)**

| Census Year | Cultivators | Agri Laborers | Household Industries | Other Workers |
|---|---|---|---|---|
| 1991 | 36.81 | 20.50 | 6.70 | 9.36 |
| 2001 | 38.51 | 28.26 | 4.15 | 29.00 |
| 2011 | 29.35 | 18.16 | 3.65 | 48.82 |

Above tables are presenting two pictures of different moods through which we can understand the scenario of working force condition in state.

Temporal growth of workforce reveals that during the census of 1991 to 2011 number of total working population has raised from 36.81 % to 39.70%. Number of Main workers has seen a significant growth from 31.24% to 52.05%. Marginal workers have registered their growth by only 5.57% to a total of 47.94%. On the other hand, non-working population have registered a decline in growth from 63.19% to 60.29%. This decrease in the population of Non workers indicates either they have shifted to Main working group or Marginal workers group or have migrated from the state.

Scenario of main workers group shows that the temporal growth of cultivator has decreased from 36.81 % in the census of 1991 to 29.35% in the census of 2011. Similarly, agricultural laborers percentage has been decreased from 20.50 % in 1991 to 18.16 % in 2011.

Number of household workers has decreased during 1991 to 2011 from 6.70% to 3.65%. Whereas the percentage of other workers has increased during 1991 (9.36%) to 2011 (48.82%). The above numbers reveal that cultivators, Agri laborers and household industrial workers are decreasing decade by decade whereas other workers percentage is increasing rapidly during each decade which expresses directly that all workers related to agriculture and household industries are shifting to other works. Other workers mean all government servants, municipal employees, teachers, factory workers, plantation workers, engaged in trade, commerce, business, transport, banking, mining, construction, political workers, social workers, priests, entertainment artists etc. In fact, all those workers other than cultivators, Agri laborers and household workers are classified as other workers. Again, I would like to mention that working population related to agriculture in state is decreasing decade by decade which shows a dull future of Agriculture in the state. Here we do not discuss other workers category in detail as those are not relevant to agriculture.

## Land Utilization

Land use and agriculture being associated with mankind since the beginning of the human civilization have been fundamental assets all over the world. Their importance still persists in developing parts where the agriculture is the chief source of livelihood. (Sinha, 2003, P.58). In this region all the land has not been properly used. During the British period, a proper utilization of land was neither realized nor it was deemed necessary. The Britishers did not take much initiative in this region because this area was poorly accessible. Even the people of this region were not much inclined towards using the land in proper way and developing agriculture because they were partly unaware of agricultural practices and they were easily getting sources of livelihood from forest gathering (Sinha, 2003, P.58).

Land utilization statistics provides the area figures, showing distribution of the total geographical area of the region inti its various uses. The detailed statistics of land utilization in India which mainly gives area of land put to different use is almost continuously available since 1884 (Hussain, 1996, P.187).

There are eleven land use categories in Jharkhand. This is 7 as broader division and eleven with sub-divisions. This classification has been used by directorate of agriculture, Govt. of India, Bihar through figures, Statistics Dept. Patna, Bihar and Jharkhand governments etc.

**Land use Classification**

| Broader Divisions | Sub Divisions |
|---|---|
| 1. Forest | |
| 2. Not Available for Cultivation | i. Land put to non-agricultural use<br>ii. Barren land |
| 3. Other Uncultivated Land Excluding Fallow Land | i. Cultivable waste land<br>ii. Permanent pasture and other grazing land.<br>iii. Land under misc. Trees. |
| 4. Fallow Land | i. Other than current fallow.<br>ii. Current Fallow |
| 5. Net Area Sown | |
| 6. Total Cropped Area | |
| 7. Area Sown More Than Once | |

For Discussion we consider only four as they are directly associated to the agricultural districts regarding land use, these are - Net Area Sown, Cultivable waste land, Fallow land and Current Fallow.

**Net Sown Area:**

Net sown are represents the actual land sown with crops during the particular year. It is a known geographical fact that areas having considerable homogeneous land in the form of flat plateaus, broad valleys or very gentle sloping land have higher percentage of land under cultivation than the areas having comparatively rough terrain, dense forest, sparse population etc. When we study the temporal land use data of Jharkhand from 1956 to 2015, it reveals that it was highest (30.5% or 2432000 Hectare) in 1965 and since then, the percentage in this category is decreasing gradually. Now it's only 17.3% or 1385000 hectares during the period 2014-15. Decreasing of Net Area Sown percentage is not a good sign of agriculture in the state.

District wise scenario of net area sown depicted on highest area in Gumla district followed by Ranchi, W. Singhbhum, Palamu, Simdega, E.Singhbhum etc. and minimum percentage have been recorded in Ramgarh, Dhanbad, Koderma and Chatra.

**Cultivable Waste Land:**

It is also called culturable waste land. Cultivable waste land is more significant than culturable waste. This represents all land available for cultivation, whether not taken for cultivation or abandoned after a few years for one reason or another. In this category there was a 5.8% or 461000 hect. land in Jharkhand during 1956 but now 2014-15 there is only 4.4% or 353000 hect. area has left under this category. This is a good sign that this waste land is getting utilized now.

District wise study shows that still highest area of this category is lying with E. Singhbhum district followed by West Singhbhum, Gumla, Giridih and Dumka etc. Whereas lowest portion is in Ramgarh, Koderma, Chatra and Palamu etc.

**Other Than Current Fallow:**

These include cropped land which are kept fallow for a period of between one and five years. Such lands are caused due to loss of fertility, disputes, lack of irrigation and poverty also. In Jharkhand these were 6.8 %- or 548000-hectare land during 1956 and has been increasing decade by decade. Now 14% or 1122000 hect of land under this category is recorded in Jharkhand during 2015. Increase in this category is a matter of anxiety for Jharkhand. In my opinion this has contributed to the decrease in the net sown area.

District wise study shows that highest Fallow Land- Other than Fallow has been recorded in W. Singhbhum district followed by Ranchi, Gumla, Hazaribagh and Simdega. Whereas lowest areas in this category have been recorded in Sarikela, Godda, Lohardaga and Ramgarh etc.

**Current Fallow:**

Current Fallow land include cropped lands which are kept fallow for a period of up to one year. In other words, these lands remain uncultivated during the year. Current fallows are caused due to unfavorable weather conditions and sometimes disputes or other reasons. In Jharkhand

**Table 1.11**
**Land Utilization in Jharkhand Area (000 Hectares) 1956-2015**

| Area | 1955-56 | 1964-65 | 1989-90 | 1995-96 | 1999-2000 | 2014-15 | Growth in Percent 1964-65 to 1989-90 | Growth in Percent 1989-90 to 1995-96 | Growth in Percent 1995-96 to 2000-01 | Growth in Percent 1999-2000 to 2014-15 |
|---|---|---|---|---|---|---|---|---|---|---|
| Total Geographical Area | 7971 | 7971 | 7971 | 7971 | 7971 | 7971 | - | - | - | - |
| Forest Area | 3186<br>40% | 3019<br>38% | 2314<br>29% | 2332<br>29.2% | 2332<br>29.2% | 2239<br>28.0% | 23.3% | 0.7% | 0.0% | -1.2% |
| Land Put to Non-Agri use | 389<br>4.8% | 464<br>5.8% | 609<br>7.6% | 657<br>8.2% | 683<br>8.6% | 565<br>7% | 31.2% | 7.8% | 3.9% | -1.6% |
| Barren Land | 402<br>5% | 385<br>4.8% | 574<br>7.2% | 574<br>7.2% | 573<br>7.2% | 568<br>7.1% | 49.0% | 0.0% | 0.2% | -0.1% |
| Cultivable Waste Land | 461<br>5.8% | 260<br>3.3% | 328<br>4.1% | 312<br>3.9% | 274<br>3.4% | 353<br>4.4% | 26.2% | -4.8% | -12.2% | +1% |
| Permanent Pasture & other Grazing Land | 148<br>1.8% | 146<br>1.8% | 105<br>1.3% | 93<br>1.1% | 88<br>1.1% | 114<br>1.4% | -28.0% | -11.4% | -5.4% | +0.3% |

| Area | 1955-56 | 1964-65 | 1989-90 | 1995-96 | 1999-2000 | 2014-15 | Growth in Percent 1964-65 to 1989-90 | Growth in Percent 1989-90 to 1995-96 | Growth in Percent 1995-96 to 2000-01 | Growth in Percent 1999-2000 to 2014-15 |
|---|---|---|---|---|---|---|---|---|---|---|
| Land Under Misc. Tree | 136<br>1.7% | 52<br>0.6% | 67<br>0.8% | 106<br>1.3% | 113<br>1.4% | 96<br>1.2% | 28.8% | 58.2% | 6.6% | -0.2% |
| Other Than Current Fallow | 548<br>6.8% | 573<br>7.2% | 857<br>10.7% | 804<br>10.0% | 779<br>9.8% | 1122<br>14% | 49.5% | -6.2% | -3.1% | +4.2% |
| Current Fallow | 694<br>8.7% | 597<br>7.5% | 1046<br>13.1% | 1154<br>14.5% | 1319<br>16.5% | 1396<br>17.3% | 75.2% | 10.3% | 14.3% | +1.2% |
| Net Area Sown | 1870<br>23.5% | 2432<br>30.5% | 2046<br>25.7% | 1830<br>23.0% | 1809<br>22.7% | 1385<br>17.3% | -15.8% | -10.55% | -1.2% | -5.3% |
| Total Cropped Area | 2444<br>30.6% | 2739<br>34.4% | 2183<br>27.4% | 2072<br>25.9% | 2068<br>25.9% | 1554<br>19.4% | -20.2% | -5.1% | -0.2% | -6.5% |
| Area Sown more than once | 373<br>4.6% | 319<br>4.0% | 139<br>1.7% | 241<br>3.0% | 261<br>3.3% | 169<br>2.1% | -56.4% | 73.3% | 8.3% | -1.2% |

**Source:** Bihar Through Figures 1991 and 99 Directorate of Statistics, Bihar, Patna, Chhotanagpur survey, Ranchi 1961 and Jharkhand Ki Rooprekha (Ram Kumar Tiwari)

**Table 1.12**
**District wise Land Use of Jharkhand 2014-15**

| Name of The District | Area | Forest | Area Under Non-Agricultural Use | Barren and Uncultured Land | Net area Sown | Other Uncultivated Land Excluding Fallow Land | Land Under Misc Tree Crops & Groves Not included in Net area Sown | Cultural Waste Land | Fallow Lands Other Than Current Fallow | Current Fallows |
|---|---|---|---|---|---|---|---|---|---|---|
| Bokaro | 288.0 | 74 | 310 | 40 | 34 | 21 | 05 | 11 | 44 | 45 |
| Chatra | 370.6 | 217 | 19 | 10 | 27 | 07 | 01 | 04 | 45 | 57 |
| Deoghar | 247.9 | 19 | 22 | 18 | 39 | 33 | 02 | 20 | 48 | 64 |
| Dhanbad | 208.9 | 19 | 49 | 31 | 13 | 14 | 02 | 11 | 45 | 34 |
| Dumka | 441.0 | 49 | 32 | 26 | 72 | 47 | 06 | 22 | 51 | 101 |
| East SinghBhum | 353.3 | 125 | 139 | 47 | 78 | 48 | 05 | 40 | 37 | 83 |
| Garhwa | 404.4 | 191 | 24 | 27 | 59 | 16 | 03 | 10 | 52 | 60 |
| Giridih | 491.9 | 159 | 33 | 38 | 61 | 44 | 10 | 22 | 57 | 102 |
| Godda | 211.0 | 24 | 21 | 18 | 57 | 24 | 09 | 09 | 22 | 67 |
| Gumla | 532.1 | 135 | 24 | 37 | 124 | 36 | 03 | 31 | 71 | 107 |
| Hazaribagh | 457.8 | 204 | 24 | 20 | 48 | 16 | 04 | 06 | 59 | 62 |
| Jamtara | 180.2 | 14 | 22 | 08 | 42 | 31 | 01 | 12 | 41 | 24 |
| Koderma | 149.4 | 59 | 09 | 10 | 19 | 06 | 01 | 03 | 27 | 26 |
| Latehar | 362.2 | 159 | 13 | 25 | 49 | 18 | 03 | 13 | 40 | 44 |
| Lohardaga | 149.1 | 44 | 11 | 10 | 36 | 09 | 01 | 07 | 22 | 21 |

| Name of The District | Area | Forest | Area Under Non-Agricultural Use | Barren and Uncultured Land | Net area Sown | Other Uncultivated Land Excluding Fallow Land | Land Under Misc Tree Crops & Groves Not included in Net area Sown | Cultural Waste Land | Fallow Lands Other Than Current Fallow | Current Fallows |
|---|---|---|---|---|---|---|---|---|---|---|
| Pakur | 180.0 | 15 | 17 | 11 | 34 | 02 | 04 | 11 | 39 | 43 |
| Palamu | 506.3 | 170 | 31 | 27 | 96 | 12 | 03 | 06 | 44 | 79 |
| Ranchi | 508.7 | 100 | 41 | 29 | 120 | 25 | 04 | 19 | 72 | 111 |
| Sahebganj | 159.9 | 28 | 21 | 10 | 47 | 16 | 03 | 08 | 37 | 40 |
| Saraikela-Kharsawan | 272.5 | 65 | 22 | 18 | 69 | 18 | 03 | 11 | 15 | 30 |
| Simdega | 375.6 | 104 | 21 | 25 | 85 | 38 | 16 | 18 | 56 | 50 |
| West Singhbhum | 718.2 | 142 | 45 | 55 | 111 | 46 | 05 | 38 | 78 | 88 |
| Khunti | 261.1 | 46 | 18 | 20 | 63 | 15 | 0.3 | 13 | 50 | 51 |
| Ramgarh | 138.7 | 42 | 18 | 13 | 09 | 05 | 0.9 | 02 | 23 | 29 |
| Jharkhand | 7971.4 | 2239 | 706 | 568 | 1395 | 565 | 96 | 353 | 1122 | 1396 |

**Source:** Jharkhand Ka Bhoogol - Ram Kumar Tiwari 2019 P. 190-191

current fallow land was only 8.7 percent or 694000 hectares during 1956 but it raised to 17.3 percent or 1386 thousand hectares in the year of 2015. It has shown a gradual raising scenario in the state. In my opinion, this is the reason that the net sown area of Jharkhand has been decreasing in the state.

Share of current fallow land is highest in the district of Ranchi followed by Gumla, Giridih and Dumka etc. whereas, lowest share has been recorded in Lohardaga district followed by Jamtara, Koderma and Ramgarh etc.

Again, I would like to produce here a comparative land use pattern for a short period of Jharkhand studied by Dr. VNP Sinha in his book "Jharkhand: Land and People" (2003, P.68)

**Table 1.13**
**Comparative Land Use pattern: Jharkhand**

| Classification of Land | 1990-91 (000 hect) | 2008-09 (000 hect) |
|---|---|---|
| 1. Net Sown Area | 1996 (25.07%) | 1504 (18.8%) |
| 2. Forest | 2332 (29.29%) | 2339(28.09%) |
| 3. Unculturable Waste Land | 573(7.20%) | 336(4.22%) |
| 4. Non Agri Use | 658(8.27%) | 1332(16.71%) |
| 5. Land Under misc tree, crops and groves | 87(1.09%) | 93(1.13%) |
| 6. Permanent Pasture & other grazing land | 100(1.26%) | 110(1.38%) |
| 7. Culturable Waste Land | 311(3.90%) | 336(4.22%) |
| 8. Fallow Land | 1904(23.92%) | 2356(29.56%) |
| Total | 7961 (100%) | 7970(100%) |

**Source:** Statistical office, Bihar 1990-91. Land Use Statistics, Ministry of Agriculture GOI 2008-09

## Transport

Jharkhand has Roads, Railways and Airways and waterways as means of transport.

### Roads:

Spanning a total distance of 2391 kms the state has 12 National Highways (NH). The length of state highways (including the urban roads and MDR) is around 6877 kms. Rural roads and minor district roads are 24300 kms (approx.).

Road density is measured as the length of roads per 1000 sq. kms. The present road density (Excluding NH and rural roads) in the state is 86.26 Km which is almost half of the national figure of 182.4 Kms, which amounts to a gap of 96.14 kms. In order to fill this gap around 7660 Kms of roads would have to be included in Road Construction Department (RCD) and would be required to be brought to the level of state highway Major District Road. This would require around Rs 19150 crores (@ 2.5 crores per km) if it is phased to 5 years, around 1530 Kms of roads have to be upgraded each year requiring an additional budget of Rs. 3825 crores each year.

### Railways:

The total railway route length in Jharkhand is 1984 Kms (2010-11). The density of railway line is 26.8 Km per 1000 sq kms. Ranchi, Dhanbad, Jamshedpur, Bokaro, Daltonganj, Barkakana are the major railway stations in Jharkhand. Most developed network of railways lie in lower Damodar basin. The Subarnarekha Basin also provides dense pattern of railways. This region is also on the main rail route of Howrah -Mumbai. This rail route touches several mineral exploiting centers as well as Jamshedpur Industrial complex. The north Koel Basin covering the areas of two districts has rail linkage. The area of upper and lower Hazaribagh plateau is also poorly accessible in terms of railways. The vast area of South-Western portions covering the pat region of Gumla- Simdega plateau lacks rail facilities. Similarly, the north eastern portion of the region consisting of the areas of Santhal Pargana has rail facilities mainly in peripheral part and the vast interior areas having hilly landscape with sparse population and a few isolated mines are devoid of rail connection. This portion is transverse on the north-east by the loop line and south- west by the chord line of eastern railways. A short branch connects (12 KMs) Rajmahal with Tinpahar on the loop line. Another branch also runs from Madhupur on the main line to Giridih, a distance of 36 km. Again, there is a branch line connecting Jasidih to Deoghar. From Sahibganj a short line runs to Sakrigali Ghat from where ferry service is available for crossing the Ganga.

**Airport:**

Jharkhand has domestic airport located in the state capital Ranchi. From here direct links are available to prominent cities of India like Delhi, Mumbai, Kolkata, Patna. Chartered flights are also available between the steel city of Jamshedpur-Kolkata. Jamshedpur, Dumka, Bokaro, Giridih, Deoghar Hazaribagh, Daltonganj and Noamundi also have airstrips. The department of civil aviation of Jharkhand plans to acquire land to develop 12 airstrips in Chatra, Simdega, Lohardaga, Godda, Pakur, Latehar, Sahebganj, Jamtara, Saraikela, Garhwa, Kodarma and Khunti district.

**Waterways:**

This is recently developed in Jharkhand and is known as Multi model terminal on National Waterways-I. This is located at south bank of Ganga in Sahebganj. This will provide an open gateway to the trading routes of Bay of Bengal and other seas to the land locked state of Jharkhand. This is a connected by road with NH-80 and Sankrigali railway station as rail route. Through this ferry services can be provided to 6 cities i.e., Patna, Varanasi, Kolkata, Haldia, Bhagalpur, Munger. It also provides river cruise services, cargo transport and tourist facilities.

## Irrigation

Uneven and irregular distribution of rainfall during agricultural seasons make the necessity of irrigation. These days production of various crops during different agricultural harvests has further accelerated the requirements of irrigation. It may be noted that previously people used to grow a few crops depending upon the natural precipitation and seasons lacking rainfall used to be without farming activities. Presently the production of crops even during seasons having negligible rain has compelled farmers to develop irrigation means. Virtually production of Rabi and Garma crops and a lot of vegetables to a considerable amount depends upon irrigation and these crops have gained significance due to changing food habits and consumption. Moreover, the dominance of undulating terrains in the state hardly allow rainfall to stagnate and drain off quickly. The region also lacks perennial rivers. Most of the rivers and rivulets are rainfed. Even when rivers and rivulets containing water hardly help in irrigating the fields away from the river course because of higher terraced fields on both sides.

Based on above conditions use of irrigation water becomes necessary wherever agricultural activities have intensified. Surface and underground water have enabled cultivators to adapt different methods of irrigation. Because of rough terrain natural water bodies are negligible. Certainly, the region provides conditions for the development of artificial reservoirs or tanks etc. It may be noted that people block the flow of the rivulet or river valley and store rain water for irrigation. In homogeneous areas Nala, Ahar, Pyne or smaller water bodies also provide the facility of surface water. In case of underground water, the region has some limitations. Because of rocky structure the development of tube well becomes difficult. Even wells dry up during summer season and hardly facilitate irrigation on large size of land. A few rivers have also enabled the facility of canal irrigation.

In the absence of sufficient rainfall, the agriculture of the Jharkhand state is primarily depended on irrigation. Since, this state is water starved and deficient state, so the managing and utilizing of available water resources is the prime concern of the state. As per the water resource department, the availability of water resource in the state is only 287810 lakh m cube out of which 237890 lakh m cube is from surface water and remaining 49920 lakh m cube is from ground water. The total utilization of surface and ground water for irrigation purposes in the state is so far only 47360 lakh m cube of which 39640 lakh m cube is from surface water and 7720 lakh m cube is from ground water.

**Table 1.14**
**Sources of Irrigation in Jharkhand (2012-13)**

| Sources | % Irrigation |
|---|---|
| Canal | 17.53 |
| Pond | 19.07 |
| Tubewell | 8.25 |
| Well | 29.38 |
| Others | 25.77 |

The Jharkhand state has a total agricultural land of 29.74 lakh ha. Against the total cultivable area of 29.74 lakh hectare, irrigation potential of 24.25 lakh hectare has been identified so far. About 1276 thousand hectares irrigation potential has to be created through major and medium irrigation projects and 1184 thousand hectares irrigation potential has to be provided by minor irrigation schemes and other sources.

**Table 1.15**
**Irrigated Area in Jharkhand (000 Hectare)**

| Year | Total Irrigated Area | Net Sown Area | Percentage of Irrigated Area to Net Sown Area |
|---|---|---|---|
| 1955-56 | 187.0 | 1870 | 10.00 |
| 1964-65 | 223.7 | 2432 | 9.20 |
| 1985-86 | 138.0 | 2071 | 7.00 |
| 1986-87 | 147.0 | 2029 | 7.24 |
| 1988-89 | 124.0 | 1985 | 6.25 |
| 1989-90 | 152.0 | 2046 | 6.7 |
| 1995-96 | 177.4 | 1830 | 9.70 |
| 1997-98 | 157.0 | 1808 | 11.30 |
| 1999-2000 | 204.3 | 1808 | 11.30 |

| 2014-15 | 207 | 1385 | 14.95 |
|---|---|---|---|

**Source:** Directorate of Statistics, GOI, Bihar ka Bhougolik Swarup (BP Rao), Wasteland management in Bihar and Chhotanagpur Survey

**Table 1.16**
**District wise Irrigated area by various means in Jharkhand 2015-16 (Hectares)**

| Districts | Canal | Pond | Wells Bore-well | Wells Other Wells | Wells Total Wells | Other Resources | Total Irrigated Area |
|---|---|---|---|---|---|---|---|
| Bokaro | - | 729 | 329 | 1785 | 2114 | 330 | 3173 |
| Chatra | 676 | 1288 | 777 | 2512 | 3289 | 1904 | 7157 |
| Deoghar | - | 2932 | 3140 | 1780 | 4920 | 1142 | 8994 |
| Dhanbad | - | 265 | 23 | 253 | 276 | 242 | 783 |
| Dumka | 221 | 1596 | 581 | 1627 | 2208 | 2540 | 6565 |
| E.Singhbhum | 945 | 1034 | 463 | 342 | 805 | 74 | 2858 |
| Garhwa | - | 2175 | 3886 | 1836 | 5722 | 4161 | 12058 |
| Giridih | - | 5695 | 727 | 19307 | 20034 | 7689 | 33418 |
| Godda | - | 1278 | 7868 | 4459 | 12327 | 1235 | 14840 |
| Gumla | 284 | 2226 | - | 2968 | 2968 | 3394 | 8872 |
| Hazaribagh | 976 | 2451 | 2574 | 2589 | 5163 | 2951 | 11541 |
| Jamtara | - | 1052 | 29 | 1769 | 1798 | 904 | 3754 |
| Khunti | 671 | 4645 | 3291 | 4742 | 8033 | 6023 | 19372 |
| Koderma | - | 622 | 1324 | 808 | 2132 | 249 | 3003 |
| Latehar | 200 | 2345 | 3179 | 3028 | 6207 | 1933 | 10685 |
| Lohardaga | 672 | 2411 | 783 | 3549 | 4332 | 4468 | 11883 |
| Pakur | - | 1079 | 1155 | 1690 | 2845 | 2301 | 6225 |
| Palamu | - | 6521 | 8028 | 6005 | 14033 | 14824 | 35378 |
| Ramgarh | - | 1141 | - | 1581 | 1581 | 304 | 3026 |
| Ranchi | - | 1682 | 2748 | 8285 | 11033 | 2904 | 15619 |
| Sahebganj | - | 1065 | 1703 | 427 | 2130 | 415 | 3610 |
| Saraikela-Kharsawan | 956 | 2263 | - | 1229 | 1229 | 2063 | 6511 |
| Simdega | 521 | 246 | 402 | 1058 | 1460 | 1517 | 3744 |
| W.Singhbhum | 497 | 499 | - | 602 | 602 | 461 | 2059 |
| Jharkhand | 6619 | 47240 | 43010 | 74231 | 117241 | 64028 | 235128 |
| Total Cropped Area is 1554000 Hectares | 0.4% | 3.0% | 2.7% | 4.7% | 7.54% | 4.1% | 15.13% |

**Source:** Directorate of Economics and statistics, GOI

The analysis of above table reveals source wise net irrigated area in the year 2015-16. In terms of canal irrigation six districts namely Deoghar, Sahibganj, Pakur, Koderma, Giridih and Bokaro are devoid of irrigation by canal. Even the districts of Godda and Dumka have less than 1% irrigated are by canal. Districts having net canal irrigated area less than 10% and more than 1% consisted of Hazaribagh, Chatra, Dhanbad, Gumla and Purbi Singhbhum. The remaining districts like Ranchi, Lohardaga, Paschimi Singhbhum, Palamu and Garhwa consisted of net irrigated area by canal more than 10%.

Tank irrigation is relatively better in position in this state. It irrigates 15.75% of the net irrigated area. The highest percentage of tank irrigated area has been in the district of Pakur (55.30%) being followed by Godda (48.40%), Deoghar (36.63%), Lohardaga (35.26%), Purbi Singhbhum (29.27%), Dumka (23.59%), Bokaro (17.7%), Chatra (17.52%) etc. Districts having tank irrigated area less than the average of the state included Garhwa, Pashimi Singhbhum, Gumla, Ranchi, Giridih, Kodarma etc.

It has been mentioned earlier that the structure of the area is not suitable for development of tube well. It is however developed in some districts. Districts having tube well irrigation even less than 1% included Dumka, Godda, Deoghar, Dhanbad, Lohardaga and Gumla. The state of Jharkhand has less than 13.40% tube well irrigated area. The districts having more than 30% irrigated area by tube well included Bokaro (40.67%), Giridih (40.13%) and Koderma (34.87%). Districts having tube well irrigation area between the regional average (13.40%) and 30% included Chatra (26.91%), Ranchi (21.22%) and Hazaribagh (21.20). Only a few districts consisted of tube well irrigated area more than 1% and less than the regional average. These districts comprised of Palamu, Garhwa, W. Singhbhum, E. Singhbhum, Sahibganj etc.

Well irrigation is comparatively better developed and irrigates 31.39% of net irrigated area. All districts have some areas under well irrigation. The highest percentage has been in the district of Lohardaga (45.20), Dhanbad (43.34), Godda (41.70), Bokaro (40.53), Ranchi (35.99), Hazaribagh (34.80) etc. Districts having well irrigated area less than the regional average included Garhwa, W. Singhbhum and E. Singhbhum, Sahibganj, Pakur etc. Even in terms of other sources of irrigation the highest percentage has been recorded in the district of Sahibganj (80.38%). This district is being followed by E. Singhbhum (57.22%), Gumla (46.65%), Dhanbad (39.26%), Deoghar (31.55%), W. Singhbhum (30.23%) etc. Other sources of irrigation irrigating less area included Bokaro, Lohardaga, Palamu, Giridih, Chatra, Hazaribagh etc.

The percentage of Net Irrigated Area to Total Net sown Area remained less than 10 percent in the districts of Dumka, Deoghar, Pakur, Bokaro, Dhanbad, Lohardaga, Gumla, E.Singhbhum and W. Singhbhum. Districts having moderate Net Sown irrigated Area included Garhwa, Palamu, Koderma, Chatra, Sahibganj, Ranchi etc.

As regards source wise gross irrigated area in the year 2012-13 remarkable spatial variation has been noted. The above table reveals gross irrigated area by different sources of irrigation in the percentage. The state of Jharkhand has 29.38% well irrigated area and 25.77% gross irrigated area by other sources. Canal, Pond and Tube Well irrigated area consists of 17.53%, 19.07% and 8.25% respectively. Canal gross irrigated area is nonexistent in the districts of Deoghar, Sahibganj, Pakur, Koderma and Bokaro. The percentage of gross irrigated area by tank less than 5% included Dumka, Godda, Gumla and Dhanbad. Districts having canal irrigated area between 5% and 14.02% (regional average) consisted of Hazaribagh and East Singhbhum. The remaining districts namely W. Singhbhum, Palamu, Garhwa, Lohardaga, Ranchi and Chatra consisted of higher percentage more than the regional average.

Tank irrigated area remained important in more districts. The highest percentage has been noted in the districts of Pakur (58.76%) being followed by Godda (57.99), Lohardaga

(34.68), Deoghar (32.94), Dumka(23.73) East Singhbhum (25.71) Chatra (22.87) etc. Districts having low percentage included Giridih, Ranchi, Gumla, W. Singhbhum, Garhwa etc. Tube well Gross Irrigated area remained non-existent in the districts of Dumka, Deoghar, Lohardaga and Gumla. This irrigation remained less important in the districts of Godda, Pakur, Sahibganj, E. Singhbhum, W. Singhbhum, Dhanbad etc. Only a few districts consisting of Hazaribagh, Kodarma, Chatra, Giridih, Ranchi recorded a better position.

Grossa Irrigated area by well remained almost prevalent in all districts. The state of Jharkhand consisted of 35.38% gross irrigated area by well. The highest percentage has been in the districts of Dumka (57.40) being followed by Giridih (55.51), Gumla (48.46), Bokaro (46.44), Dhanbad (45.66), Ranchi (44.59), Lohardaga (43.91), Hazaribagh (39.71), Godda (38.69) etc. The remaining districts consisted of gross irrigated area by well less than the regional average (35.38%). These districts included Palamu, Garhwa, E. Singhbhum, Sahibganj, Pakur, Chatra etc.

In case of other sources of irrigation all districts of the state recorded some areas. This state consisted of 24.08% gross irrigated area by other sources. The highest percentage has been in the district of Sahibganj (82.51) being followed by E. Singhbhum (59.60), Gumla (48.11), Dhanbad (38.48), Garhwa (36.05) etc. Districts having lower percentage of gross irrigated area by other sources included Godda, Hazaribagh, Giridih, Bokaro, Lohardaga etc.

From the point of view of the percentage of irrigated area to total cropped land it is only 9.32% in the state of Jharkhand. Certainly, the distribution is uneven. The highest percentage has been in the district of Garhwa (27%). It is followed by Palamu (23.34%), Hazaribagh (16.69%), Chatra (15.93%), Godda (13.96%), Koderma (12.53%) etc. Districts having moderate irrigated area included Dumka, Giridih, Ranchi, Bokaro, Pakur etc. Districts having lower percentage of irrigated land included East and West Singhbhum, Pakur, Gumla etc.

## References

Ahmad, E. (1951); Geomorphic outline of Chotanagpur, Geographical Outlook, Vol-2, Department of Geography, Ranchi University, Ranchi.

Arther L. Bloom (1978); Geomorphology: A Systematic analysis of late Cenozoic landforms. Prentice Hall of India Pvt. Ltd. New Delhi-1.

Buchanan, F. (1922); Quoted by Macpherson, H. Survey and settlement report of Santhal Pargana, Patna.

Chatterjee, S.C. (1949); The Physiographic Evolution of Chotanagpur, Calcutta University, Dept. of Geography.

Dayal, P. (1953); Bihar in Maps, Kushum Prakashan, Patna.

District Gazetteers (1938); D.G. Santhal Pargana, Govt Press.

Dunn, J.A. (1941); Economic Geology and Mineral Resources of Bihar Memoirs, G.S.I. Volume-78, Calcutta.

Ghosh, B.N. (1987); Fundamentals of Population Geography, Sterling Publishers Pvt. Ltd. New Delhi/Bangalore.

Hussain, Mazid (1996); Systematic Agriculture Geography, Rawat Publications, Jaipur/New Delhi.

Kendrew, W.G. (1961); The Climates of the continents, Clarendon Press, Oxford.

Klages, K.H.W. (1958); Ecological Crop Geography, Macmillan Co. New York.

Pearson, F.A. and Harper, F.A. (1945); The world's Hunger. New York, Cornel University Press, Ithaca.

Prasad, A. (1973); Chotanagpur: Geography of rural Settlements, Ranchi University, Ranchi.

Singh, K.N. & R.A. Siddiqui (2012); Economic Geography, Prayag Pustak Bhawan, Allahabad.

Singh, Jasbir (1974); An Agricultural Atlas of India: A Geographical Analysis. Kurukshetra, Vishal Publications.

Singh, Jasbir (1976); An Agricultural Geography of Haryana, Vishal Publications, University Campus, Kurukshetra (Haryana).

Singh, R.K. (2003); Socio economic Dynamics of Rice Production system in Jharkhand State. (ICAR) Central Rainfed Upland Rice Research Station (CRRI), Hazaribagh, Jharkhand.

Singh, R.P. (1970); Monograph of Bihar, Bharti Bhawan, Patna.

Singh, T.N.B & Purty, S.S. (2016); Ground water year book: Jharkhand, Central Ground Water Board, Patna/Ranchi.

Sinha, V.N.P. (2003); Jharkhand: Land and People, Rajesh Publications, New Delhi.

Tiwari, Ram Kumar (2019); Jharkhand Ka Bhoogol, Bihar Hindi Granth Academy, Patna

Tiwari, Ram Kumar (2015); Jansankhya Bhoogol, Prawalika Publications, Allahabad.

Wadia, D.N. (1953); Geology of India, Macmillan, London

Wadia, D.N. and Others (1935); Records, G.S.I. Vol. LCVIII, Calcutta.

# 2

## Land Tenure and Size of Holding

- Size of Land Holdings
- Bases to determine the farm size
- Types of holdings
- Pattern of farm holding in India and Jharkhand
- Size of Operational Holdings
  - Fragmentation of Land
  - Effect of Fragmentation
- A comparative study of land holding in Jharkhand between two censuses (2010-11 and 2015-16)
- Holdings in different groups
  - Marginal Holdings
  - Small Holdings
  - Semi-Medium Holdings
  - Medium Holdings
  - Large Holdings
- Holdings by size in districts
- Holding Density
- Size of Holdings

It is a geographical fact that the physical determinants which we have discussed in Chapter-1 influence the agricultural pattern of any region through out the world significantly. There are numerous sociocultural, economic, political, technological and infrastructural factors which also determine the agricultural processes. Of these factors, Land Tenancy, system of ownership, size of holdings, availability of labor and capital, religion, level of technological development, accessibility to the market, irrigation facilities, agricultural research and extension service, price incentives, government plans and international policies have a close impact on agricultural activities. In this chapter among all the factors only two factors will be studied in detail - 1. Land Tenancy System in Jharkhand and 2. Size of holdings in Jharkhand. (Husain, 1996, P.122)

## Land Tenancy

Productivity in agriculture is mainly dependent on two sets of factors, they are technological and institutional. Among the technological factors are the uses of agricultural inputs and methods such as improved seeds, fertilizers, improved ploughs, tractors, harvesters, irrigations etc. All these factors help to raise productivity, even if no land reforms are introduced. On the other hand, the institutional reforms include the redistribution of land ownership in favor of the cultivating classes so as to provide them a sense of participation in rural life, improving the size of farms, providing security of tenure, regulation of rent etc. Also, these institutional factors, such as the existence of feudal relations, small size of farms, sub-division and fragmentation, insecurity of tenancy rights, high rents, etc. help the peasantry to raise production.

When the socialist school of thought emerged, they believe that the existence of feudal or semi-feudal relations was the real cause of backwardness and poverty in the Indian rural communities. The other school of thought that emerged believes that agricultural productivity is purely a technological phenomenon and can be raised by the application of superior agricultural methods. Thus, whereas the key to higher productivity lies in technological change according to one school, it lies in institutional reform according to the other. Quite recently, both the schools of thought are converging and opinion has come to the idea that land reforms and technological change are not mutually exclusive factors but are complementary in the process of agricultural development.

The purpose of land reform and thus the tenancy reform is twofold. On one hand, it aims to make more rational use of the scarce land resources by affecting conditions of holdings, imposing ceilings and floors on holdings so that cultivation can be done in the most economical manner, i.e., without any waste of labor and capital. On the other hand, it is a means of redistributing agricultural land in the favor of less privileged classes, and of improving the terms and conditions on which the land is held for cultivation by the actual tillers, with a view to end exploitation.

## Scope of Land Reforms:

Land reforms in general and tenancy reforms in particular aim at redistributing ownership holding from the view point of social justice, and reorganizing operational holdings from the viewpoint of optimum utilization of land. Apart from this, the problem of tenancy i.e., the rights and conditions of holding land also come under the scope of land reforms.

The entire concept aims at the abolition of intermediaries and bringing the actual cultivator in direct contact with the state. The provisions of security of tenancy and rent regulation provide a congenial atmosphere in which the agriculturist feels sure of reaping the fruits of his labor. The scope of land reforms therefore entails abolition of intermediaries, tenancy reforms, i.e., regulation of rent, security of tenure for tenants and conferment of ownership on them. It also focuses on land ceiling and land holdings, agrarian reorganization including consolidation of holdings and preventing of sub-division and fragmentation and organization of cooperative farms.

## Land Tenure:

At the time of independence, there existed three types of proprietary land tenures in the country. The term land tenure is used to refer to the terms and conditions on which land is held and used.

1. **The Zamindari or Landlord Tenure:** Lord Cornwallis is considered to be the father of Zamindari system in India.

He introduced this system for the first time in 1793 in West Bengal and was later adopted in other states as well. Under this system, the land was held by a person who was responsible for the payment of land revenue. They could acquire the land mostly free of charge from the government during the British rule and it is called estate. Landlords never cultivated the land they owned and rented them out to the cultivators. The amount of land revenue may either be fixed once one for all when it was called permanent settlement, or settlement with regard to land revenue may only be temporary and may, therefore, be revised after every 30-40 years, as the practice may be. The Zamindari system is known as absentee landlordism. In this system between the actual state and the tiller there grew an intermediary who was interested in the land only to the extent of extraction of exorbitant rent. The Zamindari tenure covered about 57 percent of the area of the country.

2. **Ryotwari System:** It took birth in 1792 in Madras at the hands of Caption Read and Thomas Menro and was later extended to other states. Under this system, the responsibility of paying land revenue to the Government was of the cultivator himself and there was no intermediary between him and the state. The ryot had full right regarding sale, transfer and leasing of land and could not be evicted from the land as long as he pays the land revenue. But the settlement of land revenue under ryotwari system was done on a temporary basis and are periodic, after 20, 30 or 40 years.

3. **The Mahalwari or The Joint Village Tenure:** This system was introduced by William Bentinck in Agra and Oudh and was later extended to Madhya Pradesh and Punjab. Under this system, the village communities held the village lands commonly and it was joint responsibility of these communities to make payments of the land revenue.

Thus, the overall system of collection of revenue was based on exploitation. The British government snatched away whatever surplus above the minimum subsistence the cultivator produced. The latter were forced to lead a wretched life of slavery and deprivation. Under the above-mentioned systems the practice of cultivation by tenants became widely prevalent. These tenants were also exploited in a number of ways. It was basically to stop the exploitation of the actual tillers of the soil and pass on the ownership of the land to them that land reforms introduced in the post-independence period in India.

## Tenancy reforms:

The land reforms refer to the reforming of defective structure of the land holdings and are a planned and institutional organization of the relation between man and land. The agriculture system that existed at the time of independence consisted of several defects including that of tenancy i.e., the insecurity of tenure and high rents charged by the landlord. The reforms aimed to eliminate all forms of exploitation and social injustice within the agrarian system, to provide security for the tiller of the soil and to remove such impediments to increase in agricultural production as arise from the agrarian structure inherited from the past. One of the major aspects of the land reforms in India has been the tenancy reform.

Under the Zamindari and ryotwari systems, tenancy cultivation had been quite common in India. Tenancy cultivation may be done by small proprietors who find that they have an insufficient quantity of land or it may be carried on by landless laborer. Sometimes, the tenants holding land from an intermediary may sub-let it for cultivation. Broadly speaking tenants are divided into three categories:

1. Occupancy tenants-enjoyed permanent and heritable rights on land. They had security of tenure and could claim compensation from the landlords for any improvement affected on the land.
2. Tenants at will-did not have security of tenure and could be evicted from the land whenever landlord so desired. There is no security of tenure for them and they are also made to pay exorbitant rent to the landlords.
3. The Sub tenants-were appointed by the occupancy tenants.

The rights of tenancy are permanent and heritable. The tenants can also receive some compensation from the landlords in case they make some improvements on the land. The tenants enjoy a fixity and security of tenure, which makes them the virtual owners of the land. It can be said that the only difference between the occupancy tenant and the peasant proprietor is that the former is required to pay rent to the landlord and the latter to pay the land revenue to the state. So, for all practical purposes, occupancy tenants are treated as land owners. The position of tenants at will and that of sub-tenants is extremely weak and mostly they are subjected to ruthless exploitation. Frequent enhancement of rent, eviction of minor pretexts of several kinds, extractions and beggar are some of the popular ways of exploitation. In a country like India where the demand of land is more than its supply on account of its growing population, exploitation of weak and unprotected is a widespread evil. Fifty percent of the produce was the normal rent under Batai or sharecropping. On several occasions, the peasant had to forgo even two thirds of the produce as rent.

The tenancy cultivation suffers from three main defects; they are insecurity of tenure, rack-renting and lack of incentives of the actual cultivator. The National Sample Survey (8th round) had estimated that in 1953-54 about 90 percent of agriculture land was under tenancy system. The percentage of area leased out varied from 11 to 26 percent, though the all India average was 20 percent. It showed that about one-fifth of the total area was held under tenancy and thus it was not possible to ignore a problem affecting such a wide area. According to the 1961 census, 77 percent of the total-cultivating households were in the nature of ownership holdings, 8 percent of pure tenancy and 15 percent in mixed tenancy.

Besides this open tenancy, there is a considerable amount of land leased out on the basis of oral or hidden tenancy that accounts for anything between 35-40 percent of the total cultivated area. The informal or the oral tenancy has been a common feature of traditional agricultural societies.

Although attempts have been made to provide security of tenure, redistribution of land and fixation of fair rents, yet informal or oral tenancy has continued to exist even to this day. The term informal tenancy is referred to as oral tenancy which refers to tenancy without legal sanctions and permissions, or without any written agreement. The principal of shifting to informal tenancy is to extract higher land rents from the tenants. This is primarily done so as to get high yielding varieties program that has brought a realization among the landlords that land is a very valuable asset and promises high rates of return. India, which is marked by land hunger, it is possible here to take advantage of the situation by charging higher rents. Also, informal tenancy arrangements are a convenient device with the landlords for nullifying tenancy reforms. Thus, unrecorded or clandestine tenancy perpetuates a semi-feudal land system that was sought to be abolished by measures of land reforms.

## Measures of Tenancy Reforms:

The legislation for abolition of intermediaries was aimed at providing land to the tiller but it did not put an end to the problem of tenancy. Moreover, even with the limit of ceiling, it may not be possible for a family to cultivate the entire land and so some sub-letting is unavoidable. Besides, in order to induce agricultural population to take over non-agricultural occupations, some sub-letting to tenants may be allowed. A total ban on letting or sub-letting land would neither be socially desirable nor administratively practicable. That is why measures were taken up so as to minimize the evils of tenancy cultivation and to safeguard the interest of the tenants.

Regulation of rents: during the pre-independence period, rents were fixed either by the custom or were the result of the market forces of demand and supply. Supply of land being fixed, the demand of land rowing with an increasing population, there has been a continuous tendency for rents to rise. The decay of handicrafts increased the dependence of land further and thus pushed up the rents. Rack-renting was a common feature of the Indian agrarian structure.

It was, therefore, imperative that rents should be fixed by enacting legislation. The rates of rent prevalent were one half of the produce or more. Considering the return on investment in other sectors of economy, these rents were excessive by any standard of social justice.

Consequently, the First and the Second five-year plan recommended that rents should not exceed one fourth or one fifth of the gross produce. Various states have passed necessary legislation regulating rents, but there are large variations in the rents fixed in different states. In Gujarat, Maharashtra, and Rajasthan, one-sixth of gross produce is fixed as maximum rent. In Assam, Karnataka, Manipur and Tripura, maximum rents vary between one-fourth to one-fifth of the gross produce. In Punjab, one-third of produce has been considered as fair rent, while in Tamil Nadu it is between 33.3 and 40 percent of the gross produce. In Andhra Pradesh one-fourth of the gross produce for irrigated land and one-fifth in other cases has been fixed as rent.

Owing to the weak position of the tenants and the prevalence of the widespread land hunger, the law regulating rent is observed more in its breach then in its compliance. Another suggestion in this regard is to fix rents in cash rather than in kind. Historically, rents have been paid in kind in India but in view of the fact that the peasants have to make a good many payments in money, while purchasing seeds, fertilizers, implements and other necessaries of life, it would be desirable to switch over to cash payment of rents. This is in fitness with the requirements of a rural economy changing rapidly from barter to money economy.

Security of Tenure: the security of tenure is aimed to provide some incentives to tenants to make certain improvements of permanent nature on the land they cultivate. Many states have, therefore, enacted legislation providing for the security of tenure so long as they continue paying the rent. In some states like Maharashtra, Gujarat etc., the landlord for his personal cultivation has granted tenants security of tenure in respect of areas not resumable. In some states, the minimum period of lease of land has been prescribed, e.g., five years in Punjab, Haryana, etc., and in Andhra Pradesh four years for one category of tenants and six years for another.

Ownership rights: experience of the implementation of Zamindari abolition showed that, on the plea of resumption for personal cultivation, eviction of tenants took place on a massive scale. There is no doubt that in certain cases and categories on holder's resumption should be allowed, but the plea of resumption should not lead to large-scale ejectment of tenants. During the Second plan, states formed provisions for resumption broadly on the following patterns:

1. All tenants have been given full security of tenure, without giving the owners the right of personal cultivation.
2. Owners have been given the right to resume a limited area (not more than family holding in any case) subject, however, to the condition that a minimum area is left with a tenant.
3. A limit has been placed on the extent of land which a land-owner may resume, but the tenant is not entitled to retain minimum area for cultivation in all cases.

Thus, the tenancy reforms aim at conferring the right of ownership of land on the tenants. To encourage hard-working tenants to become owned- cultivators, many state governments have given tenants the right to purchase their holdings. This right however, is subject to the condition that the total holdings of the landlords are not thereby reduced below the limit laid down for maximum holding of a landlord.

Right of Ownership of Tenants: A very important feature of the land reform is the provision of the right of ownership of tenants. The Second plan considered it very desirable to bring tenants in non-resumable area in direct contact with the state. Earlier the right to purchase was optional to the tenants but this did not prove to be effective. Thus, the third plan suggested the optional clause to be removed and peasants be required to purchase land. Legislation for this purpose was enacted in various states. For instance, in West

Bengal the tenants and sub-tenants have been brought into direct relationship with the state by the conferment of full ownership rights. In Punjab, the right to purchase is optional. Legislation has been enacted in Gujarat, Kerala, Madhya Pradesh, Maharashtra, Karnataka, Orissa, Rajasthan, Uttar Pradesh, West Bengal and the Union Territories. Whereas in Assam, Jammu and Kashmir and Tamil Nadu, no provisions exist even for an optional right of purchase. While the state can facilitate the transfer of ownership rights from the landlords to the tenants, no financial burden is imposed on the state.

In case of voluntary surrenders, to check its evil two recommendations were made in the Third Plan. First, the volunteer surrenders of land by tenants should not be considered valid unless they were duly registered by revenue authorities, and second in case of voluntary surrenders, the land-owner should be entitled to undertake cultivation only to extent of his right resumption by law. There is much leeway to be covered in implementation in this regard so as to save the poor tenants-the most vulnerable, though the most important section in rural India. Legal Protection to Tenants unable to bring about redistribution of ownership of land, the legislation attempted to provide security of tenure to tenants, to fix land rents and condition of tenancy. Also, in order to enable tenants to obtain institutional credits, tenants should have the right to pledge their interests in land taken on rent by them. Some states have taken steps in this area of tenancy legislation. In few states, creation of a tenancy in future has been prohibited, i.e., only in special cases that also for a limited period of three years at a time. The steps for the Remission of rent for the tenants in times of famines or natural calamities when the landlord is given relief through the remission of land revenue have also been taken. Begar and other unauthorized levies have been declared illegal.

Moreover, the tenancy legislation being not comprehensive has failed to grasp the interdependence of fixation of ceiling on rents and security of occupancy rights. Myrdal focuses attention on this problem in the following words: In the absence of limits on rent, all rules about security of tenure can be nullified, the landlord can simply raise the rent beyond the tenant's capacity to pay and legally evict him for non-payment. By the same token, legislation on maximum rentals is meaningless if not buttressed by security of tenancy. Besides this, legislation aimed to provide security to a minority of tenants who paid fixed rentals and left out the majority of sharecroppers who represented the more vulnerable section of the Indian peasantry.

The most important beneficial result of the reform is that it put an end to the system of parasitic intermediaries. On the other hand, it has not put an end to absentee ownership of land nor has it led to the disappearance of tenancies. All in all, although the contribution of tenancy reforms could not be totally neglected but the programs including these reforms since independence did not lead into any significant redistribution of land, or the removal of all the obstacles to increasing agricultural production. The policies adopted in case were ambivalent and there were large gaps between policy and legislation and implementation.

## Cultural Stratigraphy of Jharkhand

Jharkhand has been an area of cultural isolation. A few cultures flowed into, a few established themselves and fewer could survive in a recognizable form. Such processes as cross-fertilization of ideas, innovations and renovation do not appear to have been operative over the greater part of the plateau during the historical past. An analysis of the ethnic composition, linguistic groups, historical and pre-historical migrations and present cultural forms enables us to discern at least seven cultural strata. The inter-relationship among them varies. In relation to one or more cultures some of them are interposed, others extraposed, while a good many of them are transposed over one another. The important cultural strata relate to

1. The Asur, Proto-Aryan and Proto Dravidians,
2. Munda
3. The Oraons

4. Chero and Kharwar
5. Later Hindus
6. Muslims and
7. Christians.

The Asurs appear to be the first people to colonize Jharkhand. Whether the Asurs were a aborigines or early Aryans is a most controversial question, but it is definite that a very advanced Asur culture flourished in the middle Ganga Plains from the pre-vedic period to the Mahabharata periods. Asur culture forms the subterranean level, the basement over which subsequent cultures developed. The Asurs were the worshippers of snake-god and Yakshas, the remnants of which still survive in this region. Mahavir or Hanuman worship is most popular in Jharkhand. Certain authorities believe that Mahavir worship was originally Yaksha worship which was adapted to Hindu way of life and subsequently assumed its present form. The Nagbansi rulers of Jharkhand might have had some link with some Nag-worshipper families of ancient Magadh. The small number of Asurs confined to the western part of the Ranchi plateau, might have been the descendants of the ancient Asurs who being defeated at the hand of the Mundas took refuge in the highest and most inaccessible areas. The Munda tradition suggests that the Asurs were the exploiters of metallic ores, particularly iron copper and gold. Reference has been made in Kautilya's Arthshastra to the supply of metals and minerals from this region to the Mauryan Empire.

The Mundas who migrated from the northern plains, must have engendered a higher level of culture, though it is doubtful that they were culturally superior to the Asurs in the beginning. Mundas immigration led to the clearance of forests and extension of cultivation on a much larger scale than was done ever before. A real revolution in the ancient and primitive culture of the Ranchi plateau appears to have started with the entry of the Oraons. The Oraons, through long wanderings and long association with the Aryans in the plains have improved their cultural acquisitions. They entered the plateau with a superior culture. Their cultural influences have, therefore, been of sweeping and lasting nature. Most of the present agricultural practices and general economy in the countryside of the Ranchi district have distinct Oraon impact. On the other hand, the Oraons adopted village administration and social organization from the Mundas.

The Cheros and Kharwars who succeeded the Oraons in Palamu were not in the primitive stage. They were largely Aryanized from the very beginning. Consequently, few of them remained in the primitive stage of culture. With the acquisition of power and wealth, they rose to the higher ranks of the Hindu society. Most of them are accepted today as Rajputs. The Cheros and Kharwars did not, therefore develop a culture of their own. They simply helped the spread of Hindu culture.

In the northern and eastern parts, the cultural succession appears to be a bit more confused. Even in these areas, Asur culture was perhaps first to flourish. Asur culture was perhaps, directly succeeded by Aryan culture. Relics of Aryan, particularly Jain settlements dating back 500 to 1000 BC have been found in various parts of Hazaribagh, Dhanbad and Singhbhum districts. The primitive cultures might have had a sojourn in this peripheral belt, but in wake of continued Aryan advance they must have had any respite to establish themselves and flourish. Consequently, the entire area north of the Ranchi scarp, except the Rajmahal hilly tract, underwent complete Aryanization. The tribals were detribalized and were absorbed in the Hindu society. With Aryanization, cultural landscape changed fast and enormously. Advanced agriculture, superior architecture and technology, trade and commerce were introduced, and the cultural landscape became a plateau version of the cultural landscape of the southern Ganga Plain of Bihar. Remnants of tribal culture that survived, went to increase the variety, diversity and complexity of the cultural landscape. The most notable contribution of Aryan culture to the landscape of the whole region was the emergence of towns. The primitive culture has no urban component and as such no tribal towns and cities were found in Jharkhand.

From the Ganga Bend in the north-east to the Subarnarekha valley in the south is a narrow-detribalized belt. This marginal area of the plateau, being contiguous with Bengal, forms zone of contact. The area has received in full the Bengali version of Aryan Culture. Several features which are typical of Bengal, are strongly represented in the cultural landscape of this area. The Bengali traits are most faithfully represented in the house types, clothing and dialects. A large number of non-tribal population are either migrants from Bengal or have matrimonial or other social ties in Bengal. The Subarnarekha valley and the Chaibasa plain that open to Bengal and Orissa received a voluminous and continuous inflow of people and culture from the two regions. In the Subarnarekha valley, Bengali influences are dominant, but in the Chaibasa plain, Oriya influences are more promptly marked.

South-Western portion of Ranchi district which forms a tripartite contact zone between tribal Jharkhand, Orissa and Madhya Pradesh, has a more intricate cultural pattern with Oriya a dominant element. A sizeable Gond and Kundahit population are found in this area. Gonds original habitat lies mostly in Madhya Pradesh whereas Kundahit is essentially an Oriya tribe. A greater degree of Oriya influences in this part is also due to matrimonial and other exchanges between the ruling families of the two regions. The cultural influences of Oriya in the south-western part and those of Bengalis in the eastern zone have been great civilizing force and have considerably molded the culture and economy on Oriya and Bengali patterns. Most of the temples in Singhbhum and in the southern part of the Ranchi district are devoted to Lord Jagannath. Rathayatra is the most highly celebrated and attended festival of the tribal and non-tribal communities alike. Both the Jagannath worship and Rathayatra are transplanted from Orissa to this area.

In the west, Jharkhand forms a common boundary with Madhya Pradesh. Though the Pat scarps and rugged surface along the border do not encourage frequent exchanges, slow and scanty movements of people across the border have taken place throughout the historical times. Half a dozen small tribes, distributed along the border area of Jharkhand, appear to have entered this region in the process of spreading out from their base in Madhya Pradesh. These minor tribes do not appear to have influenced the culture of this area to any significant extent. They rather appear to have exchanged their own culture for the dominant Oraon tribe. This portion of the plateau has also borne the brunt of Maratha invasions and raids. Impact of Maratha culture is not easily recognizable. The pony which is widely used in this area, might have been the remnant of the Maratha impact.

## Land Tenure and Tenancy System

Though the culture of Jharkhand represents today a good deal of mixture of several living and dead cultures, there are broadly two cultures in Jharkhand. They are Tribal culture and non-tribal culture. The non-tribal culture is for the most part, Hindu culture. Some of the most notable aspects of tribal culture are the tribal village organizations and peculiar tenancy system. Most of the cultural characteristics of tribal society are related to these two institutions.

The aboriginal tribes, particularly the Mundas, had in the beginning, no idea of individual ownership of landed property. They cleared the jungle and took possession of the land jointly. All the lands thus obtained belonged to a family or a group of agnate families. Clearances made by each such family came to be known as Khuntkatti Hatu or the village of the family of original settlers. The Mundas had a Patriarchal tribal organization. The oldest member of the village became a chief in religious as well as secular matters and came to be known as a Pahan. Gradually the religious and secular functions were separated. The most important man in the village became responsible for secular matters as was known as Munda. The offices were made hereditary, but neither the pahan nor the munda could enjoy any rights superior to those of the other descendants of the founders of the village. As the Mundas grew in number new villages were founded. The inhabitants of the new villages subsequently established their own burial grounds and saranas and formed independent village communities. The new villages continued to have link with the parent village. A group of such villages along with the parent village came to be known as parha or patti. In the beginning the parha was just a wider brotherhood than a village, Subsequently the Mundas came to have tribal organization of villages. Munda of the parent village or the most influential munda became manki of the patti (Chief of the group of allied villages). He was to preside over the assembly of the Mundas and Pahans of the Parha. A similar organization was adopted by the Oraons also.

In the beginning the Mundari Tenancy system was simple and free from administrative burdens. The whole Munda country was divided into patti or parha. Each patti consisted of 10 to 15 villages. Broadly speaking, three types of land tenures were prevalent among Mundas, Oraons and Hos. They were Khuntkatti, Bhuinhari - raiyatwari and tenancy systems. Khuntkatti system originated from the possession of the natural earth of the first settlers. They were the real proprietors of the land and were not required to pay any rent to anybody. New settlers were gradually attracted to the Khuntkatti villages and their neighborhood. They could not have the same rights and privileges as enjoyed by the descendants of the original settlers. Thus, was started differentiation in land tenure and tenancy system of the tribals.

A Mundari Khuntkatti village generally contains three elements. They are 1. The Khuntkattidars (owners of the title of original settlers), 2. Tenants and 3. Subsidiary artisan class. The Khuntkattidars are the owner of the whole area including in the village boundary. Raiyats are just tenure holders. Their tenures were first fixed by Khuntkattidars and later by zamindars. Bhuinhari system had a wider connotation than the term Khuntkatti. The system includes six highly complicated cognate tenures. Some of them are function tenures held by village functionaries. Raiyatwari or ordinary tenancy system was introduced by the Hindu immigrants on the pattern of the land tenures of the Ganga Plains.

The Mundari Khuntkatti system was once prevalent over greater part of Ranchi and Singhbhum districts. Several causes contributed to its break down. As the time went on, village head became conscious of his power and property and tribal chiefs arose. The chiefs began to exercise an overlordship over village communities and started receiving tributes and personal services from them. The emergence of chiefs marked the beginning of social and economic stratification among the tribal communities. This facilitated the process of Hinduization. The chiefs were first to be Hinduized. They joined hands with Hindu immigrants and collaborated with the Nagabansis in the establishment of Chotanagpur Raj. The Nagbansi rulers and chiefs, in order to establish their supremacy on firmer footing, encouraged Hindu immigration. The immigrants were granted power to collect rent and receive tributes from the cultivators. Gradually they extended their overlordship over the Khuntkattidars and reduced them to position of ordinary tenants. They took possession of forests and wastes and deprived a large number of tribal people of their cultivated land. These lands were resettled on fresh tenures. All these led to agrarian troubles and tribal rebellion. The suppression of the rebellion broke down the tribal tenancy system. The aborigines were obliged to move to new areas and establish new villages. The process led to the furtherance of human occupancy to less tractable and poorer soils in less accessible and remote areas.

## Prohibitive Laws and the Stagnation of Cultural and Economic Processes

The tribal rebellion against the greedy immigrants proved a long-drawn battle. The British administration felt it necessary to protect the rights and privileges of the aborigines. The Mundari tenancy act was passed in 1903. The act provided that Mundari Tenancies were not transferable by sale, whether in execution of decree or otherwise. Subsequently, Chotanagpur Tenancy came into force which provided greater safeguards against the alienation of lands from the tribes. In post-Independence period, by fresh legislations, tribal lands have been made almost non-transferable to non-tribals. All these legal protections to the tribal communities acted in too different ways. The non-transferability of tribal lands granted a greater degree of permanency to the tribal settlements. Formerly, being dispossessed of their lands they were to move to unoccupied areas to reclaim barren or forest lands for cultivation, thus extending the frontiers of occupancy. Some of them were obliged to migrate to towns, industrial or mining centers and tea gardens in Assam. This led to the desertion of several villages. A large number of abandoned village sites are found widely distributed in Ranchi, Palamu and Hazaribagh districts. The protective legislations have some negative effects also. The non-transferability of tribal lands to non-tribal communities has reduced the value of land in open market. There are few purchases in the tribal communities and therefore, the prices offered are very low. Those who want to dispose off their lands to invest in some other jobs are thus discouraged to do so. This has affected adversely the process of occupational change. Further, the transferability even among the tribal community is limited to the residents of an anchal, the smallest administrative unit. Most adversely affected are the non-tribal immigrants. They are almost debarred from acquiring land even for residential purposes in tribal areas. Quantum of immigration to the rural areas have been very much reduced and the process of human occupancy to non-settled areas have been slowed down. The protective legislations are, in fact, working against the natural laws of the social and economic dynamics. It has made the people immobile, both in terms of occupation and in terms of area. Redistribution of population has been retarded and both intra-regional and extra-regional socio-economic balances are being disturbed. The expansion of non-tribal settlements in rural areas has come to almost a halt. On the other hand, tribal settlements continue to expand mostly because of a higher birth rate among the tribal communities. The difference between non-tribal and tribal population is widening and the prospects of detribalization of Jharkhand are becoming dimmer. The process of culture-change has also been affected.

## Size of Land Holdings

The size of a farm is a matter of great importance to success in agriculture and accelerating agricultural production by applying modern farm technology. It is very important for farm management. All the activities connected with agriculture depend on the size of the farm. Generally, in a layman language, if there is large area of the farm, we call it a large farm and if the area is small, it will be called a small farm. But in terminology of agriculture science it has been divided in different categories which will be discussed in coming pages.

In theory, the size of a standard operational holding is fixed by socio- economic conditions in accordance with type of farming practiced. In practice it is found that in different countries the standard varies enormously in accordance with the pressure of cultivators on the cultivated area and the laws of inheritance. Even granting homogenous physical conditions, it is clear that the variation in capacity, resources and ambitions of farmers will create pressure on farms of corresponding size. (Singh, 1974, P.70).

## Bases to determine the farm size

In Geography, Economics and Agriculture Science scholars have decided different bases to determine the size of holdings or farms; among them some are very simple whereas some are quite complex to take in account.

**Area of Farm:** The first important measure to judge the size of farm is the absolute area. It is only a consistent measure used by agricultural census across many countries of the world. A 20 Hectare farm is said to be a large farm rather than a 10-hectare farm.

**Nature of Farming:** Another method to judge the size of the farm is nature of farming. Accordingly, if farming is specialized, for instance, the production of fruits or vegetables and other high value crops, the farm can be considered large. But on the other hand, an equal size with general crops is considered as a small farm.

**Location of the Farm:** If the farm is near a big city or near the market, small area can be considered as a large-scale farm. But on the other hand, If the farm is far away from city or the river bank, the large area of the farm is required for large scale farming.

**Capital Employed:** Yet another measure to know the size of the farm is the total capital employed. In order to determine the total capital all inputs viz. land, machinery etc. are summed up together. But it is not a satisfactory measure since capital is one of the inputs used in agricultural production. For instance, a 10 hectare of tobacco farm has more production with high of low capital investment as compared to 20 hectares used for rearing livestock.

**Labor Employed:** If there is a small number of labourers on a farm, they cannot get more production as all the resources cannot be exploited well. Therefore, sufficient number of labors must be for a large-scale farm.

**Volume of Output:** The volume of output is probably the most useful measure to determine the size of the farm. If there is good farm management and much of production is obtained that farm shall be a large-scale farm. Contrary to this production is less irrespective of the other conditions that shall be a small scale of farm. The output is preferably taken in value terms rather than in physical terms.

## Types of Holdings

To describe the suitable unit of cultivation it becomes pre-requisite to have a glance over the various terms viz. family holding, optimum holding as well as economic holding.

**Family Holding:** The term holding was used by the committee of land reforms taking income as the base. Assuming 2 to 3 earners in the family of an agriculturist, the annual income of the family should be Rs. 1200. It means no unit of cultivation which yielded an income less than Rs. 1200 could legitimately be considered economic.

**Optimum Holding:** Optimum holding refers to the maximum size of holding which is a family should possess. Three times the family holding was considered to be an adequate size of optimum holding. The whole idea behind imposing this ceiling on the size of holding is to fix upper limit up to which a family can be allowed to possess by law. In other words, an optimum holding in that case, would be one which permits the most efficient use of land, labor and capital with given technique of production.

**Economic Holding:** Economic holding is that holding which provides full employment to the farmers family and allow him to enjoy a reasonable standard of living. However, the term economic holding has been differently defined by different authors. According to Keatings, "An economic holding is one which allows a man a chance of producing sufficient to support himself and his family in reasonable comfort after paying for his necessary expenses." But in contrast to this definition, economic holding to Dr. Mann is "one which will provide for an average family at the minimum standards of life considered satisfactory." But the word reasonably standard and high standard are all vague and do not therefore, give a definite idea about the size of an economic holding. In reality an economic holding should mean a holding which provides the fullest and most remunerative employment of all factors of production at the disposal of the farmers.

## Pattern of Farm Holding in India

Data regarding the farm holding is collected by Agricultural census of India. Agriculture census is a large-scale statistical operation for collection of agricultural data. The first World Agricultural Census was conducted in the year 1924 by the International institute of Agriculture (IIA) in Rome. In 1930 63 countries participated. In India, Govt of India has conducted the first Agricultural census in the country with agricultural year 1970-71 (June-July). Since then it is conducted after every five-year duration. i.e. 1970-71, 1976-77, 1980-81, 1985-86, 1990-91, 1995-96, 2000-01, 2005-06, 2010-11 and presently 2015-16.

Before studying Jharkhand, we would like to study the scenario of land holding and their changing pattern in India in brief. When we study the agricultural census of 2010-11 and 2015-16 for the size of land holdings of India, results reveal that the average size of an Indian farmland shrunk by over six percent. The land holding size in country has changed its size from 1.15 hectare in 2010-11 to 1.08 hectare during 2015-16. With land holding getting smaller, the share of small and marginal holdings in the country (between zero and two hectare) has risen to 86.21 percent of total operational holding in 2015-16 which comes to around 126 million as against 84.97 percent in 2010-11. On the other hand, the share of semi- medium and medium operational holdings (2-10 hectares) in total land holdings dropped from 14.29% to 13.22%, while that of large holding (10 hectare and above) fell down from 0.71 % in 20110-11 to 0.57 % in 2015- 16. This also means the number of small holdings in the country has grown in five years, due to fragmentation of land, while that of medium and large holdings have gone down. Operational holding has been defined as all land used wholly or partly for agricultural production and is operated as one technical unit by one person alone or with others, without regard to title, legal form, size or location. Total operated area which includes both cultivated and uncultivated, provided part of it is put to farm use, fell from million hectares in 2010-11 to 157.14 million hectares in 2015-16.

**Table 2.1**
**Distribution of Personal holdings in India**

| Category | Size (Hectares) | Number in Millions | | Area (In Million Hectares) | |
|---|---|---|---|---|---|
| | | 2010-11 | 2015-16 | 2010-11 | 2015-16 |
| Marginal Holdings | 0-1 | 928 | 1002 | 359 | 379 |
| Small Holdings | 1-2 | 247 | 258 | 352 | 362 |
| Semi Medium Holdings | 2-4 | 138 | 140 | 377 | 376 |
| Medium Holdings | 4-10 | 58 | 55 | 338 | 318 |
| Large Holdings | 10 and above | 9 | 8 | 169 | 143 |

| India | 1383 | 1464 | 1595 | 1578 |
|---|---|---|---|---|

Due to round off total may not tally
**Source:** Agricultural Census of India

**Table 2.2**
**Number of Operational Holdings in Jharkhand (000)**

| Census | All Groups | Marginal | Small | Semi-Medium | Medium | Large |
|---|---|---|---|---|---|---|
| 2010-11 | 2709 | 1848 | 429 | 283 | 129 | 20 |
| 2015-16 | 2803 | 1962 | 419 | 277 | 126 | 20 |

**Source:** Agricultural Census of India

**Table 2.3**
**Area of Operational Holdings in Jharkhand (000 Hectares)**

| Census | All Groups | Marginal | Small | Semi-Medium | Medium | Large |
|---|---|---|---|---|---|---|
| 2010-11 | 3165 | 764 | 591 | 775 | 725 | 311 |
| 2015-16 | 3091 | 754 | 569 | 755 | 708 | 304 |

**Source:** Agricultural Census of India

**Table 2.4**
**Number of Operational Holdings in Jharkhand for SC (000)**

| Census | All Groups | Marginal | Small | Semi-Medium | Medium | Large |
|---|---|---|---|---|---|---|
| 2010-11 | 375 | 294 | 47 | 24 | 10 | 1 |
| 2015-16 | 379 | 301 | 45 | 23 | 9 | 1 |

**Source:** Agricultural Census of India

**Tablc 2.5**
**Area of Operational Holdings in Jharkhand for SC (000 Hectares)**

| Census | All Groups | Marginal | Small | Semi-Medium | Medium | Large |
|---|---|---|---|---|---|---|
| 2010-11 | 322 | 124 | 63 | 66 | 52 | 17 |
| 2015-16 | 310 | 119 | 61 | 63 | 50 | 17 |

**Source:** Agricultural Census of India

**Table 2.6**
**Number of Operational Holdings in Jharkhand for ST (000)**

| Census | All Groups | Marginal | Small | Semi-Medium | Medium | Large |
|---|---|---|---|---|---|---|
| 2010-11 | 967 | 588 | 172 | 127 | 67 | 12 |
| 2015-16 | 979 | 608 | 168 | 125 | 66 | 12 |

**Source:** Agricultural Census of India

**Table 2.7**
**Area of Operational Holdings in Jharkhand for ST (000 Hectares)**

| Census | All Groups | Marginal | Small | Semi-Medium | Medium | Large |
|---|---|---|---|---|---|---|
| 2010-11 | 1431 | 262 | 239 | 354 | 385 | 190 |
| 2015-16 | 1388 | 253 | 229 | 348 | 376 | 182 |

**Source:** Agricultural Census of India

**Table 2.8**
**Percentage Distribution of number of Operational Holdings in Jharkhand 2010-11 & 2015-16 (All Social Groups)**

| Census | Marginal | Small | Semi-Medium | Medium | Large |
|---|---|---|---|---|---|
| 2010-11 | 68.23 | 15.83 | 10.44 | 4.75 | 0.75 |
| 2015-16 | 69.98 | 14.94 | 9.89 | 4.48 | 0.71 |

**Source:** Agricultural Census of India

**Table 2.9**
**Percentage Distribution of Area Operated by Operational Holdings in Jharkhand 2010-11 & 2015-16 (All Social Groups)**

| Census | Marginal | Small | Semi-Medium | Medium | Large |
|---|---|---|---|---|---|
| 2010-11 | 24.13 | 18.66 | 24.49 | 22.90 | 9.82 |
| 2015-16 | 24.40 | 18.43 | 24.4 | 22.90 | 9.83 |

**Source:** Agricultural Census of India

### Table 2.10
### Percentage Distribution of number of Operational Holdings in Jharkhand 2010-11 & 2015-16 (Scheduled Caste)

| Census | Marginal | Small | Semi-Medium | Medium | Large |
|---|---|---|---|---|---|
| 2010-11 | 78.37 | 12.41 | 6.34 | 2.56 | 0.32 |
| 2015-16 | 79.28 | 11.90 | 6.07 | 2.44 | 0.31 |

**Source:** Agricultural Census of India

### Table 2.11
### Percentage Distribution of Area Operated by Operational Holdings in Jharkhand 2010-11 & 2015-16 (Scheduled Caste)

| Census | Marginal | Small | Semi-Medium | Medium | Large |
|---|---|---|---|---|---|
| 2010-11 | 38.62 | 19.49 | 20.38 | 16.11 | 5.41 |
| 2015-16 | 38.51 | 19.55 | 20.34 | 16.22 | 5.39 |

**Source:** Agricultural Census of India

### Table 2.12
### Percentage Distribution of number of Operational Holdings in Jharkhand 2010-11 & 2015-16 (Scheduled Tribes)

| Census | Marginal | Small | Semi-Medium | Medium | Large |
|---|---|---|---|---|---|
| 2010-11 | 60.86 | 17.75 | 13.15 | 6.97 | 1.27 |
| 2015-16 | 62.12 | 17.14 | 12.79 | 6.74 | 1.20 |

**Source:** Agricultural Census of India

### Table 2.13
### Percentage Distribution of Area Operated by Operational Holdings in Jharkhand 2010-11 & 2015-16 (Scheduled Tribes)

| Census | Marginal | Small | Semi-Medium | Medium | Large |
|---|---|---|---|---|---|
| 2010-11 | 18.31 | 16.74 | 24.75 | 26.91 | 13.30 |
| 2015-16 | 18.22 | 16.48 | 25.04 | 27.12 | 13.14 |

**Source:** Agricultural Census of India

## Size of Operational Holdings

An analysis of mean size of operational holding shows the enormity and the seriousness of the land problems in the predominantly agrarian economy of India. The size of the farm decides the degree of risk that a farm operator may bear, i.e. larger the size of the farm, greater the capacity of the farmer to take the risk and vice versa. This in turn would affect the extent of specialization and also the quantity of equipment and power to be used. Moreover, size is linked with the pressure of population, the economic requirements and the fertility of land. In India the standard size of a farm that would be most suitable to agriculture cannot be maintained because of the increasing hold of the rural community on arable land and the operation of the Hindu Law of Inheritance. These two factors lead to division of holding into small, often widely separated fragments which fail to conform to any reasonable economic standard from the point of view of operations.

The size of the farm is one of the important factors which influence the farmer's decisions regarding agricultural work, and determines his income from farming. This explains why analysis of such a relationship had been the perpetual theme of study in agricultural economics in the past. In recent times, however, the size of farm has been given greater importance, as on account of high prices, land has become scarce as compared to family labor. Therefore, any increase in size of the farm is not possible except by adding leased arable land. Consequently, the size of the farm includes the net tilled area of the land owned and the land leased in. In practice, each operational holding does not constitute a single farm. But as it is well known, it may comprise several plots of land lying at varying distances from each other. In India this is a matter of no small importance from the viewpoint of managing efficiency. The operational holdings, whether large or small, are also fragmented.

To conclude, the size of a farm, is an important consideration in determining the volume of agricultural yield. In recent times, land values have gone up, and land ceiling legislation has also been implemented. Agricultural land has become relatively scarce as compared to family labor and capital- hence the former has assumed more significance. The possibilities of increase in size of an operated holding in terms of hectarage are limited. The only way to increase agricultural productivity and volume of farm income is by making more intensive use of the existing land resources. A liberal provision of credit facilities to the sub-marginal, marginal and medium-size farms should be able to meet their economic needs. Provision of credit at low rates of interest to small farmers and of mechanical power to large farmers could be the two main elements in agricultural growth as a size-oriented growth policy. The stupendous agricultural progress achieved by Japan, despite the small size of her farms should dispel wrong notions in this regard and our agriculture development programs may be formulated and implemented on the same pattern.

## Fragmentation of Land and Mode of Cultivation

Knowledge of the distribution of land in different size groups alone is not enough. The fragmentation of cultivated land and the mode under which it is cultivated are the other two important aspects which further helps us to understand the operation of holdings (Dantwala and Donde, 1965). Besides as Iqbozurike(1976) has rightly said, other farm factors, besides the size of the farm, also need to be emphasized in analytical discussions of farmland. These are fragmentation and sharp regional and local variations in skill, opportunities and socio-economic aspirations of the farmers. The extent and limit of fragmentation should be particularly stressed in studies on farm management. Fragmentation and land tenurial systems are the crucial problems faced by farmers in India. These problems have caused considerable wastage of resources.

**Fragmentation of Land:** Fragmentation here refers to division or segmentation, and may be defined from an agricultural viewpoint as a situation under which a cultivator owns an operational holding consisting of more than one plot of land. Therefore, the fragmented farm has been regarded as comprising two or more plots of land which may be separated from one another. This does not include plots which constitute a single block of land including those that are contiguous to each other having separate survey (Khasra) numbers written in the village revenue records.

In his study, Chishom (1962) has clearly stated that several plots of land lying at varying distances from each other in an operational holding are a matter of great significance from the standpoint of managerial efficiency. Further, in many studies on economics of farm management, it has been established that fragmentation is a great obstacle and one of the major deterrents to economically viable cultivation. It results in wastage of land, labor and material inputs. It is responsible for increased overhead costs, including even the cost of production resulting in low returns from agriculture.

Thus, repeated subdivision and fragmentation of holdings of land are major problems adversely affecting agriculture. Division of land into equal lots among settlers is followed by further subdivision. Unhampered, continuous growth of population has further aggravated the problem as it became necessary to break up and distribute common lands. Sale of a few high value farmlands and purchase of low value ones took place in some areas. Systems of inheritance made such problems more acute. The principle of equal sharing among heirs has a historical background both in the Oriental and the Occidental worlds. It is rare for a peasant farmer to have enough land to leave each heir with an adequate holding. More often than not heirs have no option but to sell their inheritance in order to share its value (Naylon, 1959). In a country where the local custom favors 'gavelkind' rather than the right of 'primogeniture' all landed property is divided equally among all children. The law of succession in India also favors the gavelkind system with the result that the average size of a farm has gone down rapidly. When the father dies, the land is divided equally among his collateral heirs. Each successor insists on having a share from each location and piece of land, resulting in further fragmentation. It is a wasteful method of land utilization and many improved agricultural practices cannot be adopted (Randhawa, 1974).

Therefore, the most common characteristic of the Indian rural scene is that it is rare to find average size holdings. Farms are still either very large and generally enclosed within a single boundary or very small, split up into scattered fragments (At times of a few square meters). A bird's eye view of countryside will show a complicated pattern of an infinite number of boundaries, footpaths, cart-tracks, bushy fences, ditches, ponds etc. In this sense, agrarian property is a property in tatters and that cultivation in many regions is an absurdity. It is often difficult to fathom the intensity of the problem because of lack of cadastral maps; yet it is certain that the average size of holdings is declining in many parts of India. From the social viewpoint, the division of rural landed property may seem to be justified, but economically it is very harmful to the use of modern inputs. Besides, it has been observed that the problem of fragmentation is less severe in higher size-classes and comparatively acute and rampant in the small size operational holdings. This has a greater adverse effect on the utilization of agricultural land resources and the introduction of modern methods of cultivation. The input-wastage should not be ignored by agricultural planners while preparing the agricultural developmental schemes for peasant proprietors.

## Effects of Fragmentation

The disadvantages of fragmentation and scattered holdings are not unknown. They are the main obstacles in rationalizing agricultural activities. The economic disabilities of small holdings are indeed so great that a large proportion of land cannot be properly cultivated. The most common and widely known effects of fragmentation to the detriment of agricultural development are:

i. Most of the labor of the farmer and his animals is expended in journey from one plot to another.
ii. Considerable time of the farmer is lost in shifting animals, fertilizers, seeds and many other farming equipment and

in supervising the establishment.

iii. Cultivation of many of the scattered plots is perforce given up for two reasons- first, they are located at rather uneconomic distances; second, it is difficult for a pair of yoked bullocks to turn round since these plots are too small in size.

iv. A good proportion of cultivated land is wasted in boundaries, hedges, ditches, footpaths, cart-tracks, drove ways and multiplication of amenities such as staying and watering places.

v. Excessive fragmentation raises the cost of agricultural produce because of duplication of invested capital.

vi. No irrigation development is possible unless there is consolidation

vii. Excessively high costs of applying manure, limitations on introducing crop rotation and maintaining breeding stock slows down agricultural progress.

viii. The menace of soil erosion becomes too unwieldy to check, and crop diseases are not easy to combat and

ix. The use of modern farm machinery in small fragments is virtually impossible despite the favorablee3d conditions for all round mechanization.

It has been estimated that on a holding of one hectare a tractor can spend 32 percent of its working time in simply turning around. (Naylon, 1959). Credit facilities cannot be put to proper use because of the farmers inability to do so; nor can crediting agencies be turned into convenient bases to establish credit facilities to farmers (without any creditworthiness). Lastly, never-ending litigations over boundaries, inheritance etc. dis hearted farmers and reduce them to penury.

## A Comparative Study of Land Holdings in Jharkhand Between two Censuses (2010-11 and 2015-16)

At first, we shall compare the number and area of holdings according to their group i.e. Marginal, Small, Semi-Medium, Medium and Large and their district wise study will compare their size and class so that the changes in groups regarding number and area can help us understand the situation of state and the districts.

On the basis of Agricultural survey of India, number of operational holdings in Jharkhand was 27.09 lakh during the census of 2010-11 which has increased during the census of 2015-16 to 28.03 lakhs. This shows that the number of land holdings in Jharkhand has grown by around 0.94 Lakhs during five years. In percent we can say that the number of operational holdings in Jharkhand was 0.19 percent to the total of India which has grown to 0.20 percent during 2015-16 or after 5 years of span.

At the same time the study of the data of the holdings reveal that it was 3165000 hectares in all groups during 2010-11 census which reduced to 3091000 hectares in the census of 2015-16. Accordingly, the average size of holding was 1.16 hectares per holding which became 1.10 hectares per holding in between the census of 2010 and 2015. The prominent cause of reduction in size of the holdings in the state is fragmentation of land due to division of family and therefore resulting in the growth of tenants.

As scheduled caste and scheduled tribe have been enumerated separately in each census, the study of both are essential separately. This study will clear the situation of the society of the state that in which conditions a deprived class performs their agricultural practices.

Number of operational holding of scheduled caste was 375000 during 2010-11 census in all groups and it raised to 379000 during the census of 2015-16. This was a growth of 4000 holdings during the period of five years. This is the growth of 1.055 percent which is very less with respect to the state growth of 3.35 percent. So far, the study of the area of operational holdings of scheduled caste is concerned it was 322000 hectares during the census of 2010-11 and it became 310000 during 2015- 16 census. It was a reduction of 3.87 percent in the area of holding among the SC of the state. With the help of this data when we calculate the size of holdings of SC, the result shows that the per-holding size was 0.85 hectare during the 2010-11 census and it became 0.81 hectare during 2015-16 census. Size of the holding was reduced to about 4.7% during the five year.

Now the study of Scheduled Tribe data reveals that the number of operational holdings in all groups was 967000 during 2010-11 and it raised to 979000 in the census of 2015-16. This was a growth of 12000 holdings in number within a period of five years. This is equal to a growth of 1.22 percent in number of holdings within a time period of 5 years. At the same time the area of operational holdings of ST was enumerated 14311000 hectares during 2010-11 census and this became 1388000 hectares during 2015-16 census. This was a reduction of 3.00 percent in between the period of both censuses.

When we compare the growth of number and area among the SC and ST, we find the result that the growth in number of holding was 1.05 percent whereas among the ST it was 1.22 percent. Similarly, the growth of area of the holding of SC was -3.87 percent where as among the ST it was -3.00 percent. Size of holding of SC was 0.85 and it reduced to 0.81 but the size of holding was 1.48 hectare per holding and it reduced to 1.42 hectare per holding in between 2010-11 to 2015-16.

Now it is clear that the number of holding among ST is growing more than the SC which reveal that the rate of fragmentation of holding is higher among ST than SC. So far size is concerned, the land holding size is bigger with ST than SC. It is 0.81/hectare with SC and 1.42/hectare with ST. Now the result force to think that how it is possible with ST. They have rapid fragmentation of land but size is still bigger than the size of the land of SC. No doubt the ST of the state have a better position I respect of land holding to the SC of the state.

## Holdings in Different Groups

Agriculture census has divided the land holding into five groups - marginal, small, semi-medium, medium and large.

All groups will be studied separately to understand the situation of holding in respect of their comparative growth, number and area or size.

1. **Marginal Holdings:** If the size of the holding is less than 1 hectare, it is known as marginal holding. In Jharkhand number of marginal holding is highest among all other groups. Thus, it is obvious that maximum land holding in the state is in a size of less than one hectare. As per the census of 2010-11 out of 2709 thousand holdings in the state 1848 thousand were marginal category of less than one hectare. During 2015-16 census 1962 thousand out of a total of 2803 thousand holding belongs to the size of less than one hectare. These holdings were accounted for 764 thousand hectares during 2015-16 census in that area. Here picture is very clear that the number of marginal holdings i.e. less than one hectare has jumped to 114 thousand in number within five year while the area of the marginal holding reduced to 10 thousand hectares in between the census of 2010- 11 and 2015-16. This is a decrease of 1.33 percent in between these censuses. Growing number of holding indicating the rapid fragmentation but decreasing the area shows either transfer of holding in other group or converted the land use or purpose of the land.

Number of operational holdings of SC was 294 thousand during 2010-11 and it became 301 thousand in number during 2015-15. But the area of the operational holding has been decreased from 124 thousand hectare to 119 thousand hectares. This was a decrease of 4.2 percent which is greater than the average of state (1.3 percent decrease).

Number of operational holdings of ST was 588 thousand during 2010-11 and it goes to 608 thousand in number during 2015-16. But the area of the operational holdings has been decreased from 262 thousand hectare in between these two censuses. This was a decrease of 3.56 percent which is also greater than the state average. When we compare the situation of SC and ST, we find that Area of operational holdings of marginal group of scheduled tribes is reducing more than the scheduled caste.

2. **Small Holdings:** When the size of the land holding is in between 1 to 2 hectares it is known as small holding. In Jharkhand number of small holding has second position among different groups of holdings. As per the census of 2010- 11 out of 2709 thousand holdings in all the state, 429 thousand were small category of the area of 1 to 2 hectares. During 2015-16 census 419 thousand out of a total of 2803 thousand holdings belong to the size of 1 to 2 hectares. In between these two census small holdings registered a decrease of percentage from 15.83 to 14.94. This category of land holding was accounted for 591 thousand hectares area during 2010-11 census and 569 thousand hectares area in 2015-16 census. Small holding group of land has decreased its area from 18.66 percent to 18.43 percent during these two censuses of 2010-11 and 2015-16.

Here picture of small land holding is very clear that there is a decrease in number of 0.89 percent and in area a decrease of 0.23 percent. It is noticeable that in marginal group the number of holding was increased and area was decreased but in small holding group both have recorded their decrease. This indicates a great pressure over this small holding group and shifting to other land holding category.

In small land holding group of SC the number of operational holding was 47 thousand during census of 2010-11 and it was enumerated 45 thousand during the census of 2015-16. This was a decrease of 0.51 percent within five years of time period. On the other hand, percentage of the area of small holding group was recorded 19.49 hectare in 2010-11 and it became 19.55 in the census of 2015-16. It was an increase of 0.31 percent.

Number of operational holdings of Scheduled Tribe was enumerated under small land holding group was 17.2 thousand in number during 2010-11 census and it decreased to 168 thousand during the census of 2015-16. This was a decrease from 17.75 percent to 17.14 percent in between 2010-11 to 2015-16. This was a decrease of 0.61 percent in number of holding of ST. Similarly, area under small holding was enumerated 239 thousand hectares during 2010-11 and again 229 thousand hectares during 2015-16 which were 16.74 percent and 16.48 percent respectively during said census years. This was a decrease of 10 thousand hectares of land under small holding which is similar to the decrease of 0.26 percent.

If we compare the situation of small holdings of SC and ST for the census year of 2010-11 and 2015-16 the result reveals that small holdings of SC in number has decreasing tendency, but the area of holding represent an increase while for the ST holding number and area both are in decreasing tendency.

3. **Semi-Medium Holding:** Semi-medium holding size has been decided as the holding size in between 2 to 4 hectares. In Jharkhand the number of semi-medium holding has third position among the different groups of land holding. As per the census of 2010-11 out of 2709 thousand holdings in all of the state, 283 thousand holdings were in the category of semi-medium holding of the area of 2 to 4 hectares. During the 2015-16 census 277 thousand out of a total of 2803 thousand holdings belong to the size of 2 to 4 hectares. In between these two censuses semi-medium holdings registered a decrease of percentage from 10.44 to 9.89 percent. This category of Land Holding was accounted for 775 thousand hectares area during the census of 2015-16. Just like small holding, the semi-medium holding also recorded decreasing in their percentage of area from 10.44 percent during census of 2010-11 to 9.89 percent during the census of 2015-16. This is a decrease of 0.55 percent within a period of five years. Similarly, there was a decrease from 24.49 percent in 2010-11 census to percent in 2015-16. This was a reduction of 0.05 percent in between five years. Now it is noticeable that the nature of semi-medium holding groups was very much similar to the nature of small holding group.

The semi-medium land holding was 24 thousand during the 2010-11 census and it became 23 thousand during the 2015-16 census. This was a reduction from 6.34 percent to

6.07 percent in between the two censuses. This amount to a decrease of 0.27 percent in five years. On the other hand, semi-medium area of the holding of SC was registered 66 thousand hectares during 2010-11 and 66 thousand hectares during 2015-16 which were 20.38 percent and 20.34 percent during the respective census years. This was a decrease of only 0.04 percent in the semi-medium land holding area of the scheduled caste in the state.

The number of operational lands holding of semi-medium of ST was recorded 127 thousand during the census of 2010-11 and 125 thousand during the census of 2015-16. These were 13.15 percent and 12.77 percent during the respective census years. This was a decrease of 0.36 percent in the semi-medium holding of scheduled caste. Similarly, the area of the semi-medium land holding of scheduled tribe was registered 354 thousand hectares during 2010-11 and 348 thousand hectares during the census of 2015-16. Which was respectively 24.75 percent and 25.04 percent of the respective area. This was an increase of 0.4 percent of the area of semi-medium land holding of ST. This was a different nature of semi-medium holding among the tribal people.

4. **Medium Holdings:** Medium holding has been defined by holding size of 4 to 10 hectares. In Jharkhand the number of medium holdings has fourth position among their different size groups. As per the census of 2010-11 out of 2709 thousand holdings in all of the state, 129 thousand holdings have been recorded in the category of medium holdings with an area between 4 to 10 hectares. In between these two census medium holdings recorded a decrease of 3 thousand land holdings in number. The same data shows that medium holding enumerated 4.75 percent and 4.48 percent during their respective census year. This shows a decrease of about 0.27 percent in medium holdings. When we study the area of this category, we find a decrease in holding size. Area of the medium holding has enumerated 725 thousand hectares during 2010-11 census and again 708 thousand hectares during the census of 2015-16. Here area has also registered a decreased number of holdings. There is a decrease of 17000-

hectare area in the medium holding category during the period of five years. In percent the same data reveals the same position during both censuses i.e. 22.90 percent. We can say that there is no change in the holding area of medium holding in percent.

Among the Scheduled Caste agricultural population share of the number of land holding was registered 10 thousand under medium land holding in the census of 2015-16. Figure of percentage of the same data reveal that the medium land holding was recorded 2.56 percent during 2010-11 and 2.44 percent during 2015-16 census for SC population. This shows a decrease of 0.12 percent in numbers of medium land holdings of SC. When we study the area of the medium holding of SC it was recorded 52 thousand hectares during 2010-11 and 50 thousand hectares during 2015- 16 census under the medium land holding group of SC. This shows a decrease of two thousand hectares. Figures show it was 16.11 percent during 2010-11 and 16.22 percent during 2015-16. Thus, it was an increase of 0.11 percent.

Scheduled Tribe agricultural population have their share of 67 thousand land holdings under medium holding group during 2010-11 census and it was enumerated 66 thousand during the census of 2015-16. This data clearly shows a decrease from 6.97 percent of 2010-11 to 2015-16 population. This is a reduction of 0.23 percent of land holding under medium group of ST. When we study the medium group data of ST it enumerates 385 thousand hectares during 2010-11 census and 376 thousand hectares during 2015-16 census. It was a change of 0.09 thousand hectares under medium holdings of ST. But the figure of percentage clearly shows an increase in holdings in the area. It was 26.91 percent during the 2010-11 census which increased to 27.12 percent during 2015-16. It was a change of only 0.21 percent of the area.

Scheduled Tribes agricultural population have recorded the decreasing share of the number of holding of medium group and area also which shows an unhealthy situation in big holding. They are losing the number and area of medium holdings.

5. **Large Holdings:** Large holding has been defined as the holding that consists of the size of 10 hectare and above. There were 20 thousand holdings in this category in Jharkhand during the census of 2010-11 and the numbers remained same during the subsequent census of 2015-16 as well. So far, the area is concerned, there were 311 thousand hectares are under the large holdings. Census of 2015-16 has recorded 304 thousand hectares are under large holding in the state. This is a loss of 7 thousand hectares of area under this category. Large holding has recorded 0.75 percent of the total operational number of holdings during 2010-11 and it was decreased to 0.71 percent during the next census of 2015-16.

Area under this category has recorded 311 thousand hectares during 2010-11 and it was 304 thousand hectares during 2015-16. Data shows that the percentage remained same during both the censuses.

Number of large holdings of Scheduled Caste was 1000 in number during 2010-11 census and it remained same during the next 2015-16 census as well. The area recorded under large holdings for Scheduled Caste was 176 thousand hectares during the 2010-11 census and it remained the same during next census as well. Percentage distribution also remains the same during both censuses.

Number of Large holdings of Scheduled Tribe enumerated as 12 thousand during the census of 2010-11 and 2015-16 as well. Thus, the number of Large Holdings has remained same for Scheduled Tribes as well for both censuses. However, the percentage shows a decrease from 1.27 percent during 2010-11 to 1.20 percent in 2015-16 census. Area of large holdings among tribal agricultural population was 190 thousand during 2010-11 census and it was registered 182 thousand hectares during 2015-16. It was a decrease of 8 thousand hectares in this category. Figure of percentage also shows a decrease from 13.30 percent to 13.14 percent during the census of 2010-11 and 2015-16.

Hence about the large holding we may conclude that in

Jharkhand the Number and Area both show neither a decrease or increase in numbers over two censuses. It implies ST are in a better position than the SC population in Jharkhand state as far as Large holdings are concerned.

## Holdings by Size in Districts

Prof. Jasbir Singh has rightly observed in his book "An agricultural Geography of Haryana" (1976, P. 159) during the analysis of Land Holdings, Land Tenure and size of Operational Holdings that, "Launching a direct attack on technical problem alone i.e. extension work aimed at the use of improved technical methods and modern agricultural requisites, will by itself be not sufficient to ensure success for the proposed agricultural development plans and package programs. Simultaneous attention must be given to other problems which at the moment are impediments to the rapid development of agricultural production. Among these problems some are sociological. The most important sociological impediments in oriental agrarian society are: 1. Regime of ownership and problems of fragmented holdings, 2. Land Tenure problems and 3. Limitations imposed by the size of operational holdings."

No doubt the ownership of the land and land tenure systems are more directly related to the socio-judicial organization of the rural population of the state but the influence of the size of an individual holding is even more powerful impediment in agricultural development i.e. tiny sized fragmented sub-marginal and marginal uneconomic holdings. Here we shall concentrate on discussing the holding by their size in districts of the state in detail. Hence, data has been arranged as 'Holding Density' and the amount of land per hectare ratio, calculated from the tables of Agricultural Census of India of 2010-11 and 2015-16.

There is a common formula to calculate the average size of the holding. It is a very simple equation that divides the number of cultivators to the cultivated area. This equation shows only the pressure of cultivators per unit of cultivated area. This equation shows only the pressure of cultivators over per unit of cultivated area. In other words, it will be called average cultivated area per cultivator. But in particular the size of land holding may be calculated by relating the area of land holding to the number of holdings. This will be an exact size of the holding of the area. At the same time if we calculate by dividing number of holding to the total area of holding, it will be the density of holdings for the area. This is perhaps a new concept because this is no where in the entire references of thesis. This idea was introduced by my professor Dr. Ram Kumar Tiwari during his teaching of Agricultural Geography in the classes of Masters Degree. Concept was so useful that I still remember and could use in this proper place.

## Holding Density

Using the technique mentioned above, the land holding density has been computed and table was prepared for the census of 2010-11 illustration. The table/map is showing the average holding density of the state is 0.86 Holdings per hectare area of the state. The idea generates that the high density of holding has concentrated in the district of Chatra, Dhanbad, E. Singhbhum, Garhwa, Giridih, Hazaribagh, Pakur, Palamu and Ranchi. These are nine in number where density of holding is more than 1 holding per hectare. Medium density has been enumerated in the district of Bokaro, Deoghar, Gumla, Jamtara, Khunti, Koderma, Latehar, Lohardaga, Ranchi, Saraikela-kharsawan. These are ten in number where density of holding is in between 1 to 0.50 holding per hectare. Now the low density of holding has been concentrated in the districts of Dumka, Godda, Ramgarh, Simdega and W. Singhbhum. These are five in number where the land holding density is below 0.50 holdings per hectare.

When we study the same thing for the census of 2015-16 the high density of land holding has concentrated in Chatra, Dhanbad, E. Singhbhum, Garhwa, Giridih, Hazaribagh, Koderma, Pakur, Palamu and Sahebganj. These are ten in number where holding density is more than 1 holding per hectare. The medium density of holdings has been found in the districts of Bokaro, Deoghar, Dumka, Godda, Gumla,

Jamtara, Khunti, Latehar, Lohardaga, Ramgarh, Ranchi, Saraikela-Kharsawan, Simdega and W. Singhbhum. These are fourteen in number where the density of holding is in between 1 to 0.50 holding per hectare.

We did not find any districts in the state with Low density of holding.

When we compare the changes recorded between these

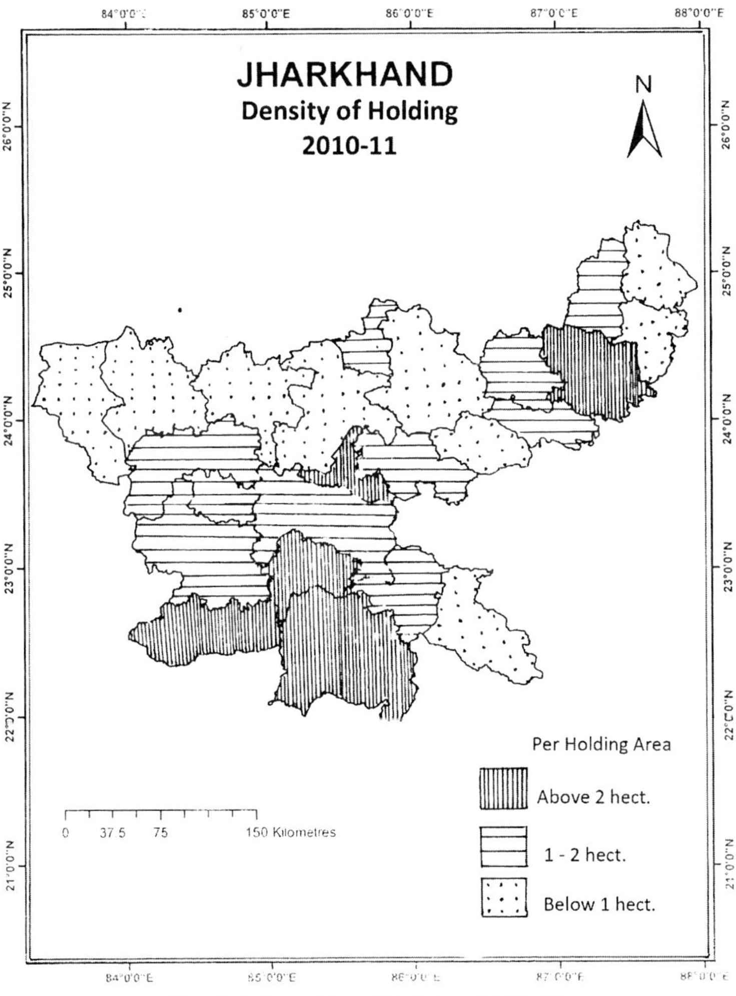


**Map 2.1: Density of Holding (2010-11)**

two censuses 2010-11 and 2015-16, it clearly shows that during the census of 2010-11 high density was recorded only in nine districts while in 2015-16 it was recorded in ten districts. District Koderma was under the category of Medium density during 2010-11 which has changed its category to high in 2015-16. Similarly, under medium category there were only ten districts in 2010-11 which increased to fourteen districts during the census of 2015-16. Four new districts were added

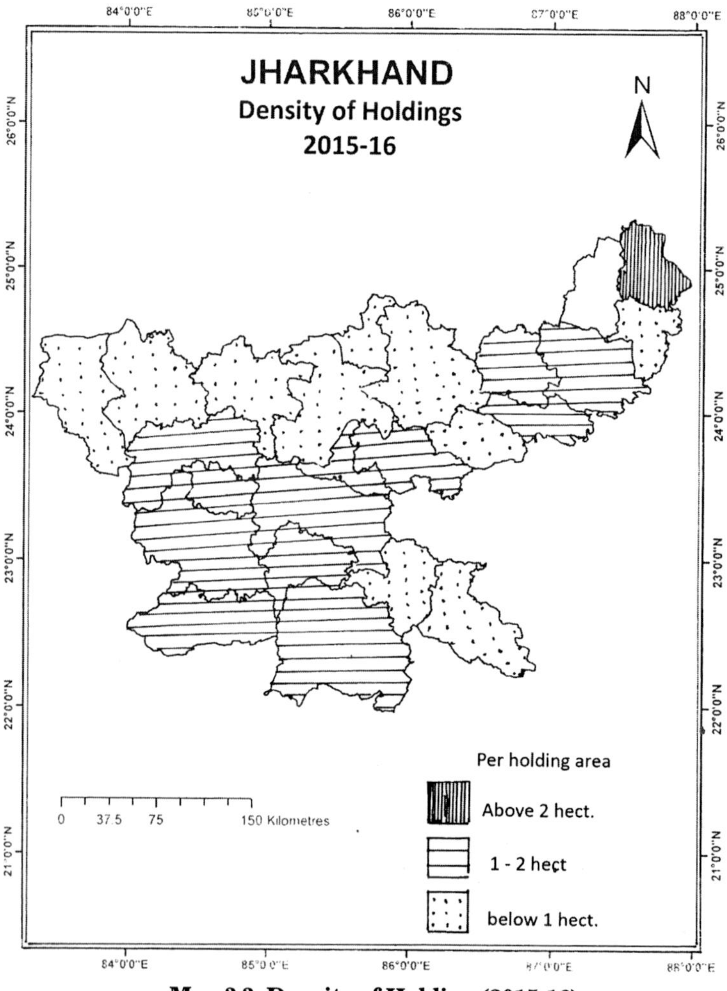


**Map 2.2: Density of Holding (2015-16)**

to this category which were listed under low density holding in the previous census. This is the reason there is not a single district listed under low density category.

Changing in density in medium category is high that shows a large transformation of holdings under this category. This indicates that small size holdings are being

transferred to medium size holdings. We can further imagine that poor cultivators who were possessing small holdings, transferred their holdings to richer more capable cultivators which then changed their size of holdings.

## Size of Holdings

Size of holding generally is the measure of the average area per holding. This can be calculated by dividing the area of land holding to the number of holding in the district. Here table has been prepared through the data obtained from Agricultural census of 2010-11 and 2015-16. There were 2803 thousand holdings in number during the census of 2015-16. It was 2709 thousand during the census of 2010-11. On the other hand, the area of these holdings was 3091 thousand hectare and 3165 thousand hectares during their respective census years. The figure shows their holding size during 2010-11 that the average of state was 1.17 hectare per holding. This was reduced to 1.10 hectares per holding during the next census of 2015-16. Now it is clear that the size of holding is reducing year by year in the state. Certainly, this is the result of fragmentation of land.

When we compare to the holding size of the country it was 1.15 hectare per holding during the census of 2010-11 and it reduced to 1.07 hectare per holding during the next census of 2015-16. Thus, the average land holding size of Jharkhand is always higher than the average holding size of the country. So far, the reduction in land is concerned, the country has recorded a reduction of 6.95 percent in the holding size whereas, Jharkhand has recorded a reduction of only 5.98 percent. This shows Jharkhand in a better position than the national average.

**Census:2010-11:**

District wise analysis of Jharkhand in land holding size depicted above 2 hectare per holding in the district of Simdega (2.16 hect/holding), Ramgarh (2,04 hect/holding), West Singhbhum (2.03 hect/holding) and Khunti (2.01 hect/holding) respectively during the census of 2010-11. Land holding size in between 1 to 2 hectares/holding has been enumerated in the districts of Gumla (1.63 hect/holding), Lohardaga (1.68 hect/holding), Latehar (1.54 hect/holding), Jamtara (1.50 hect/holding), Ranchi (1.42 hect/holding), Deoghar (1.33 hect/holding), Godda (1.32 hect/holding), Koderma (1.24 hect/holding), Bokaro (1.19 hect/holding) and Saraikela-Kharsawan (1.16 hect/holding) respectively.

Land holding size below 1 hect. per holding during the census of 2010-11 has been recorded in the district of Chatra (0.95 hect), Dhanbad (0.84 hect), E. Singhbhum (0.86 hect), Garhwa (0.78 hect), Giridih (0.63 hect), Hazaribagh (0.79 hect), Pakur (0.74 hect), Palamu (0.65 hect) and Sahebganj (0.95 hect)

**Census: 2015-16:**

District wise analysis of land holding size during the census of 2015-16 depicted a picture that the holding size above 2 hectares per holding containing districts is only one i.e. Sahebganj (2.15 hect/holding).

Holding size in between 1 to 2 hectares belong to the district of Khunti (1.83 hect), Dumka (1.77 hect), Simdega (1.67 hect), Lohardaga (1.62 hect), Gumla (1.60 hect), Jamtara (1.56 hect), Ramgarh (1.54 hect), W. Singhbhum (1.49 hect), Latehar (1.43 hect), Ranchi (1.36 hect), Deoghar (1.32 hect), Godda (1.19 hect) and Bokaro (1.08 hect).

Holding size below 1 hectare per holding has includes the districts Chatra (0.85 hect), Dhanbad (0.77 hect), E. Singhbhum (0.89 hect), Garhwa (0.83 hect), Giridih (0.61 hect), Hazaribagh (0.76 hect), Koderma (0.97 hect), Pakur (0.87 hect), Palamu (0.68 hect) and Saraikela - Kharsawan (0.45 hect).

When we compare the result of the data analysis of the census of 2010- 11 and 2015-16 at first all attention goes to the greater change in the holding size of above 2 hectare per holding has decreased. Five districts were having above 2 hect/per holding size and after five years it concentrated to only one district. This shows a large number of fragmentations of the holdings. Again, we notice an increase in number of districts belonging to the holding size less than 1 hectare per holding. Maximum concentration of land holding size between 1 and 2 hectares has been observed in maximum number of districts. All these changes indicate towards the rapid fragmentation of the land holding.

Situation of Sahebganj district has a different dimension. During the census of 2010-11 Sahebganj was under the category of below 1 hectare per holding and within a time period of five years it reached to the category of 2 hectare per holding category. This seems a very strange phenomenon and a question mark before agriculture scientists. When we take a deep dive into the data for the district, we notice a great decrease in number of holdings as well as reduction in area too. Area has decreased perhaps due to transfer of land category from cultivated to uncultivated or non-agricultural use. While decrease in number of holdings indicate transfer of holdings from small cultivators to rich ones as the larger holdings belong to rich cultivators without a doubt.

**Table 2.14**
**District wise Number and Size of Holdings in Jharkhand (2010-11)**

| Districts | Number of Holdings | Holding Density (Number of Holding per hectare/ Area of Holding) | Area of Holdings (Hectare) | Area/ Holding = Per holding Area (Hectare) |
|---|---|---|---|---|
| Bokaro | 68389 | 0.84 | 81275 | 1.19 |
| Chatra | 123117 | 1.05 | 117020 | 0.95 |
| Deoghar | 110698 | 0.75 | 147212 | 1.33 |
| Dhanbad | 49127 | 1.20 | 41081 | 0.84 |
| Dumka | 55554 | 0.48 | 115529 | 2.08 |

| | | | | |
|---|---|---|---|---|
| E. Singhbhum | 167393 | 1.16 | 143888 | 0.86 |
| Garhwa | 180500 | 1.29 | 140053 | 0.78 |
| Giridih | 236921 | 1.59 | 148729 | 0.63 |
| Godda | 83406 | 0.03 | 110452 | 1.32 |
| Gumla | 146930 | 0.61 | 238941 | 1.63 |
| Hazaribagh | 127887 | 1.27 | 100502 | 0.79 |
| Jamtara | 53126 | 0.66 | 79612 | 1.50 |
| Khunti | 60593 | 0.50 | 122028 | 2.01 |
| Koderma | 14580 | 0.81 | 18086 | 1.24 |
| Latehar | 71078 | 0.65 | 109100 | 1.54 |
| Lohardaga | 42021 | 0.61 | 68593 | 1.63 |
| Pakur | 117074 | 1.35 | 86720 | 0.74 |
| Palamu | 270628 | 1.53 | 176534 | 0.65 |
| Ramgarh | 30382 | 0.49 | 62009 | 2.04 |
| Ranchi | 202784 | 0.70 | 287891 | 1.42 |
| Sahibganj | 94200 | 1.05 | 89639 | 0.95 |
| Saraikela-Kharsawan | 138126 | 0.86 | 160236 | 1.16 |
| Simdega | 70393 | 0.46 | 152007 | 2.16 |
| W. Singhbhum | 140800 | 0.49 | 285548 | 2.03 |
| Jharkhand | 2709000 | 0.86 | 3165000 | 1.17 |

**Source:** Agricultural Census of India 2010-11

**Table 2.15**
**District wise Number and Size of Holdings in Jharkhand (2015-16)**

| Districts | Number of Holdings | Holding Density (Number of Holding per hectare/ Area of Holding) | Area of Holdings (Hectare) | Area/ Holding = Per holding Area (Hectare) |
|---|---|---|---|---|
| Bokaro | 74398 | 0.93 | 80218 | 1.08 |
| Chatra | 144127 | 1.18 | 121991 | 0.85 |
| Deoghar | 99035 | 0.76 | 130871 | 1.32 |
| Dhanbad | 50653 | 1.30 | 39113 | 0.77 |
| Dumka | 108103 | 0.57 | 19112 | 1.77 |
| E. Singhbhum | 157689 | 1.23 | 139725 | 0.89 |
| Garhwa | 183394 | 1.21 | 151698 | 0.83 |
| Giridih | 239893 | 1.64 | 146112 | 0.61 |
| Godda | 85148 | 0.84 | 101638 | 1.19 |
| Gumla | 151756 | 0.63 | 242232 | 1.60 |
| Hazaribagh | 151258 | 1.31 | 115592 | 0.76 |

| | | | | |
|---|---|---|---|---|
| Jamtara | 54408 | 0.64 | 84887 | 1.56 |
| Khunti | 60491 | 0.55 | 110928 | 1.83 |
| Koderma | 35750 | 1.04 | 34502 | 0.97 |
| Latehar | 75190 | 0.70 | 107132 | 1.43 |
| Lohardaga | 47168 | 0.62 | 76224 | 1.62 |
| Pakur | 100685 | 1.15 | 87657 | 0.87 |
| Palamu | 274040 | 1.48 | 185215 | 0.68 |
| Ramgarh | 33632 | 0.65 | 51934 | 1.54 |
| Ranchi | 201971 | 0.73 | 275427 | 1.36 |
| Sahibganj | 88337 | 1.07 | 82614 | 2.15 |
| Saraikela-Kharsawan | 132512 | 0.94 | 140860 | 0.45 |
| Simdega | 84593 | 0.60 | 141380 | 1.67 |
| W. Singhbhum | 168625 | 0.67 | 251593 | 1.49 |
| Jharkhand | 2803000 | 0.91 | 3091000 | 1.10 |

**Source:** Agricultural Census of India 2015-16

**Table 2.16**
**Number and Area of Holdings by Size Class in Jharkhand Agriculture Census 2010-11**

| District | Category | Size (Hectare) | Number of Holdings | % to Total | Area of Holding | % to Total |
|---|---|---|---|---|---|---|
| Bokaro | Marginal | 0-1 | 43671 | 63.86 | 19631 | 24.15 |
| | Small | 1-2 | 19157 | 28.01 | 32324 | 39.77 |
| | Semi-Medium | 2-4 | 3739 | 5.46 | 13990 | 17.21 |
| | Medium | 4-10 | 1785 | 2.61 | 14297 | 5.28 |
| | Large | Above 10 | 37 | 0.05 | 1035 | 1.27 |
| | **Total** | | **68389** | | **81275** | |
| Chatra | Marginal | 0-1 | 84272 | 68.44 | 36051 | 30.80 |
| | Small | 1-2 | 31157 | 25.30 | 45025 | 38.47 |
| | Semi-Medium | 2-4 | 6279 | 5.10 | 25057 | 21.41 |
| | Medium | 4-10 | 1403 | 1.13 | 10673 | 9.12 |
| | Large | Above 10 | 6 | 0.0048 | 214 | 0.18 |
| | **Total** | | **123117** | | **117020** | |
| Deoghar | Marginal | 0-1 | 66068 | 59.68 | 34922 | 23.72 |
| | Small | 1-2 | 32781 | 29.61 | 53458 | 36.31 |
| | Semi-Medium | 2-4 | 8529 | 7.70 | 30627 | 20.80 |
| | Medium | 4-10 | 3213 | 2.90 | 24629 | 16.73 |

| | | | | | | |
|---|---|---|---|---|---|---|
| | Large | Above 10 | 107 | 0.096 | 3577 | 2.42 |
| | **Total** | | **110698** | | **147212** | |
| Dhanbad | Marginal | 0-1 | 38016 | 77.38 | 15779 | 38.40 |
| | Small | 1-2 | 8770 | 17.85 | 14038 | 34.17 |
| | Semi-Medium | 2-4 | 1638 | 3.33 | 6083 | 14.80 |
| | Medium | 4-10 | 693 | 1.41 | 4914 | 11.96 |
| | Large | Above 10 | 10 | 0.020 | 267 | 0.64 |
| | **Total** | | **49127** | | **41081** | |
| Dumka | Marginal | 0-1 | 23082 | 41.44 | 12063 | 10.44 |
| | Small | 1-2 | 18709 | 33.67 | 35788 | 30.97 |
| | Semi-Medium | 2-4 | 9931 | 17.87 | 39129 | 33.86 |
| | Medium | 4-10 | 3829 | 6.89 | 28488 | 24.65 |
| | Large | Above 10 | 3 | 0.0054 | 60 | 0.051 |
| | **Total** | | **55554** | | **115529** | |
| East Singhbhum | Marginal | 0-1 | 125001 | 74.67 | 48323 | 33.58 |
| | Small | 1-2 | 34947 | 20.87 | 60107 | 41.77 |
| | Semi-Medium | 2-4 | 5753 | 3.43 | 21184 | 14.72 |
| | Medium | 4-10 | 1670 | 0.99 | 11872 | 8.25 |
| | Large | Above 10 | 22 | 0.013 | 2402 | 1.66 |
| | **Total** | | **167393** | | **143888** | |
| Garhwa | Marginal | 0-1 | 147560 | 81.75 | 62782 | 44.89 |
| | Small | 1-2 | 25321 | 14.02 | 42118 | 30.07 |
| | Semi-Medium | 2-4 | 5953 | 3.29 | 22225 | 15.86 |
| | Medium | 4-10 | 1621 | 0.89 | 11605 | 8.28 |
| | Large | Above 10 | 51 | 0.028 | 1232 | 0.87 |
| | **Total** | | **180500** | | **140053** | |
| Giridih | Marginal | 0-1 | 191681 | 80.90 | 67783 | 45.57 |
| | Small | 1-2 | 40187 | 16.96 | 59757 | 40.16 |
| | Semi-Medium | 2-4 | 4565 | 1.92 | 16152 | 10.86 |
| | Medium | 4-10 | 688 | 0.29 | 5037 | 3.38 |
| | Large | Above 10 | 00 | 00 | 00 | |
| | **Total** | | **236921** | | **198729** | |
| Godda | Marginal | 0-1 | 49553 | 59.41 | 22625 | 20.48 |
| | Small | 1-2 | 24079 | 28.86 | 41149 | 37.25 |
| | Semi-Medium | 2-4 | 7014 | 8.40 | 25418 | 23.01 |

| | | | | | | |
|---|---|---|---|---|---|---|
| | Medium | 4-10 | 2743 | 3.28 | 20863 | 18.88 |
| | Large | Above 10 | 17 | 0.020 | 395 | 0.35 |
| | **Total** | | **83406** | | **110452** | |
| Gumla | Marginal | 0-1 | 86451 | 59.03 | 42010 | 17.58 |
| | Small | 1-2 | 37296 | 25.47 | 65385 | 27.86 |
| | Semi-Medium | 2-4 | 14632 | 9.99 | 56166 | 23.50 |
| | Medium | 4-10 | 7614 | 5.199 | 65473 | 27.40 |
| | Large | Above 10 | 437 | 0.2984 | 9907 | 4.14 |
| | **Total** | | **146430** | | **238941** | |
| Hazaribagh | Marginal | 0-1 | 102762 | 80.35 | 42921 | 42.70 |
| | Small | 1-2 | 20644 | 16.14 | 35440 | 35.26 |
| | Semi-Medium | 2-4 | 3331 | 2.60 | 12357 | 12.29 |
| | Medium | 4-10 | 1110 | 0.867 | 8617 | 8.573 |
| | Large | Above 10 | 40 | 0.031 | 1166 | 0.015 |
| | **Total** | | **127887** | | **100502** | |
| Jamtara | Marginal | 0-1 | 32203 | 60.61 | 16826 | 21.13 |
| | Small | 1-2 | 13849 | 26.06 | 27264 | 34.24 |
| | Semi-Medium | 2-4 | 5731 | 9.658 | 19686 | 24.72 |
| | Medium | 4-10 | 1876 | 3.531 | 14424 | 18.117 |
| | Large | Above 10 | 67 | 0.126 | 1413 | 1.77 |
| | **Total** | | **53126** | | **79612** | |
| Khunti | Marginal | 0-1 | 35023 | 57.8004 | 16095 | 13.189 |
| | Small | 1-2 | 14551 | 24.014 | 27618 | 22.63 |
| | Semi-Medium | 2-4 | 5257 | 8.675 | 20466 | 16.77 |
| | Medium | 4-10 | 5317 | 8.77 | 44355 | 36.34 |
| | Large | Above 10 | 67 | 0.126 | 1413 | 1.77 |
| | **Total** | | **60593** | | **122028** | |
| Koderma | Marginal | 0-1 | 9727 | 66.71 | 4269 | 23.60 |
| | Small | 1-2 | 3256 | 22.331 | 5895 | 32.59 |
| | Semi-Medium | 2-4 | 1084 | 7.434 | 4243 | 23.46 |
| | Medium | 4-10 | 511 | 3.504 | 3629 | 20.06 |
| | Large | Above 10 | 2 | 0.0137 | 50 | 0.276 |
| | **Total** | | **14580** | | **18086** | |
| Latehar | Marginal | 0-1 | 45223 | 63.63 | 18505 | 16.36 |
| | Small | 1-2 | 15274 | 2.181 | 24716 | 22.65 |

| | | | | | | |
|---|---|---|---|---|---|---|
| | Semi-Medium | 2-4 | 5442 | 7.656 | 20352 | 18.65 |
| | Medium | 4-10 | 4928 | 0.07039 | 40335 | 36.97 |
| | Large | Above 10 | 211 | 0.2968 | 5193 | 4.759 |
| | **Total** | | **71078** | | **109100** | |
| Lohardaga | Marginal | 0-1 | 24801 | 59.020 | 10514 | 14.80 |
| | Small | 1-2 | 10283 | 24.471 | 18583 | 27.091 |
| | Semi-Medium | 2-4 | 3881 | 9.235 | 15131 | 22.059 |
| | Medium | 4-10 | 3004 | 701488 | 23375 | 34.077 |
| | Large | Above 10 | 52 | 0.123 | 1344 | 1.95 |
| | **Total** | | **42021** | | **63593** | |
| Pakur | Marginal | 0-1 | 94723 | 80.90 | 36719 | 42.34 |
| | Small | 1-2 | 18738 | 15.70 | 33101 | 38.17 |
| | Semi-Medium | 2-4 | 3493 | 2.98 | 12859 | 14.82 |
| | Medium | 4-10 | 473 | 0.40 | 3881 | 4.48 |
| | Large | Above 10 | 7 | 0.0059 | 160 | 0.18 |
| | **Total** | | **117074** | | **86720** | |
| Palamu | Marginal | 0-1 | 222877 | 82.35 | 72111 | 40.84 |
| | Small | 1-2 | 38209 | 13.010 | 59856 | 33.90 |
| | Semi-Medium | 2-4 | 7254 | 2.6804 | 26433 | 14.97 |
| | Medium | 4-10 | 2190 | 0.8092 | 15899 | 9.0061 |
| | Large | Above 10 | 93 | 0.034 | 2236 | 1.266 |
| | **Total** | | **270628** | | **176534** | |
| Ramgarh | Marginal | 0-1 | 15176 | 49.95 | 7824 | 12.61 |
| | Small | 1-2 | 9108 | 29.97 | 16537 | 26.668 |
| | Semi-Medium | 2-4 | 3700 | 12.178 | 15149 | 24.43 |
| | Medium | 4-10 | 2321 | 7.639 | 20548 | 33.137 |
| | Large | Above 10 | 77 | 2.27 | 1952 | 3.147 |
| | **Total** | | **30382** | | **62009** | |
| Ranchi | Marginal | 0-1 | 106897 | 52.71 | 52324 | 18.21 |
| | Small | 1-2 | 74166 | 36.57 | 116174 | 40.43 |
| | Semi-Medium | 2-4 | 13126 | 6.4728 | 50734 | 17.66 |
| | Medium | 4-10 | 8250 | 4.068 | 59476 | 20.70 |
| | Large | Above 10 | 345 | 0.0701 | 8574 | 2.98 |

| | | | | | | |
|---|---|---|---|---|---|---|
| | **Total** | | **202784** | | **287291** | |
| Sahebganj | Marginal | 0-1 | 69219 | 73.41 | 29650 | 30.06 |
| | Small | 1-2 | 19152 | 20.31 | 31762 | 32.20 |
| | Semi-Medium | 2-4 | 4236 | 4.49 | 15925 | 17.77 |
| | Medium | 4-10 | 1678 | 1.78 | 18188 | 20.29 |
| | Large | Above 10 | 5 | 0.005 | 115 | 0.13 |
| | **Total** | | **94290** | | **89639** | |
| Saraikela-Kharsawan | Marginal | 0-1 | 92224 | 66.76 | 37084 | 23.143 |
| | Small | 1-2 | 35626 | 25.79 | 65134 | 40.64 |
| | Semi-Medium | 2-4 | 8027 | 5.811 | 36852 | 22.83 |
| | Medium | 4-10 | 2234 | 1.6173 | 20540 | 12.818 |
| | Large | Above 10 | 15 | 0.010 | 623 | 0.3888 |
| | **Total** | | **138126** | | **160236** | |
| Simdega | Marginal | 0-1 | 34143 | 48.503 | 15060 | 9.907 |
| | Small | 1-2 | 21722 | 30.858 | 39801 | 26.183 |
| | Semi-Medium | 2-4 | 7461 | 10.599 | 29311 | 19.28 |
| | Medium | 4-10 | 6701 | 9.519 | 57511 | 37.83 |
| | Large | Above 10 | 366 | 0.5199 | 10324 | 6.7919 |
| | **Total** | | **70393** | | **152007** | |
| West-Singhbhum | Marginal | 0-1 | 77079 | 54.743 | 27432 | 9.2817 |
| | Small | 1-2 | 34848 | 24.75 | 59489 | 20.833 |
| | Semi-Medium | 2-4 | 13831 | 9.823 | 51200 | 17.93 |
| | Medium | 4-10 | 13811 | 9.8089 | 11646 | 40.77 |
| | Large | Above 10 | 1231 | 0.874 | 30981 | 10.84 |
| | **Total** | | **140800** | | **285548** | |
| **Total/ Jharkhand** | **Marginal** | **0-1** | **1848000** | | **764000** | **24.13** |
| | **Small** | **1-2** | **429000** | | **591000** | **18.67** |
| | **Semi-Medium** | **2-4** | **283000** | | **775000** | **24.48** |
| | **Medium** | **4-10** | **129000** | | **725000** | **22.90** |
| | **Large** | **Above 10** | **20000** | | **311000** | **9.82** |
| **Total** | **All Groups** | **2709000** | **2709000** | **0.20** | **3165000** | **0.20** |

**Source:** Agriculture Census 2010-11

**Table 2.17**
**Number and Area of Holdings by Size Class in Jharkhand Agriculture Census 2015-16**

| District | Category | Size (Hectare) | Number of Holdings | % to Total | Area of Holding | % to Total |
|---|---|---|---|---|---|---|
| Bokaro | Marginal | 0-1 | 51031 | 68.59 | 19238 | 23.98 |
| | Small | 1-2 | 17549 | 23.58 | 29592 | 36.88 |
| | Semi-Medium | 2-4 | 3864 | 5.19 | 14560 | 18.15 |
| | Medium | 4-10 | 1885 | 2.53 | 15051 | 18.76 |
| | Large | Above 10 | 69 | 0.092 | 1777 | 2.21 |
| | **Total** | | **74398** | | **80218** | |
| Chatra | Marginal | 0-1 | 107742 | 74.75 | 41625 | 34.12 |
| | Small | 1-2 | 29449 | 20.43 | 45895 | 37.62 |
| | Semi-Medium | 2-4 | 5431 | 3.76 | 21468 | 17.59 |
| | Medium | 4-10 | 1395 | 0.96 | 10370 | 8.50 |
| | Large | Above 10 | 110 | 0.076 | 2633 | 2.15 |
| | **Total** | | **144127** | | **121991** | |
| Deoghar | Marginal | 0-1 | 60612 | 61.20 | 29168 | 22.28 |
| | Small | 1-2 | 27939 | 28.21 | 45730 | 39.94 |
| | Semi-Medium | 2-4 | 7476 | 7.54 | 28693 | 21.92 |
| | Medium | 4-10 | 2840 | 2.86 | 23321 | 17.81 |
| | Large | Above 10 | 168 | 0.16 | 3959 | 3.02 |
| | **Total** | | **99035** | | **130871** | |
| Dhanbad | Marginal | 0-1 | 31090 | 61.37 | 17764 | 45.41 |
| | Small | 1-2 | 7760 | 15.31 | 12851 | 32.85 |
| | Semi-Medium | 2-4 | 1411 | 2.78 | 5136 | 13.13 |
| | Medium | 4-10 | 382 | 0.75 | 3086 | 7.88 |
| | Large | Above 10 | 10 | 0.019 | 275 | 0.70 |
| | **Total** | | **50653** | | **39113** | |
| Dumka | Marginal | 0-1 | 55526 | 51.36 | 25355 | 13.26 |
| | Small | 1-2 | 32487 | 30.05 | 57890 | 30.27 |
| | Semi-Medium | 2-4 | 14509 | 13.421 | 56073 | 23.32 |
| | Medium | 4-10 | 5986 | 5.537 | 47483 | 24.83 |
| | Large | Above 10 | 195 | 0.183 | 4392 | 2.29 |

| | | | | | | |
|---|---|---|---|---|---|---|
| | **Total** | | **108103** | | **191192** | |
| East Singhbhum | Marginal | 0-1 | 119533 | 75.83 | 43920 | 31.43 |
| | Small | 1-2 | 30217 | 19.162 | 52484 | 37.56 |
| | Semi-Medium | 2-4 | 5346 | 3.390 | 20108 | 14.07 |
| | Medium | 4-10 | 2524 | 1.6006 | 19667 | 14.07 |
| | Large | Above 10 | 69 | 0.043 | 3545 | 2.53 |
| | **Total** | | **108103** | | **191192** | |
| Garhwa | Marginal | 0-1 | 145880 | 79.54 | 56640 | 37.33 |
| | Small | 1-2 | 28262 | 15.41 | 45484 | 29.98 |
| | Semi-Medium | 2-4 | 6308 | 4.55 | 23898 | 15.75 |
| | Medium | 4-10 | 2824 | 1.53 | 22743 | 14.99 |
| | Large | Above 10 | 120 | 0.064 | 2933 | 1.933 |
| | **Total** | | **183394** | | **151698** | |
| Giridih | Marginal | 0-1 | 195270 | 81.39 | 64027 | 43.82 |
| | Small | 1-2 | 39908 | 16.63 | 61380 | 42.00 |
| | Semi-Medium | 2-4 | 4049 | 1.687 | 15130 | 10.35 |
| | Medium | 4-10 | 641 | 0.26 | 4924 | 3.37 |
| | Large | Above 10 | 25 | 0.010 | 651 | 0.42 |
| | **Total** | | **239893** | | **146112** | |
| Godda | Marginal | 0-1 | 53670 | 63.03 | 22826 | 22.45 |
| | Small | 1-2 | 23161 | 27.20 | 36738 | 36.14 |
| | Semi-Medium | 2-4 | 5809 | 6.82 | 21454 | 21.10 |
| | Medium | 4-10 | 2413 | 2.83 | 18447 | 18.14 |
| | Large | Above 10 | 95 | 0.11 | 2174 | 2.13 |
| | **Total** | | **85148** | | **101638** | |
| Gumla | Marginal | 0-1 | 95336 | 62.82 | 42270 | 17.45 |
| | Small | 1-2 | 33361 | 21.98 | 60759 | 25.08 |
| | Semi-Medium | 2-4 | 13644 | 8.99 | 52140 | 21.52 |
| | Medium | 4-10 | 8853 | 5.83 | 72931 | 30.10 |
| | Large | Above 10 | 562 | 0.37 | 14133 | 5.83 |
| | **Total** | | **151756** | | **242232** | |
| Hazaribagh | Marginal | 0-1 | 120572 | 79.71 | 44508 | 38.50 |
| | Small | 1-2 | 24246 | 16.02 | 37661 | 32.58 |
| | Semi-Medium | 2-4 | 4636 | 3.06 | 17601 | 15.22 |
| | Medium | 4-10 | 1713 | 1.13 | 13180 | 11.40 |

| | | | | | | |
|---|---|---|---|---|---|---|
| | Large | Above 10 | 91 | 0.060 | 2642 | 2.28 |
| | **Total** | | **151258** | | **115592** | |
| Jamtara | Marginal | 0-1 | 32590 | 59.89 | 14885 | 17.53 |
| | Small | 1-2 | 13965 | 25.66 | 26464 | 31.17 |
| | Semi-Medium | 2-4 | 4735 | 8.70 | 17741 | 20.89 |
| | Medium | 4-10 | 3025 | 5.55 | 23504 | 27.68 |
| | Large | Above 10 | 93 | 0.17 | 2081 | 2.45 |
| | **Total** | | **54408** | | **84887** | |
| Khunti | Marginal | 0-1 | 35357 | 58.45 | 15848 | 14.28 |
| | Small | 1-2 | 15230 | 25.17 | 27569 | 24.85 |
| | Semi-Medium | 2-4 | 5137 | 8.49 | 19818 | 17.86 |
| | Medium | 4-10 | 4450 | 7.35 | 38396 | 34.61 |
| | Large | Above 10 | 320 | 0.52 | 9303 | 8.38 |
| | **Total** | | **60491** | | **110928** | |
| Koderma | Marginal | 0-1 | 25389 | 71.01 | 12387 | 35.90 |
| | Small | 1-2 | 8892 | 24.87 | 15663 | 45.39 |
| | Semi-Medium | 2-4 | 1245 | 3.48 | 4584 | 13.28 |
| | Medium | 4-10 | 219 | 0.612 | 1662 | 4.81 |
| | Large | Above 10 | 5 | 0.013 | 205 | 0.59 |
| | **Total** | | **35750** | | **34502** | |
| Latehar | Marginal | 0-1 | 49288 | 65.55 | 17295 | 16.14 |
| | Small | 1-2 | 16272 | 21.64 | 28229 | 26.34 |
| | Semi-Medium | 2-4 | 4719 | 6.27 | 18211 | 16.99 |
| | Medium | 4-10 | 4736 | 6.29 | 39141 | 36.53 |
| | Large | Above 10 | 175 | 0.23 | 4256 | 3.97 |
| | **Total** | | **75190** | | **107132** | |
| Lohardaga | Marginal | 0-1 | 28794 | 61.045 | 10468 | 13.73 |
| | Small | 1-2 | 10605 | 22.48 | 18861 | 24.74 |
| | Semi-Medium | 2-4 | 4124 | 8.74 | 16030 | 2.1030 |
| | Medium | 4-10 | 3549 | 7.51 | 28409 | 37.27 |
| | Large | Above 10 | 96 | 0.20 | 2456 | 3.222 |
| | **Total** | | **47168** | | **76224** | |
| Pakur | Marginal | 0-1 | 76342 | 75.822 | 28171 | 32.13 |
| | Small | 1-2 | 18983 | 18.85 | 33016 | 37.66 |
| | Semi-Medium | 2-4 | 3894 | 3.867 | 14751 | 16.82 |

| | | | | | | |
|---|---|---|---|---|---|---|
| | Medium | 4-10 | 1432 | 1.422 | 10758 | 12.27 |
| | Large | Above 10 | 34 | 0.3376 | 961 | 1.096 |
| | **Total** | | **100685** | | **87657** | |
| Palamu | Marginal | 0-1 | 222904 | 81.33 | 71559 | 38.63 |
| | Small | 1-2 | 41081 | 14.99 | 62806 | 33.909 |
| | Semi-Medium | 2-4 | 7440 | 2.714 | 27722 | 14.96 |
| | Medium | 4-10 | 2493 | 0.909 | 20113 | 10.85 |
| | Large | Above 10 | 122 | 0.044 | 3015 | 1.627 |
| | **Total** | | **274040** | | **18215** | |
| Ramgarh | Marginal | 0-1 | 19844 | 59 | 8485 | 16.33 |
| | Small | 1-2 | 9230 | 27.44 | 15695 | 30.22 |
| | Semi-Medium | 2-4 | 2619 | 7.78 | 10091 | 19.4304 |
| | Medium | 4-10 | 1887 | 5.61 | 16146 | 31.08 |
| | Large | Above 10 | 52 | 0.15 | 1518 | 2.922 |
| | **Total** | | **33632** | | **51934** | |
| Ranchi | Marginal | 0-1 | 116323 | 57.59 | 51102 | 18.55 |
| | Small | 1-2 | 64251 | 31.81 | 104246 | 37.84 |
| | Semi-Medium | 2-4 | 12696 | 6.286 | 48080 | 17.45 |
| | Medium | 4-10 | 8375 | 4.14 | 62969 | 22.86 |
| | Large | Above 10 | 326 | 0.1614 | 9030 | 3.2785 |
| | **Total** | | **201971** | | **275427** | |
| Sahebganj | Marginal | 0-1 | 65860 | 74.55 | 26263 | 31.79 |
| | Small | 1-2 | 17179 | 19.44 | 29960 | 36.26 |
| | Semi-Medium | 2-4 | 3541 | 4.042 | 13313 | 16.114 |
| | Medium | 4-10 | 1721 | 1.9482 | 12244 | 14.82 |
| | Large | Above 10 | 36 | 0.0407 | 834 | 1.0095 |
| | **Total** | | **88337** | | **82614** | |
| Saraikela-Kharsawan | Marginal | 0-1 | 91996 | 69.42 | 33962 | 24.1104 |
| | Small | 1-2 | 29859 | 21.85 | 49466 | 35.11 |
| | Semi-Medium | 2-4 | 7340 | 5.539 | 29976 | 21.28 |
| | Medium | 4-10 | 3229 | 2.436 | 25336 | 17.98 |
| | Large | Above 10 | 88 | 0.0664 | 2121 | 1.5057 |
| | **Total** | | **132512** | | **140860** | |
| Simdega | Marginal | 0-1 | 50533 | 59.73 | 19980 | 14.132 |
| | Small | 1-2 | 21318 | 25.20 | 38929 | 27.53 |

| | | | | | | |
|---|---|---|---|---|---|---|
| | Semi-Medium | 2-4 | 6858 | 8.107 | 26489 | 18.73 |
| | Medium | 4-10 | 5604 | 6.62 | 48627 | 34.39 |
| | Large | Above 10 | 280 | 0.330 | 7354 | 5.201 |
| | **Total** | | **84593** | | **14130** | |
| West-Singhbhum | Marginal | 0-1 | 100133 | 59.38 | 36265 | 14.41 |
| | Small | 1-2 | 43246 | 25.64 | 73405 | 29.17 |
| | Semi Medium | 2-4 | 15715 | 9.3194 | 59656 | 23.71 |
| | Medium | 4-10 | 9108 | 5.401 | 71507 | 28.42 |
| | Large | Above 10 | 423 | 0.250 | 10760 | 4.2767 |
| | **Total** | | **168625** | | **251593** | |
| **Total/ Jharkhand** | **Marginal** | **0-1** | **1962000** | **69.98** | **754000** | **24.40** |
| | **Small** | **1-2** | **419000** | **14.94** | **569000** | **18.43** |
| | **Semi-Medium** | **2-4** | **277000** | **9.89** | **755000** | **24.44** |
| | **Medium** | **4-10** | **126000** | **4.48** | **708000** | **22.90** |
| | **Large** | **Above 10** | **20000** | **0.71** | **304000** | **9.83** |
| **Total** | **All Groups** | | **2803000** | **0.19** | **3091000** | **0.20** |

**Source:** Agriculture Census 2015-16

## References

A. Prasad (1973); Chotanagpur: Geography of Rural Settlement, Ranchi University, Ranchi.

Camil, Bulke (1963); Yaksha Puja, Ramkatha, Allahabad University, Allahabad.

Chotanagpur Tenancy Act (1931 & 1946); Patna.

Dantawala, M.L. & W. B Donde (1965); The Economic Cultivator, The Journal of Agricultural Economics, Vol-15.

Hallet, M.G & Stewart Mcpherson (1917); Bihar and Orissa district Gazetteer: Ranchi, Province Govt Publication, Superintendent, Govt. Prints, Patna.

Husain, Majid (1996); Systematic Agriculture Geography, Rawat Publications, Jaipur/Delhi.

Kautilyas; Arthshastra, Jaico Publishing House, Mumbai/ Bhopal/Kolkata.

Naylon, J (1959); Land Consolidation in Spain, Annals of the Association of American Geographers, Vol-49.

Singh, Jasbir (1974); An Agricultural Atlas of India: A Geographical Analysis.

Singh, Jasbir & SS Dhillon (1984); Agriculture Geography, Tata Mc Graw Hill Publishing Co. Ltd, New Delhi.

Singh, Jasbir (1976); An Agriculture Geography of Haryana, Vishal Publications, University Campus, Kurukshetra, Haryana.

# 3

## Farm Technology

- Meaning of Technology
- Types of Technology
- Farm Technology
  - Types of Farm Mechanization
  - Progress of Mechanization
  - Factors responsible for expanding mechanization
  - Argument in favor of farm mechanization
  - A Correct Approach
  - Scope of farm mechanization
  - Suggestions to promote farm mechanization.
- Sub-Mission on Agriculture Mechanization
- Methodology for assessment of farm power availability
- Data of Jharkhand
- Analysis of Farm Power Availability
- Mechanization Gaps and Future Requirements
- Farm Workers
- Use of Fertilizers
- High Yielding Seeds
- Plant Protection

### Meaning of Technology

Generally, technology has been defined as "The Knowledge being applied by human being to improve production or marketing process". In other words, it is the operative knowledge used to produce goods and services. A technology which is well suited for a particular industry or a firm at a particular point of time may become outdated at another time when new inventions and innovations take place. Innovation a prime mover capital accumulation take place due to various reasons which can be classified into two broad categories:

a. To reduce cost, increase output and profit.
b. To provide substitutes for the factor which are scarce and therefore priced heavily.

According to prof. Kuznets (1973), "technical progress consists of not only invention and innovations that require heavy capital investment, but also a stream of relatively cheap changes of improvements whose cultivate effect is a drastic reduction in input resources accompanied by increase in output."

## Types of Technology

Technological change may be broadly divided into two parts viz. embodied technological change and disembodied technological change. Embodied change is the introduction of change in physical capital inputs. Therefore, all mechanized changes can be regarded as embodied technological change. For instance, use of tractor, thresher etc. Disembodied change is not embodied in the physical capital. These kinds of changes are organizational. For example, better information which tends to increase the managerial ability of the farm-firm operator.

Besides, the terms like land-augmenting, labor saving, capital saving and neutral technological change are used widely in theory. So, it is necessary to know the meaning of these terms.

1. **Land Augmenting Technological Change:** As it is known land is inelastic in its supply, therefore, some techniques might help to augment land availability. Multiple cropping, fertilizers, pesticides, high yielding variety seeds are such techniques. Suppose an individual possesses 5 hectare of land and he is cultivating paddy. It takes six months to harvest because it is a traditional variety of paddy. Now assume that the farmer adopts a high yielding variety, the

duration of which is only three months, then he can cultivate twice and the total number of hectares cultivated by the individual increase to 10 hectares. These biological, chemical and technical changes introduced in agriculture are referred as land-augmenting technological changes.

2. **Labor Saving Technical Change:** All mechanized changes, like the use of tractors, threshers etc. are called labor saving. These changes are labor displacing. In other words, the utilization of labor in relation to capital becomes less when technological change is labor saving.

3. **Capital Saving Technical Changes:** In the case of capital saving technical change, the capital-labor ratio declines. This kind of change tents to increase the marginal product of capital.

4. **Neutral Technical Changes:** In this case capital-labor ratio remains constant after technical change. It means the capital-labor ratio before technological change and capital-labor ratio after change are equal.

## Farm Technology

At first, we should understand the word farm. A farm is an area of land that is devoted primarily to agriculture processes with the primary objective of producing food and other crops with a house and other building for keeping animals. It is the basic facility in food production. Farming originated independently in different parts of the world, as hunter-gatherer societies transitioned to food production rather than food capture. It may have started about 12000 years ago with the domestication of livestock in the fertile crescent in western Asia soon to be followed by cultivation of crops. In Jharkhand the agriculture may have started in between 10000-6000 years ago (Tiwari, 2019, P.79).

The next word 'Technology' is the sum of techniques, skill, method and processes used in the production of goods or services or in the accomplishment of objectives, such as scientific investigation. Hence the technology is a form of such scientific knowledge which develops machinery and equipment.

Now, in context of agriculture we can say that a "full development of the potential of the soil of any place will never be achieved only by human farm-power and traditional farming practices. Some degree of 'technification' will be essential if the existing level of agricultural production per unit area is to be increased, if new land is to be brought under multiple cropping, if the conflicting demands of food and cash crops are to be resolved by making re-adjustments in the cropping pattern and if the quality of the agricultural products is to be improved" (Singh, 1976, P.187)

In modern usages, 'technification of farm' or 'farm mechanization' covers two aspects - biological and mechanical. Eddowes (1969, P.61) has asserted that the main controllable factors affecting crop production in a given environment are: Seed, fertilizers and manure, mechanical power, chemical power and the sequence of cropping.

By mechanization of agriculture, we mean the use of power operated machines to do various agricultural operations such as irrigation, ploughing, sowing or harvesting in place of human and animal power. It implies the substitution of subsistence farming by mechanized farming. For example, ploughing is to be done by tractor; sowing and application of fertilizer by drill and threshing with combine harvest and so on. J.B. Bhattacharjee defines mechanization of agriculture and farming process as "application of medium power to work on land, usually performed by bullocks, horses and other draught animals or by human labor." Thus, mechanization of agriculture is that part of farm technology which stands for the utilization of machinery in all farming operations right from land preparation up to the marketing of the produce. Mechanization of agriculture necessitates intensive farming or farming on a relatively large scale. However, agriculture operations can be classified into two types. Firstly, traction work which includes ploughing, land levelling, sowing, weeding, and harvesting; secondly, stationary work of irrigation, threshing, crushing of sugarcane or grinding etc. This form of mechanization has tremendous impact on the whole of the economy because it normally leads to proper use of land resource, abundant agricultural surpluses and higher farm income. Farm mechanization is, therefore, bound to have considerable bearing on future development of agriculture and agro-industries in the state. As a consequence, various states especially, Punjab, Haryana, Andhra Pradesh and western part of Uttar Pradesh have experienced mechanization to a greater extent.

## Types of Farm Mechanisation

Mechanization may either be partial or complete. It is partial when only a part of farm work is done by machine. When animals or human labor is completely dispensed with by power supplying machines, it is termed as complete. However, some of the more used agricultural machines in India are as follows:

1. Tractors: Completely dispenses wooden plough.
2. Combine Drills: Performs sets of operation of sowing and putting fertilizer simultaneously.
3. Combined Harvesters: Performs the work of reaping and threshing.
4. Oil Electric Pumps: Used for irrigation purposes.
5. Aero planes: Used for dusting and saving crops from diseases.
6. Sugarcane Harvesters: Cuts the canes and loads it at a speed of 7 or 8 mounds an hour.

## Progress of Mechanisation

M L. Darling has rightly affirmed that "the plough that looks like a half open pen knife and just scratches the soil, the hand sickle made more for a child than for a man; the old-fashioned winnowing fray that woos the wind to shift the grain from the chaff; and the rude choffer with its waste of fodder are un-displaced from their primitive but immemorial functions" Thus, it has deep rooted impact on the economy of the region."

Tractor is the basic mechanical input which largely determines the extent of use of allied machinery and equipment. A perusal of the table shows that there was a rapid progress in the number of tractors in the country. In

1961, the country had only 31000 tractors which increased to 2,52000 in 1966. This increase is indeed too phenomenal to beg overlooked. This fact could be attributed to the coincidence of this period with the green revolution. Consequently, the number of tractors increased to 4,555,000 in 1990 and is expected to have further increased to over by the end of 1992. The number of tube wells operated with electricity increased from 10,6000 in 1961 to 4,55,600 in 1990. In the same manner, tube wells operated with oil engines also rose steeply. The number of tube wells further increased to 10,85,000 in 1990. The increase was much more in case of electric tube wells while the diesel operated tube wells declined, clearly this change was due to the poor economies of diesel operated tube wells as compared to electric tube wells. Recently, there was revolution in the harvesting and threshing operations of wheat and paddy crops as a result of which the number of combine harvesters increased to about 60,350 as tractor mounted and about 31,60 as self-propelled.

## Factors Responsible for Expanding Mechanisation

After analyzing the progress of mechanization in the country, now, the question that creeps up in mind is that after all what are the factors that induce the farmers to expand the use of machines in their farms. The brief description of these factors is given as under:

1. **Cheap Rate of Capital and Foreign Exchange:** Cheap rate of capital and foreign exchange is the basic factor for expanding mechanization. Today, improved farm machineries can be imported on liberal terms and were sold to farmers on liberal terms. Moreover, the establishment of multinational farms has further facilitated the domestic production of these implements.

2. **Easy Access to Credit:** Easy availability of credit on soft terms has also paved the way for expanding mechanization. Farmers are, now, not required to depend on the mercy of big landlords for their credit requirements. They can easily take loans from commercial, co-operative and land-development banks.

3. **Labor Shortage:** Labor shortage is the most important factor resulting in farm mechanization. As with the success of new technology, farmers in Haryana and Punjab had increased their income manifold and are now enjoying a better standard of living. This resulted in a sharp decline in the availability of family labor which in turn led to shortage of labor.

4. **To reduce Dependance on Tenant:** Dr. K.N. Ray pointed out that mechanization was undertaken by large farmers to reduce their dependance on tenants. In certain cases, mechanization was undertaken by owners of large sized holdings mainly as a means of reducing their dependance on tenants and casual labor.

**Magnitude of Mechanization:** The national commission on Agriculture has estimated the magnitude of mechanization in India by year 2000 as shown in the below table.

**Table 3.1**
**Extents of Mechanization in Jharkhand**

| Particulars | Total Power (million hp) |
|---|---|
| Men | 8.40 (6.5%) |
| Draught Animals | 25.60 (19.8%) |
| Land Development | 1.28 (1.0%) |
| Tractors | 18.34 (14.1%) |
| Electric Motor | 53.00 (48.9%) |
| Diesel Engine | 7.00 (5.4%) |
| Power Sprayer | 16.00 (12.3%) |
| Total | 129.62 (100%) |

Accordingly, the average power per hectare in the country as a whole would be 0.86 hp. Electric motors and diesel engine would account for 46 per cent of the total power requirements. Agricultural tractors would account for 14.1 per cent while the share of draught animals would be to the tune of 19.8 percent. Power protection equipment would account for per cent in the power spectrum.

## Argument in Favor of Farm Mechanization

Though industrialization in the country as well as in the

region has made head way to push the economy, yet the agriculture has to play strategic role in the economic development. Thus, farm mechanization can bring about a significant improvement in agricultural productivity. Thus, there is strong case for mechanization of agricultural operations. The main points favoring mechanization may be enumerated as follows:

1. **Increase in Productivity:** Farm mechanization leads to increase in agricultural productivity per acre and per man. This is the outcome of precision and timeliness in agricultural operations by mechanization for operations. For example, sowing the seed at desired distance and depth with the help of seed drill and sowing the crop at appropriate time with the help of tractor certainly has impact on yields.

2. **Reduction in the Cost of Production:** Normally, there are good chances to reduce the cost of production if farm operations are mechanized as it saves labor both human and bullock. Further, higher production means less cost on the farms. Moreover, it reduces the risk as the work could be completed well in time.

3. **Risk of Non-Availability of Labor is Reduced:** The work done by machines is faster which helps to keep time schedule of sowing intact Otherwise, there are chances of delay. For example, during sowing and harvesting period, there is spoilage of labor force. Thus, timely sowing and harvesting yields better results.

4. **Helpful to Make the Agriculture Commercial:** As production increases with mechanization of the farm operation it creates a good scope for commercialization of agriculture.

5. **Employment Oriented:** The use of farm mechanization enlarges the employment opportunities both on farms and in non-farm sectors through multicropping, agro-industries and transportation services.

6. **Technological Improvement in Agriculture:** The use of different machines at farms helps to make improvement in agriculture technology, development of agro-based

industries, and repair business.

7. Change in Occupational System: The farm machinery, being lumpy in nature, require comparatively larger farm size to operate to reap the benefits of it Due to this structural change in agriculture, the occupational distribution in the rural economy will undergo a change. In other words, less people will depend on agriculture for their livelihood as some farming population would migrate to other sectors.

8. Solution to the Problem of Labor: There is a hot controversy over the effect of farm mechanization on employment of human labor. But, actually the development of allied industries and infrastructure gives the opportunity to solve the disguised unemployment.

## Arguments against Farm Mechanization

It is true that farm mechanization has shown good results as of raising the agricultural production and improving the standard of living of cultivators within very short period. But a number of arguments have been advanced against farm mechanization. It is separate issue that some of them are controversial. They are:

1. **Small Holding:** Small size and scattered holdings of the farmers stand in the way of mechanization. As a result of this, farm machinery generally remains under Utilized

2. **Paucity of Capital:** Majority Of small cultivators are poor who are not in a position to purchase the costly machinery like tractors, combine, harvesters etc. Thus, paucity of funds is another limitation for adoption.

3. **Surplus Cattle Population:** The use of machinery at farm will turn the draft cattle population surplus which will increase the burden on the economy, Unnecessarily.

4. **Fear of Unemployment:** No doubt, the increasing farm mechanization is going to increase employment in secondary and tertiary sectors but its impact on creating unemployment in primary sector especially in the rural area is quite visible where there is already limited scope of employment opportunities in non-agricultural sector.

5. **Lack of Technical Knowledge:** Lack of proper knowledge of fanner to operate the machinery creates another problem. This demands employment of skilled labor to handle such machinery, thus, incurring heavy cost. Otherwise, poor maintenance and improper operating the machinery is also Uneconomical and risky too.

6. **Shortage of Petrol and Diesel:** There is great shortage and diesel in the country as a whole. It is evident from the long queues of farmers which are commonly observed on diesel pumps in Punjab. Thus, to use so extensive oil-based farm machinery is not desirable. This will be an extra burden to import the petrol and diesel.

7. **Poor Repair Facilities:** The State lacks repair and replacement facilities especially in the remote rural areas

8. **Idle Machinery:** Due to the seasonal nature of the agriculture the farm machinery remains idle for much of the time. Thus, idle machinery means unnecessary high costs unless proper alternate use of such machinery in the off time is made.

## A Correct Approach

The above analysis indicates that most of the ill-based arguments against farm mechanization can easily be solved without much difficulty. But there is no dispute regarding the need to boost the agricultural production to stabilize the economy. Production, on the other hand, cannot be enhanced with primitive and traditional operations. Thus, keeping in view all facts in mind, partial mechanization is the need of the hour to raise the agricultural production to meet the basic domestic food requirements at least.

## Benefit of Farm Mechanisation

The major benefits of farm mechanization are given below as it helps:

1. To ease the tough farm operations and thus to reduce drudgery for agricultural labor.
2. To increase the productivity of land and labor.
3. To improve timeliness of seasonal farm operations.

4. To reduce spoilage, wastage and other losses of products by improving accuracy in farm operations and efficient marketing of perishable farm products.
5. To preserve and properly process farm products and food supplies.
6. To increase agricultural production.
7. To convert feed and fodder production areas kept for draft animals to human consumption.
8. To improve water supplies and water control systems.
9. To develop new land for agriculture by clearing of obstructions or by draining, levelling or other reclamation operations or uncultivated.
10. To reclaim land abandoned due to primitive operations.
11. To create a greater measure of well-being for the rural masses.
12. To raise the cropping intensity and shift the cropping pattern in desirable direction.

## Scope of Farming Mechanization

The factors that justify the scope and strengthening of farm mechanization in the country can be enumerated below:

1. **Timeliness of operations:** This factor is significant in obtaining optimal fields from different crops. For instance, the sowing of wheat is done during the first fort night of November. A delay beyond this period by every one week leads to about 1.50 quintals per acre decrease in the yield. This is also correct in the case of other crops like paddy, cotton, groundnut and potato. Similarly, other farm operations like hoeing, irrigation, harvesting, threshing and marketing need to be performed at appropriate time otherwise it is going to adversely affect the yield and income of the farmer.

2. **Better Quality and Precision of Operations:** The quality and precision of the operations are equally significant for realizing higher yields. The various operations such as land levelling, irrigation, sowing and planting, use of fertilizers, plant protection, harvesting and threshing need a high degree of precision to increase the efficiency of the inputs and reduce the losses. For example, sowing of the required

quantity of seed at proper depth and applying the given dose of fertilizer at a specified depth in a specific manner can only be possible with the use of proper mechanical devices. However, when such operations are performed through indigenous methods, their efficiency is reduced.

3. **Higher Intensity of Cropping:** The time taken to perform one operation after the other is a factor determining the cropping intensity. So as to ensure timeliness of various operations, it is quite inevitable to use such mechanical equipment which have higher output capacity and which cut down the number of operations to be performed in achieving the desired degree of perfection in the task to be performed.

4. **Higher Power Availability for Higher Yields:** The factors such as timeliness, precision and better quality of operations, individually as well as collectively lead to higher yield per unit area. Thus, without adequate power availability on the farm, the operations like seed-bed preparation. Stowing and planting, plant protection, harvesting and threshing take longer time which leads to decrease in productivity.

5. **Higher Productivity of Land and Labor:** Higher productivity of land and labor is another factor which clearly justify farm mechanization. Not only the output per hour is more, the total labor requirement is also reduced. The displaced labor is of course absorbed in the other alternatives created by the increased mechanization such as manufacturing, repair and service shops and the sales services. Thus, it only results in the shifting of the labor from one vocation to the other.

## Causes for Slow Progress of Mechanization

According to the study made by Verma, Singh and Mittal, the chief bottlenecks of farm mechanization can be cited under three heads namely:

A. Research development and testing.
B. Manufacture, standardization and quality control.
C. Education, training and popularization of farm

equipment.

**A. Research Development and Testing:**

The equipment which is required to be developed include:

1. Equipment for dry farming areas.
2. Equipment for small and marginal farmers for efficient seed-bed preparation, seeding, harvesting etc.
3. Efficient hand tools such as/for weeding and hoeing.
4. Rotary tillage equipment such a rotavator and auger plough.
5. Till-plant machines, i.e., equipment which can prepare the seed- bed simultaneously.
6. Paddy transplanting equipment.
7. Efficient fertilizer broad-casters and applicator.
8. Harvesting equipment such as reaper, binders, combines etc.
9. Equipment for cotton mechanization, viz., planters, cotton picking devices, cotton stick puller etc.
10. Machines for sugarcane mechanization viz., sugarcane planter, stalk-shaker, prayers, etc.
11. Multi-purpose seeding and planting machines.
12. Multi-crop threshers with capability to thresh wheat, paddy, maize and pulses together with an optional bhusa making attachment.
13. Anhydrous Ammonia Applicator.
14. Sprayers and dusters for tall crops like cotton, sugarcane as well as for orchard spraying.
15. Equipment for vegetable and fruit mechanization like pit-hole digger, portable chain saw, universal fruit grader, orchard tiller etc.

**B. Manufacture, Standardization and Quality Control:**

Although there are about 300 firms engaged in fabrication and manufacture of farm equipment in the state, the main problems indicated are:

1. Poor quality and durability of the equipment.
2. Lack of an organization for quality making.
3. Lack of interchangeability of components due to non-adoption of standard designs.

4. Lack of incentive for adoption of standardized designs.
5. Lack of legislative measures to ban firms manufacturing sub- standard and hazardous equipment.
6. Absence of feeder industries engaged in the manufacture of vital/ critical components of farm machinery needing high degree of precision, heat treatment and superior raw-material.
7. Acute shortage of testing facilities.
8. Non-employment of qualified engineers.
9. Lack of modernization of manufacturing processes and tooling.

**C. Education, Training and Popularization of Farm Equipment:**

1. Inadequate training facilities for farmers, skilled operators, mechanics and rural artisans.
2. Lack of adequate incentive for the small and marginal farmers to adopt improved machinery by way of subsidized cost, low interest rates, etc.
3. Inadequate repair and service facilities for tractors and farm equipment in the rural areas
4. Lack of cheap and quality equipment for the small farmers.
5. Excessive excise duty and taxes especially on the tractors, diesel and lubricants.
6. Inadequate agricultural engineering staff for extension activities.
7. Inadequate custom-hiring services including agro -service centers in the state.

## Suggestions to Promote Farm Mechanisation

The problems discussed above need careful consideration at different levels i.e., by the Govt., the manufacturers, engineers and the farmers. To be specific, the following suggestions can go a long way to promote farm mechanization in the State:

1. Suitable farm machinery for small and marginal farmers, need to be developed and should be subsidized by the Govt. Since majority of the farms are small, it is the main

hurdle in the adoption of big farm machinery.

2. Adequate loans for purchase of farm machinery and for repair units should be provided.
3. Technical know-how should be provided to the farmers with respect to appropriateness of farm machinery for the situation and for its proper use.
4. A standardization and quality marking cell for farm equipment should be established in the State.
5. The regular supply of diesel, petrol and oil should be ensured.
6. The farm machinery suitable for different type of soil and operations for important crops should be manufactured.
7. The selected farm equipment should be imported to cut down time and expenditure on research.
8. Polytechnical training centers should impart more training relating to farm machinery and equipment.
9. The equipment beating certificate marking should be provided incentives.
10. The excise duty and taxes on tractors and other machinery and equipment should be reduced. Similar tax cut needs to be explored for the diesel and petrol.
11. More service units should be encouraged and mobile service units should be started.
12. Incentives should be given to the cooperative fanning societies so that the different effects and economical use of farm machinery is made.

## Mechanization v/s Employment

Generally, it is opined that new technology and mechanization are the two sides of the same coin since the adoption of new technology has proved useful for inevitably raising the agricultural production and, therefore, mechanization of agriculture has also become inevitable. But we do not agree with this view. As mechanization is not only the factor leading to increase production. There are certain other factors. Therefore, it becomes essential to make the analysis of this aspect and see the effect of new technology and mechanization on employment.

Tractor is considered to be a versatile instrument for a large number of farm operations. Therefore, technically the correlation between tractor- use and HYV of seeds is unfounded. This has been proved by the fact that even before the onset of green revolution, there were a large number of tractors in Punjab and elsewhere in India and the production was of low order. New technology has given only economic impetus to farmers to intensify mechanization and to increase their private profits.

In view of this, it becomes necessary to understand the true nature of new technology vis-a-vis demand for labor. The availability of water has helped to increase the cropping intensity. The increased cropping intensity has also increased the production and marketable surplus. This, in turn, has increased the demand of labor for various harvesting and post-harvesting operations. Thus, it is obvious that new farm technology has a tendency to increase the demand for labor.

## Sub-Mission on Agricultural Mechanization (SMAM)

Agriculture occupies the most important role in Indian economy. Earlier, Indian farmers largely relied upon the human and animal power. But with the passage of time, tractor and tractor driven agricultural implements have been introduced with the Government sponsored scheme. It has been observed that during the last 10 years, the concept of farm mechanization and its importance has been well accepted and accordingly new farm implements/equipment have been devéloped and commercialized. This has resulted in increase of productivity and production to feed the ever-increasing population besides reduction in drudgery associated with farm operations. Farm mechanization has played a positive role in increasing employment in rural areas through generation of opportunities for operators, mechanics, salesmen etc. The adoption of tractor and farm machinery/implements has been very rapid in India with the assistance of centrally sponsored schemes but its spread has been uneven. The Ministry of Agriculture and Farmers

Welfare launched a scheme Sub-Mission on Agricultural Mechanization (SMAM) for promotion of farm mechanization during the year 2014-15. To evaluate the effectiveness and status of implementation with regard to physical and financial progress, mechanization strategies, extent of adoption of technologies and bottlenecks during the implementation of scheme for the year 2014-15, 2015-16 and 2016-17, WAPCOS Ltd., Gurugram has been awarded a project to conduct monitoring, evaluation and impact assessment of SMAM.

## Sub-Mission Components

1. Promotion and Strengthening of Agricultural Mechanization through Training, Testing and Demonstration: Aims to ensure performance testing of agricultural machinery and equipment, capacity building of farmers and end users and promoting farm mechanization through demonstrations.
2. Demonstration, Training and Distribution of Post-Harvest Technology and Management (PHTM): Aims at popularizing technology for primary processing, value addition, low-cost scientific storage/transport and the crop by-product management through demonstrations, capacity building of farmers and end users. Provides financial assistance for establishing PHT units.
3. Financial Assistance for Procurement of Agriculture Machinery and Equipment: Promotes ownership of various agricultural machinery & equipment as per the norms of financial assistance.
4. Establish Farm Machinery Banks for Custom Hiring: Provides suitable financial assistance to establish Farm Machinery Banks for Custom Hiring for suitable locations and crops.
5. Establish Hi-Tech, High Productive Equipment Hub for Custom Hiring: Provides financial assistance to set up hi-tech machinery hubs for high value crops like sugarcane, cotton etc.
6. Promotion of Farm Mechanization in Selected Villages: Provides financial assistance to promote appropriate

technologies and to set up Farm Machinery Banks in identified villages in low mechanized States.

7. Financial Assistance for Promotion of Mechanized Operations per hectare carried out through Custom Hiring Centers: Provides financial assistance on per hectare basis to the beneficiaries hiring machinery/equipment from custom hiring centers in low mechanized areas.
8. Promotion of Farm Machinery and Equipment in North-Eastern Region: Extends financial assistance to beneficiaries in high- potential but low mechanized states of north-east.

### Objectives of Sub Mission on Agriculture Mechanization (SMAM)

1. Increasing the reach of farm mechanization to small and marginal farmers and to the regions where availability of farm power is less;
2. Promoting 'Custom Hiring Centers' to offset the adverse effect on Farm economy due to small landholding and high cost of individual ownership of farm machinery;
3. Creating hubs for hi-tech & high value farm equipment;
4. Creating awareness among stakeholders through demonstration and capacity building activities;
5. Ensuring performance testing and certification at designated testing centers located all over the country.

## Methodology for Assessment of Farm Power Availability at District Level

For the estimation of Farm power, the Household Questionnaire was designed to collect the source of power for farm operations right from primary tillage operations to harvesting and primary processing. The data was collected from the respondent farmers belonging to different land holding categories. The parameters to assess the power availability which was filled in questionnaire (SUB-2) during survey by the field investigators are:

Total land holding owned and hired, Tractor (HP), Power Tiller (HP), Engine (HP), Motor (HP/KW), Animal Power (No of Bullocks and others), Human Power for farm work

(Male/Female) etc. Total questionnaires based upon sampling plan was filled in each state. The whole data was fed in computer and as per standard norms for each source of farm power, the data was analyzed and calculated for actual total power available with each farmer surveyed. Accordingly, total land holding available and total farm power available with all the farmers surveyed in each district was calculated using secondary data from the reliable sources. Based upon this analysis, the average farm availability per hectare in each district was calculated. In this methodology, it was an assumption that farmers surveyed as per sampling plan represented the whole district. For the districts which had not been surveyed, suitable statistical tools were used for making interpolation /extrapolation.

The average farm power availability was worked out for the districts/States using following formula

**Farm Power Availability (kW/ha) =**

**{(Number of agricultural Worker × 0.05) + (Number of draught animal × 0.38) + (Number of Tractors × 26.1) + (Number of Power tillers × 5.6) + (Number of electric motors × 3.7) + (Number of diesel engine × 5.6)} / Available cultivated land in ha.**

## Data for Jharkhand

As stated in methodology a vast survey was done with help of ministry of Agriculture and different data were collected under different states of the country. Here only the data related to the Jharkhand state has been taken from the final report (2018) of the department of Agriculture to understand the Farm Mechanization of this state.

**Table 3.2**
**Fund Utilization Pattern in Jharkhand during 2014-17**

| Year | 2014-15 | 2015-16 | 2016-17 | 2017-18 |
|---|---|---|---|---|
| Fund Received (In Lakh) | 358 | 254 | 225 | 837 |
| Fund Utilized (In Lakh) | 49 | 38 | 64 | 151 |

**Table 3.3**
**Fund Utilization Pattern by Jharkhand Govt and number of machines distributed during 2014-17**

| Fund Utilized (in lakh) | Number of Machines Distributed | | | | Household Data Validated and Analyzed |
|---|---|---|---|---|---|
| 151 | 2014-15 | 2015-16 | 2016-17 | Total | |
| | 00 | 00 | 00 | 00 | 00 |

**Table 3.4**
**Hiring Rates for Different Implements**

| Name of Equipment | Hiring Charge in Local Market (Rs/Hour) |
|---|---|
| Tractor with Rotavators | 1200 |
| Tractor with Cultivators | 800 |
| Tractor with Seed Drill | 800 |
| Power Tiller | 600 |
| Bush Cutter | 350 per day |
| Thresher | 1000 |
| Paddy Combine | 1400 |

## Analysis of Farm Power Availability

The Farm power availability (FPA) of different state has been calculated by the Agriculture Department under the SMA-M survey program. The report categorizes Jharkhand is in the third category. The states were categorized in five categories according to power utilization.

| | |
|---|---|
| Category I | Where power availability is more than the national average i.e., 2025 KW/Hectare (according to 2016-17). |
| Category II | When Power availability is in between 2.025 and 1.726 KW/ha (according to 2013-14). |
| Category III | The states which have power availability in between 1.726 and 1.00 KW/ha |
| Category IV | States which have farm power availability less than 1.00 KW/ha |

On basis of above criteria Jharkhand is under Category III of power utilization or FPA (Farm Power Availability).

**Table 3.5**
**District wise Farm Power Availability during 2013-14 in Jharkhand**

| District | Net Area Sown | Tractor | Draught Animals | Agricultural Laborers | Electric Motor | Diesel Engine | Power Tiller | Total Farm Power Availability |
|---|---|---|---|---|---|---|---|---|
| Bokaro | 15,510 | 745 | 90,432 | 202,942 | 340 | 643 | 120 | 4.480 |
| Chatra | 45,070 | 603 | 90,919 | 102,588 | 230 | 3,004 | 62 | 1.630 |
| Deoghar | 63,680 | 538 | 135,087 | 146,836 | 151 | 1,985 | 56 | 1.330 |
| Dhanbad | 34,600 | 415 | 77,924 | 264,038 | 369 | 304 | 28 | 1.644 |
| Dumka | 119,100 | 320 | 163,536 | 130,027 | 236 | 1,283 | 37 | 0.716 |
| Garhwa | 64,770 | 926 | 97,756 | 130,154 | 1,153 | 2,707 | 164 | 1.361 |
| Giridih | 77,590 | 1,682 | 197,070 | 240,666 | 180 | 2,566 | 170 | 1.892 |
| Godda | 78,210 | 471 | 71,692 | 129,071 | 170 | 785 | 40 | 0.655 |
| Gumla | 161,750 | 367 | 151,022 | 100,949 | 194 | 1,018 | 124 | 0.489 |
| Hazaribagh | 106,140 | 1,353 | 122,794 | 170,667 | 2,238 | 5,381 | 850 | 1.259 |
| Jamtara | 55,340 | 1,807 | 86,401 | 77,775 | 98 | 484 | 34 | 1.575 |
| Koderma | 17,040 | 566 | 37,615 | 70,587 | 78 | 1,186 | 368 | 2.441 |
| Latehar | 59,770 | 155 | 92,561 | 71,424 | 29 | 955 | 15 | 0.809 |
| Lohardaga | 39,640 | 302 | 51,871 | 45,446 | 65 | 1,401 | 43 | 0.963 |
| Pakur | 57,840 | 153 | 76,116 | 88,503 | 29 | 361 | 25 | 0.685 |
| Palamau | 97,750 | 1,610 | 129,448 | 190,580 | 1,313 | 2,750 | 337 | 1.257 |
| Ranchi | 255,850 | 1,190 | 140,032 | 286,612 | 1,695 | 7,394 | 352 | 0.579 |
| Sahebganj | 41,460 | 374 | 32,692 | 113,191 | 9 | 565 | 93 | 0.761 |

| District | Net Area Sown | Tractor | Draught Animals | Agricultural Laborers | Electric Motor | Diesel Engine | Power Tiller | Total Farm Power Availability |
|---|---|---|---|---|---|---|---|---|
| Saraikela | 60,420 | 202 | 84,346 | 104,670 | 12 | 252 | 6 | 0.729 |
| Simdega | 81,670 | 313 | 118,523 | 59,036 | 38 | 557 | 62 | 0.732 |
| **Total** | **1,533,200** | **14,093** | **2,047,837** | **2,725,762** | **8,628** | **35,583** | **2,984** | **0.998** |

**Source:** Final Report on Agricultural mechanization, Ministry of Agriculture, Mechanization and Technology Division

**Table 3.6**
**District wise Farm Power Availability during 2016-17 in Jharkhand**

| District | Net Area Sown | Tractor | Draught Animals | Agricultural Laborers | Electric Motor | Diesel Engine | Power Tiller | Total Farm Power Availability |
|---|---|---|---|---|---|---|---|---|
| Bokaro | 1,375 | 85,075 | 219,195 | 367 | 741 | 178 | 5.524 | 23.3 |
| Chatra | 1,114 | 85,533 | 110,803 | 248 | 3,461 | 91 | 1.951 | 19.7 |
| Deoghar | 994 | 127,085 | 158,596 | 163 | 2,287 | 82 | 1.508 | 13.4 |
| Dhanbad | 867 | 73,308 | 285,184 | 398 | 350 | 41 | 1.977 | 20.3 |
| Dumka | 891 | 153,848 | 140,441 | 255 | 1,478 | 55 | 0.825 | 15.2 |
| Garhwa | 1,709 | 91,965 | 140,578 | 1,243 | 3,118 | 241 | 1.698 | 24.8 |
| Giridih | 3,106 | 185,396 | 259,940 | 194 | 2,956 | 250 | 2.361 | 24.8 |
| Godda | 870 | 67,445 | 139,408 | 183 | 905 | 59 | 0.785 | 19.8 |
| Gumla | 777 | 142,076 | 109,034 | 209 | 1,173 | 182 | 0.545 | 11.3 |
| Hazaribagh | 2,498 | 115,519 | 184,336 | 2,412 | 6,199 | 1,252 | 1.592 | 26.4 |
| Jamtara | 2,337 | 81,283 | 84,004 | 105 | 558 | 50 | 1.805 | 14.6 |
| Koderma | 1,045 | 35,387 | 76,240 | 84 | 1,367 | 542 | 3.259 | 33.5 |
| Latehar | 453 | 87,077 | 77,144 | 31 | 1,100 | 23 | 0.923 | 14.1 |
| Lohardaga | 624 | 48,799 | 49,086 | 71 | 1,615 | 64 | 1.184 | 22.9 |
| Pakur | 385 | 71,607 | 95,591 | 31 | 416 | 36 | 0.773 | 12.8 |
| Palamau | 2,973 | 121,780 | 205,843 | 1,415 | 3,169 | 496 | 1.636 | 30.1 |
| Ranchi | 2,198 | 131,737 | 309,566 | 1,826 | 8,518 | 519 | 0.705 | 21.6 |
| Sahebganj | 690 | 30,755 | 122,256 | 10 | 651 | 137 | 0.971 | 27.6 |

| District | Net Area Sown | Tractor | Draught Animals | Agricultural Laborers | Electric Motor | Diesel Engine | Power Tiller | Total Farm Power Availability |
|---|---|---|---|---|---|---|---|---|
| Saraikela | 487 | 79,349 | 113,052 | 13 | 290 | 9 | 0.832 | 14.1 |
| Simdega | 625 | 111,501 | 63,764 | 41 | 641 | 91 | 0.810 | 10.6 |
| **Total** | **26,018** | **1,926,524** | **2,944,059** | **9,299** | **40,992** | **4,397** | **1.205** | **20.7** |

**Source:** Final Report on Agricultural mechanization, Ministry of Agriculture, Mechanization and Technology Division

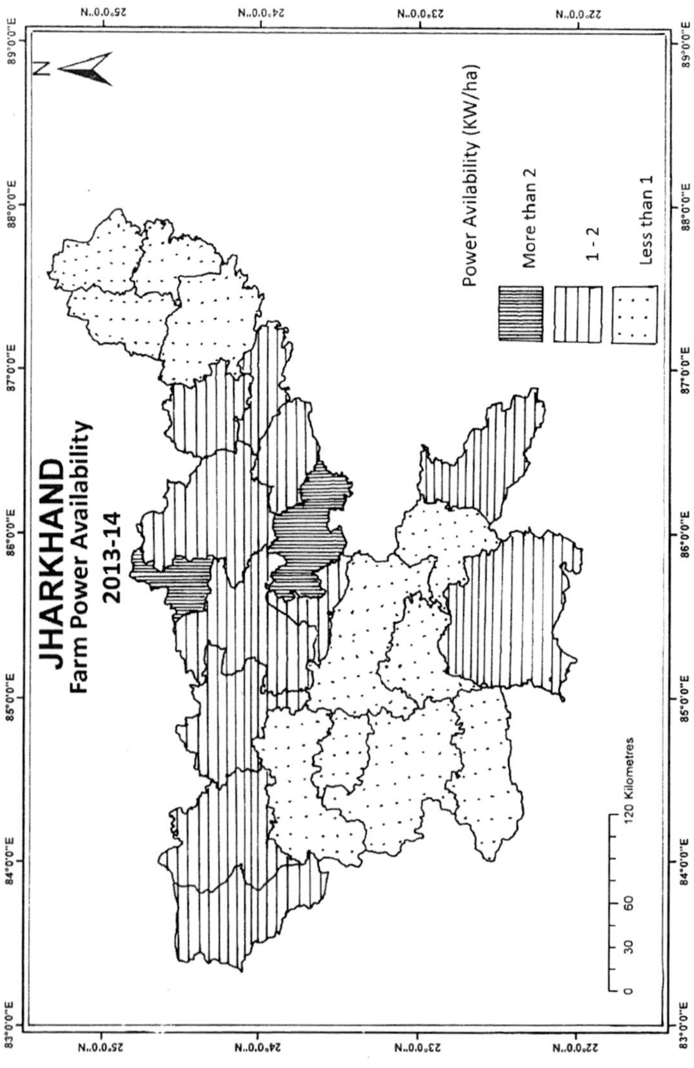


**Map 3.1: Farm Power Availability (2013-14)**

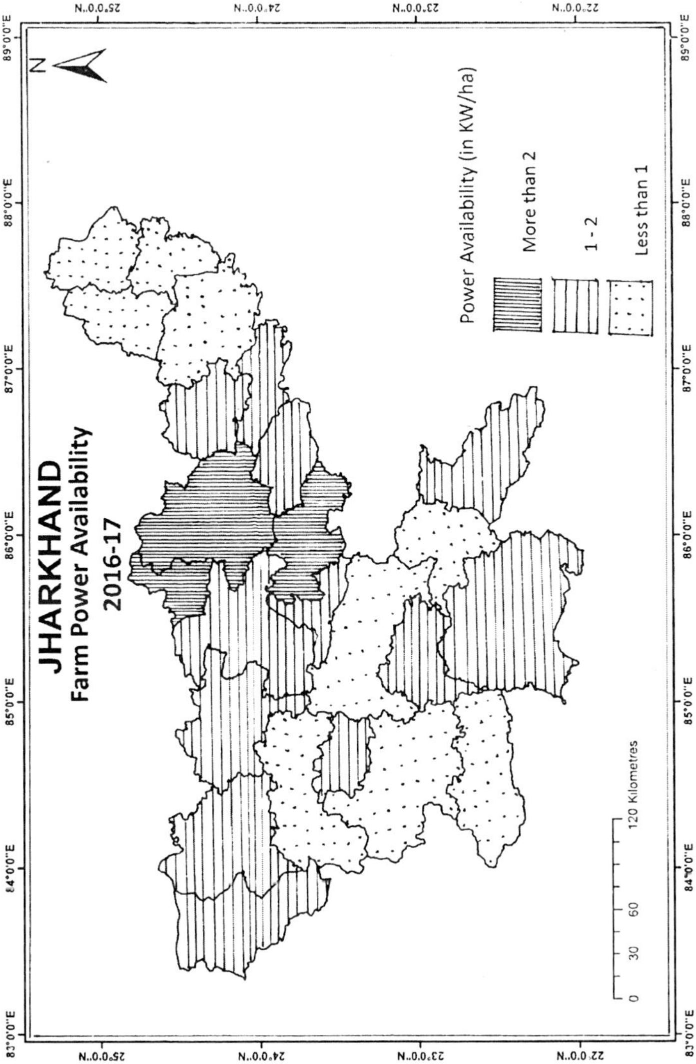


**Map 3.2: Farm Power Availability (2016-17)**

The average farm power availability in Jharkhand before the implementation of SMAM (Sub-Mission on Agricultural Mechanization) uses 0.998 KW/ha (2014) and it increased to 1.205 Kw/ha by the end of 2016-17, thus registering a 20.7 percent increase in FPA in three years. The FPA in Jharkhand is 40 percent less than the National Average i.e., Kw/ha at the end of 2016-17. The district wise farm power availability (2016-17) in state of Jharkhand is graded as given below and the complete analysis of farm power availability in Jharkhand is mentioned in the below table.

FPA (Farm Power Availability)

| **More than 2.03 KW/ha** | **Between 1.0 to 2.03 KW/ha** | **Less than 1.00 KW/ha** |
|---|---|---|
| Bokaro (5.524) | Chatra (1.951) | Dumka (0.825) |
| Koderma (3.259) | Deoghar (1.508) | Godda (0.785) |
| Giridih (2.361) | Dhanbad (1.977) | Gumla (0.545) |
| | Garhwa (1.698) | Latehar (0.923) |
| | Hazaribagh (1.592) | Pakur (0.773) |
| | Jamtara (1.805) | Ranchi (0.705) |
| | Lohardaga (1.184) | Sahebganj (0.971) |
| | Palamu (1.636) | Saraikela (0.832) |
| | | Simdega (0.810) |

According to the district wise analysis a highest growth (above 2.03 Kw/ha) has been recorded in Bokaro district (5.524 Kw/ha) followed by Koderma (3.259 KW/ha) and Giridih (2.361 Kw/ha). In second category which registered a growth between 1.00 to 2.05 KW/ha are Dhanbad (1.977 Kw/ha), Chatra (1.951), Deoghar (1.508), Garhwa (1.698), Hazaribagh (1.592), Jamtara (1.805), Lohardaga (1.184) and Palamu (1.636) districts respectively. Lowest growth (less than 1.00 KW/ha) has been recorded in Dumka (0.825), Godda (0.785), Gumla (0.545), Latehar (0.923), Pakur (0.773), Ranchi (0.705), Sahebganj (0.971), Saraikela (0.832) and Simdega (0.810) districts.

## Mechanization Gaps and Future Requirements

The efficacy of agricultural inputs and natural resources - seed, fertilizers, chemicals, land and water has been increased through adoption of appropriate agricultural equipment. Energy is the key input for agricultural, industrial and socio-economic development of a nation. The energy in agriculture is required for field operations which facilitate crop production and processing besides indirect energy in terms of seeds, fertilizer, irrigation, and chemicals. Human, animal, electrical and mechanical energy sources are extensively used in agriculture.

The farm mechanization facilitates use of inputs effectively and in reducing the drudgery in farm operations besides ensuring timeliness. The Indian farmers have adopted improved farm machinery irrespective of land holding size. Equipment for tillage, sowing, irrigation, plant protection and threshing have widely been accepted by them. Farmers with small holdings utilize selected improved farm equipment through custom hiring to increase productivity and reduce cost of production.

About 300 improved agricultural equipment and technologies have been designed and developed for various pre and post-harvest operations operated by human, animal, mechanical and electrical power for timely field operation, facilitating timeliness, removal of farm drudgery, reducing post-harvest losses and addition of value to the agro-produce. Increased productivity does not mean additional income to the growers unless it is matched with appropriate postharvest technologies that minimize post- harvest losses and add value to the produce and by-products.

Agricultural mechanization refers to interjection of improved tools, implements and machines between farm workers and materials handled by them. Independent India ushered in a process of agricultural mechanization and revival of rural agro-processing which got acceleration during post-Green Revolution period. Irrigation pump sets, power threshers, tractors, power tillers and matching implements, including for 65 million draft animals have become popular. Seed and seed-cum- fertilizer drills, planters, mechanical rice transplanters, vertical conveyor reapers, and combines soon followed. In the recent past, Zero-till Drill and Raise Bed

Planters have found good acceptance from the farmers. Currently mechanization is in increasing demand. Farmers and policy makers and developmental agencies now realize that for increasing production and productivity at reduced unit cost of production, free of arduous labor, agricultural mechanization is essential. It is brought in center stage with globalization of world markets. Introduction of electromechanical power units supplementing and substituting traditional animate sources of farm power is going to continue.

For achieving desired intensity of cropping average and productivity, farm power requirement of 2.5 kW/ha by 2022 and 4.00 kW/ha by 2030 is considered essential, currently it is 2.03 kW/ha. Shifts in agriculture leading to crop diversification towards horticulture, animal husbandry fishery, forestry and on-farm agro-processing are going to bring in greater degree of mechanization. India dominated by small and marginal land holdings may not have same trend of mechanization as the developed world but it is going to grow close to it. Though, the national average farm power availability in India is 2.03 kW/ha at the end of 2016- 17, many states still lag behind the national average.

Availability of adequate farm power for mobile and stationary farm operations should be increased from the present level of about 1.21 kW/ ha to about 1.54 and 2.65 kW/ ha in Jharkhand by 2020 and 2030, respectively. For stationary operations like water lifting, threshing, chaff cutting, cane crushing, cleaning, grading and other agro-processing and value addition activities, adequate electrical energy should be provided. The strategies needed for the mechanization of Jharkhand State are as follows:

1. Mechanization of tribal belt of Chotanagpur region using small hand tools.
2. Introduction of package of animal drawn equipment for different crops in rainfed regions of State.
3. Mechanization of rice cultivation through promotion of puddler, rice seeder, rice transplanter, cono-weeder and pedal paddy thresher.

4. Equipment for mechanization of orchard crops - pit making, transplanting of saplings, pruning, spraying in tall crops, harvesting of fruits etc. need to be identified/ imported/designed, introduced and popularized.
5. Vegetable crop production has to be mechanized for which equipment for seed bed preparation, planting, transplanting of seedlings, inter culture, irrigation, spraying harvesting, picking/digging has to be identified/ designed and introduced.
6. Conducting front line demonstrations of improved equipment for different crops.
7. Manufacturing small hand tools and animal drawn implements to meet the demand of the farmers through involving agricultural implements manufacturers.
8. Mechanization of hill agriculture in the State.
9. Promoting processing of agricultural produce for employment generation through value addition of raw material.
10. Promoting modern irrigation equipment and methods (sprinkler/drip irrigation).

It has been mentioned in presiding paragraph of this work quoting Eddowes (1969) that "the controllable factors effecting crop production in a given environment are seed, fertilizer and manures, mechanical power, chemical power and sequence of cropping". We have already examined the Farm Power Availability for the state. Now we shall examine or study the other factors.

## Farm Workers

The primary effects of improved mechanical techniques are to save labor but at the same time they do contribute to the enhancing of yield per hectare by making possible more intensive cultivation and the application of advanced biological techniques. They have multiplied the efficiency of a farm worker. In the highly developed countries less than 10 per cent of the economically active population is sufficient to provide food requirement of the entire population, while in developing countries as much as 70 to 80 percent of the working force is engaged in agriculture and yet the national food supply remain inadequate and the people are undernourished. Agriculture in many developed countries combines the high input of manual labor with heavy investment in machinery.

In 1991 78.7 percent, in 2001 77.7 percent and in 2011 75.9 percent of the total population was classified as rural in Jharkhand. This huge concentration of population in rural society confirms the deep-rooted relationship between land and people; the ultimate possibilities of mechanization may be realized only when farm workers data are placed together with information on mechanical and biological technology. In the national perspective, the proportion of the rural population was not very large than Jharkhand i.e., 74.3 percent in 1991, 72.20 percent in 2001 and 68.84 percent in 2011. Rural population ratio is always high in Jharkhand than the country.

So far as agricultural population (Cultivator + Agricultural Labor) is concerned the national proportion depicted 67.1 percent in 1991, 58.2 percent in 2001 and 48.9 percent in 2011 census whereas the proportion of Jharkhand registered 71.1 percent in 1991, 66.85 percent in 2001 and percent in 2011 census.

**Table 3.7**
**Rural and Agricultural Population**

| Census | India | | Jharkhand | |
|---|---|---|---|---|
| | Rural Population (in %) | Agri Population (in %) | Rural Population (in %) | Agri Population (in %) |
| 1991 | 74.3 | 67.1 | 78.8 | 71.1 |
| 2001 | 72.2 | 58.2 | 77.7 | 66.8 |
| 2011 | 68.8 | 48.9 | 75.9 | 62.9 |

Above table reveals that the rural and Agricultural population of the country and the state is decreasing decade by decade. But the pace of change is slow in state rather than the country. Change in rural population in the country is 7.99 percent but it is only 3.82 percent in state. The Agricultural population change is 37.9 percent in country but it is only 13.0 percent in Jharkhand. Thus, we can say that the high proportion of rural and agricultural population in Jharkhand level confirms that man-land relation has been strengthened rather than changed.

Gap between the rural and agricultural population percentage it is 7, 14 and 19 in India respectively but in Jharkhand it is only 7, 10 and 13 in respective census years. This also shows that in Jharkhand the proportion of farm

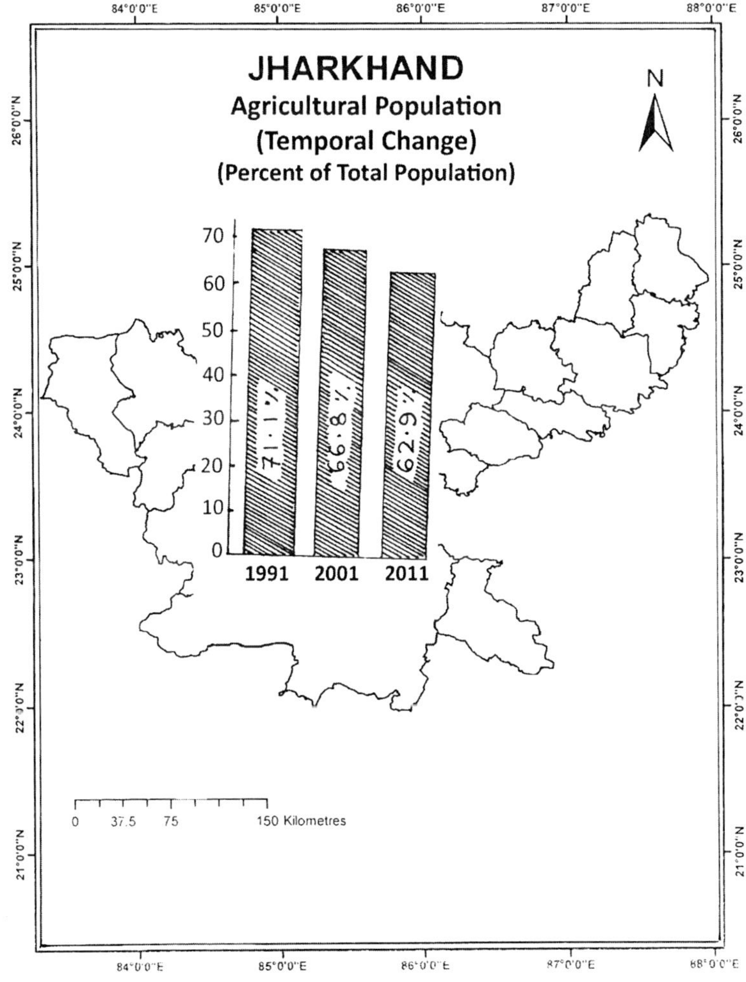


**Map 3.3: Agricultural Population (Temporal Change)**

worker dominate considerably to the rural population. On the other hand, comparatively, low change indicates a low change in Agricultural workers. In India as well as in the state of Jharkhand both have reducing percent of Agricultural population. Cause behind this fact is the increasing percent of other workers or house hold industries.

In Jharkhand tribal and non-tribal both populations form the major proportion of farming communities and lend a peculiar character to rural areas. Farm workers comprises of cultivators and hired agricultural workers, both male and female. Among tribal population Oraon, Munda, Santhal, Ho, Kharia, Khond, Kisan, Korwa, Mahli are the dominant cultivators as well as agricultural laborers. But the primitive tribes are engaged mainly as agricultural laborers and other miscellaneous work. Among the non-tribal population or Sadan, Kurmi, Kumhar, Koiry, Teli, Rajwar are the dominant cultivators and agricultural laborers. Mostly backward caste people are laborers except some exceptions.

## Animal Drawn and Specialized Implements

In Jharkhand differences exist within the hierarchy of farming between the high-status cultivator with a large farm and low status cultivator with small holdings. These differences also reflect the use of traditional and modern use of agricultural implements. Social differences within a farming community and difference between communities, however, still effect the scale and type of farming operations, the choice of enterprise and the use of agricultural implements. On the whole, sub-marginal, marginal and small farmers are conservative and their farm-equipment are as traditional that there is little possibility of change in their attitude. Further, the ease in tilling a soil, depending on its softness or hardness to a significant degree determines the density of wooden or Iron-based ploughing implements. Now a days by the efforts of agriculture universities training program the cultivators have realized, whether consciously or un-consciously, the importance of moisture conservation for some crops in soil of areas with precarious rainfall and low intensity of irrigation. This demands deep ploughing and overturning of the soil, needing iron-based soil scratchers. Due to this teaching of the cultivator the agricultural implements are changing gradually to the modern implements of farming which has changed the production ratio and economic status of the farmers in the area. But this is not true for every part of the state and there still are primitive farm-implements. Nevertheless, a data of modern agricultural implements used by the farmers in Jharkhand is useful here.

**Table 3.8**
**Number of Agricultural Implements used in Jharkhand (As recorded in Livestock census - 2003)**

| Name of Implements | Number of Implements |
|---|---|
| **Manually Operated Instruments** | |
| Seed cum fertilizer drill/ seed drill | 72167 |
| Inter Culture Equipment | 45038 |
| Rice Transplanter | 4653 |
| Thresher | 19820 |
| Winnowers | 105886 |
| **Animal Operated Instruments** | |
| Wooden Plough | 2071685 |
| Steel Plough | 107057 |
| Cultivator | 10828 |
| Disk Harrow | 47357 |
| Seed cum fertilizer drill | 26512 |
| Leveler | 916653 |
| Wctland Peddler | 137372 |
| Sugarcane Crusher | 12708 |
| Animal Cart | 362819 |
| Ghani | 5861 |
| **Plant Protection Equipment** | 55383 |
| Power Operated Sprayers | 1895 |
| Tractor/Power Tiller | 3384 |
| Irrigation Equipment | |
| Water Lifting Devices (manual) | 215306 |

| | |
|---|---|
| Water Lifting Devices (Animal Operated) | 23025 |
| Diesel Engine Pump set | 21442 |
| Electric Pump set | 21442 |
| Drip Irrigation Set | 1340 |
| Sprinkler Irrigation Set | 3288 |
| **Tractor and Other Power Operated Equipment** | |
| Power Tiller | 1051 |
| Agriculture Tractor | 6269 |
| Crawler Tractor | 1060 |
| Disc Plough | 639 |
| Cultivator | 839 |
| Disc Harrow | 657 |
| Rotavator | 78 |
| Seed cum Fertilizer Drill | 79 |
| Planter | 851 |
| Leveler | 5592 |
| Potato Digger | 375 |
| Tractor Operator | 864 |
| Self-Propelled Harvester | 801 |
| Trailer | 813 |
| **Power Operated Thresher** | |
| Paddy Thresher | 23451 |
| Wheat Thresher | 7464 |
| Multi Crop Thresher | 746 |
| Maize Sheller | 112 |
| Sugarcane Crusher | 463 |
| Rice Planter | 1531 |
| Reaper | 876 |
| **Power Operated Horticultural Equipment** | |
| Hand Tools | 3528715 |
| Power Operated Tools | 76100 |

Above mentioned data reveal that power operated Horticulture equipment which is hand tool are maximum (3528715) in number. The implement which comes second is the wooden plough (2071685). Leveler is in third position

(916653) followed by animal cart (362819), manual water lifting device (215306), Diesel Engine Pump Set (166002), Steel Plough (107057), Winnowers (105886) etc. This data shows nothing mechanical but a traditional picture of Agricultural Landscape in Jharkhand.

## Use of Fertilizers

The effect of biological techniques is primarily yield-increasing. Obviously, progress on this front can be obtained through the use of fertilizers, pesticides and insecticides and improved high-yielding strains. These are the triad pre-conditions of yield-raising means of agricultural production. Increasing food production by the area increasing method is no longer possible in the state because all land which is

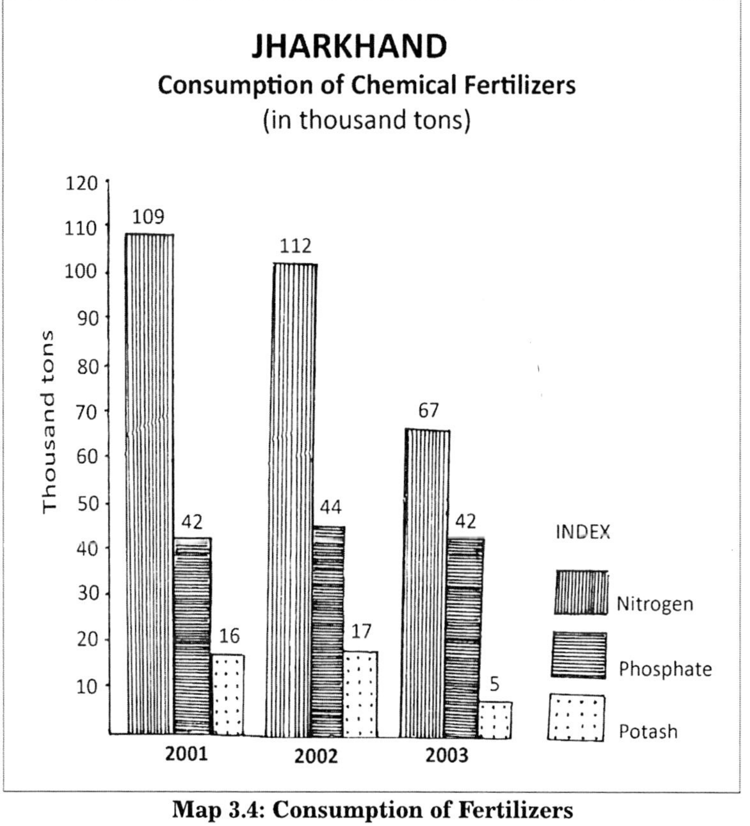


Map 3.4: Consumption of Fertilizers

suitable for cultivation has already been brought under plough. If physical area under arable farming is not possible to increase then how food production can be increased for the growing population of the state. Answer to this question is variously stated that we should have moved towards modernizing and commercializing farming. In simple terms, the solution lies in increasing the productivity of soil by augmenting the supply of plant foods through the application of balanced fertilizers and, wherever possible, supplementing the water supply by irrigation on the one hand, and on the other by using progressively high-yielding varieties of seeds through genetic research directed towards increasing the fertilizer responsiveness and the yield potential of the crops. (Singh, 1976, P216-217)

**Table 3.9**
**Consumption of Chemical Fertilizers in Jharkhand (000 Tons)**

| Types of Fertilizers | Jharkhand 2001 | 2002 | 2003 | India 2001 | 2002 | 2003 |
|---|---|---|---|---|---|---|
| Nitrogen (N) | 109 | 112 | 67 | 11310 | 10474 | 11076 |
| Phosphate (P) | 42 | 44 | 42 | 4382 | 4019 | 4124 |
| Potash (K) | 16 | 17 | 5 | 1667 | 1601 | 1598 |
| Total | 167 | 174 | 114 | 17359 | 16094 | 16797 |
| Consumption per hectare (Kg) | 54 | 67 | 44 | 90 | 85 | - |

**Source:** Agriculture Statistics, IASRI, New Delhi (2005)

There is no doubt that fertilizers used efficiently and in combination with other improved practices can be one of the most effective means of increasing production per unit area per unit time. Their use is very important because they act as a booster or a lead in securing the adoption of other improved techniques. Fertilizer use is commended as one of the primary practices to be adopted, because the response to fertilizers is usually strikingly visual even to the eye of an untrained observer. The farmers get relatively quick returns from the use of fertilizers and the capital required is much less than for many other mechanical techniques and economic improvements that may be desirable for raising the production. Hence teaching farmers how to use fertilizers advantageously can be a strong motivational factor in encouraging other changes. Perhaps the most effective method of teaching farmers the value of the fertilizers is by means of a large number of field demonstrations and trials. Such as those being carried out in the fertilizer program of the State Directorate of Agriculture and other organizations. This is needed in Jharkhand because a large number of farmers cannot read and write printed directions on fertilizer bag. The impact of these programs is evident from the dramatic increase in the use of fertilizers from 114 thousand metric tons in 2003-04 to 197 thousand metric tons in 2012-13. But after 2012- 13 Jharkhand witnessed steep decline in consumption of total fertilizers by about 48 percent in 2013-14. Total fertilizer consumption went down from 1997 thousand metric tons during 2012-13 to 104 thousand metric tons during 2013-14. The growth in consumption was negative.

**Table 3.10**

**Fertilizer Consumption in Jharkhand (000 tons)**

| Types of Fertilizer | 2010-11 | 2011-12 | 2012-13 |
|---|---|---|---|
| Nitrogen + Phosphate + Potash | 126<br>-24.59% | 171<br>35.79 % | 139<br>-19.01% |
| Total India | 28122<br>-1.18 % | 27790<br>-7.94% | 25583 |

Again, the data for 2010-11 to 2012-13 depicted an increasing pattern of the use of NPK in the study area. In total 126 thousand metric tons was used during 2010-11 followed by 171 thousand metric tons and 139 thousand metric tons was used during 2011-12 and 2012-13 respectively. Fertilizer consumption pattern in the state is being to increasing and decreasing time to time.

To sum up, it would appear that in general, the main obstacles to fertilizer use in the whole Jharkhand are firstly, inadequate irrigation facilities needed for the full potential from fertilizers to be realized. Secondly, uncertain rainfall, thirdly insufficient information on the kind and amount of fertilizer needed by particular crops under various environmental conditions and deficiencies in nutrients of soil, fourth according to the need of the farmer inadequate fertilizer distribution system, fifth, land holding arrangements and cropping system that tend to discourage the economic use of fertilizer. All these inadequacies need to be tackled immediately for making to increase the agricultural production through efficient, judicious and critical minimum fertilizer use in the state.

District wise analysis of fertilizer distribution depicted an uneven picture of the state. District wise distribution of fertilizer during 2000-01 in Jharkhand registered the highest consumption of fertilizer i.e., 160 Kg/ha in Deoghar district, followed by Ranchi district (101 Kg/ha), Sahebganj (88 Kg/ha), Hazaribagh (81 Kg/ha), Lohardaga (65 Kg/ha), Giridih (58 Kg/ha), Godda (54 Kg/ha) and so on. On the other hand, minimum consumption has been registered in the Saraikela-Kharsawan district i.e., 6 Kg/ha. Similar case with West-Singhbhum district. Gumla and Simdega districts have recorded only 8Kg/ha fertilizer consumption which is slightly more than the lowest unit. These are rice dominating area in Jharkhand. Perhaps farmers do not use fertilizers in this region. Agriculture department should try to make them trained in use of fertilizers (Mishra, R.K., 2006).

## High Yielding Seeds

We shall discuss about the high-yielding seeds later on. Firstly, a story regarding hunger of Jharkhand which is an outcome of the National Sample Survey Organization (NSSO) during their 55th round of survey. This survey very clearly shows that 10.5 percent of household in Jharkhand face seasonal food insecurity and 2.5 percent of households face chronic food shortage. Furthermore, 64 percent of the household face food shortage for two to three months while as many as 28 percent of household face the same for four to five months (Singh, 2012, P, 75-88). This is the story of a general economic condition of the general people of Jharkhand. Why is the condition so? Answer to this story

### Table 3.11
### Comparative Productivity (kg/ha) Performance in Major Crops Cultivated in Jharkhand and India

| Year | Rice | | Wheat | | Maize | | Pulses | | Oilseed | |
|---|---|---|---|---|---|---|---|---|---|---|
| | Jharkhand | India | Jharkhand | India | Jharkhand | India | Jharkhand | India | Jharkhand | India |
| 2001-02 | 1797 | 2079 | 1764 | 2762 | 1496 | 2000 | 783 | 607 | 661 | 913 |
| 2002-03 | 1497 | 1744 | 1734 | 2610 | 1708 | 1681 | 613 | 543 | 634 | 691 |
| 2003-04 | 1694 | 2078 | 1719 | 2713 | 1600 | 2041 | 556 | 635 | 352 | 1064 |
| 2004-05 | 1495 | 1984 | 1614 | 2602 | 1457 | 1907 | 653 | 577 | 564 | 885 |
| 2005-06 | 1150 | 2102 | 1340 | 2619 | 1266 | 1938 | 427 | 598 | 548 | 1004 |
| 2006-07 | 1832 | 2131 | 1528 | 2708 | 1504 | 1912 | 727 | 612 | 497 | 916 |
| 2007-08 | 2023 | 2202 | 1500 | 2802 | 1500 | 2335 | 749 | 625 | 554 | 1115 |
| 2008-09 | 2035 | 2178 | 1541 | 2907 | 1407 | 2414 | 764 | 659 | 561 | 1006 |
| 2009-10 | 1505 | 2125 | 1550 | 2839 | 1332 | 2024 | 733 | 630 | 475 | 958 |
| 2010-11 | 1451 | 2239 | 1500 | 2988 | 1235 | 2540 | 656 | 691 | 487 | 1193 |
| 2011-12 | 3197 | 2393 | 1908 | 3177 | 1446 | 2476 | 817 | 699 | 679 | 1135 |
| 2012-13 | 2833 | 2462 | 1944 | 3118 | 1812 | 2552 | 1169 | 786 | 783 | 1169 |
| 2013-14 | 2694 | 2419 | 2007 | 3059 | 1851 | 2602 | 993 | 770 | 706 | 1149 |

may be given in different ways from different sections of society or by a social scientist. One may say that Jharkhand has sufficient rainfall but deficient water availability for irrigation. Other may say that Jharkhand has an ideal topography for traditional cultivation of fruits and tuber crops with food crops but improper exploitation of natural resources (soil, biodiversity and even fertilizers), it has become a place of monoculture. Insufficient infrastructure, market facilities and poor technology adaption are the supporter for this poor and insufficient agriculture of this region. Before separation from Bihar, all above factors were a hallmark in the Agriculture of Jharkhand.

After a separate existence as a state Jharkhand began to improve its agricultural environment particularly through the infrastructure development and economic support of the farmers. Farmers need a genetically diverse portfolio of improved crop seeds varieties suited to the local environment of this plateau area. Crop improvement and seed improvement are like two wheels of the same cart, having to co-exist with uninterrupted complimentary playing supplementary role and having to move with same force and speed. (Pattnaik, 2013, P. 14-18). However, timely delivery to farmers of the seed of high-yielding varieties (HYVs) or hybrids requires big improvements in the system that connects plant germplasm collections, plant breeding and seed production and delivery (Ghosh, 2013, P. 91-98). In India and other part of the world, the so-called green revolution was due to higher-quality seed of semi-dwarf varieties of major cereal crops, particularly Rice and Wheat. Replacement processes traditional seed to a high-yielding seed is called seed replacement rate (SRR) is defined as the percentage of area sown with certified/quality seed out of the total cultivated area under a crop in a season (Singh and Chand, 2011, P.22-25). Seed is the carrier of technology and the medium for translating scientific achievements to the field. Often, Indian farmers do not distinguish between grain and seed (Chand, 2007, P. 133-148).

Now we shall study the crop productivity and seed

requirement of popular crop and status of SRRs in comparison to standard SRRs among different crops grown in Jharkhand.

As we have discussed earlier in Chapter-I that Jharkhand has 380000 hectare of cultivable land area out of its total geographical area of 7971400 hectare. Here net sown area is only 2575000 hectares (32.3 percent) only one-fourth of the net sown area is utilized during rabi season and approximately 70 percent of total cultivable area during Kharif season. In other words, any one can say that despite good rainfall, cropped area and cropping intensity is low in Jharkhand. The level of technology adaptation including quality seed is also poor leading to lower productivity (NITI Ayog, 2015). Other strategies to improve SRR in Jharkhand are - Extension Activities, Introduction of emphasis on farmers managed seed systems, strengthening seed systems, public-private partnership (PPP), (Singh & Singh, 2016, P, 6987-7007).

**Table 3.12**
**Seed Requirement of Different Categories of Crops Grown in Jharkhand**

| Crop | Area (ha) | Seed Rate (kg/ha) | SMR | Certified seed Requirement to cover Entire Area | Certified seed Requirement as per standard SRR & area required to produce the desired quantity of CS | |
|---|---|---|---|---|---|---|
| **Kharif Crops** | | | | | | |
| Rice | 1692000 | 50.00 | 50.00 | 846000 | 279180 | 11167.20 |
| Maize | 250000 | 20.00 | 50.00 | 50000 | 25000 | 1666.67 |
| Finger millet | 30000 | 7.00 | 60.00 | 2100 | 700 | 100.00 |
| Pigeon Pea | 104000 | 20.00 | 15.00 | 20800 | 6684 | 1144.00 |
| Black Gram | 84000 | 20.00 | 20.00 | 16800 | 5544 | 1386.00 |
| Green Gram | 16000 | 20.00 | 20.00 | 3200 | 1056 | 264.00 |
| Groundnut | 22000 | 100.00 | 6.00 | 22000 | 7260 | 907.50 |
| Til | 3120 | 8.00 | 50.00 | 250 | 83 | 20.75 |
| Soybean | 5000 | 75.00 | 15.00 | 3750 | 1238 | 123.80 |

| | | | | | | |
|---|---|---|---|---|---|---|
| Sunflower | 2000 | 15.00 | 15.00 | 300 | 150 | 37.50 |
| Nigar | 15000 | 5.00 | 100.00 | 750 | 375 | 93.75 |
| Horse Gram | 31900 | 25.00 | 30.00 | 7975 | 2632 | 438.67 |
| Sub Total | 2255020 | | | 973925 | 330082 | 17349.84 |
| **Rabi Crops** | | | | | | |
| Wheat | 120000 | 100.00 | 20.00 | 120000 | 39600 | 2460.00 |
| Lentil | 32000 | 25.00 | 30.00 | 8000 | 2460 | 528.00 |
| Pea | 35000 | 75.00 | 10.00 | 26250 | 8663 | 866.00 |
| Linseed | 38000 | 30.00 | 30.00 | 11400 | 3762 | 752.00 |
| Gram | 150000 | 75.00 | 10.00 | 112500 | 37125 | 5303.57 |
| Rapeseed & Mustard | 130000 | 7.00 | 80.00 | 9100 | 3003 | 600.60 |
| **Sub Total** | **505000** | | | **287250** | **94739** | **10690.87** |
| **Grand Total** | **2760020** | | | **1261175** | **424875** | **28040.71** |

Above Data Presented 1. Comparative Productivity of major crops, 2. Seed Requirements of Jharkhand and 3. Status of seed replacement rate. These data clearly show the impact of high-yielding seeds over the productivity of the crops. Only data of 3 tables itself express the situation of hybrid impact on the productivity as the productivity has been mentioned from 2001-02 to 2013-14. This data represents the picture of productivity when Jharkhand was just created or separated from Bihar. On the other hand, there is the scenario of productivity after 14 years in the same table. Hence comparison is obvious. Nevertheless, the average productivity of major food crops in Jharkhand is far below the national average, which indicates the backwardness of the agriculture practices in the state. Table 1 shows the comparative performance with regard to major food crops productivity in Jharkhand and India. In Jharkhand, the increased production and productivity of rice is very important for food security reasons. The overall productivity of rice in Jharkhand was less than the national average throughout the first decade of the new millennium, although the latest three years of data shows that productivity of rice has increased (Table 1). The wheat and maize productivity of the Jharkhand state is quite low as compared to the all-India level. In pulses the productivity of the state is high but in the case of oilseeds it is drastically low if one compares it with the national average.

**Table 3.13**

**Progressive Status of Seed Replacement Rate among different field crops in Jharkhand**

| Crop | Norms | 2007-08 | 2008-09 | 2009-10 | 2010-11 | 2011-12 | 2012-13 | 2013-14 | 2014-15 | 2015-16 |
|---|---|---|---|---|---|---|---|---|---|---|
| Rice | 33 | 7.6 | 14.6 | 23.4 | 25.1 | 17.1 | 22.4 | 15.65 | 15.41 | 28.29 |
| Maize | 50 | 2.0 | 7.6 | 19.2 | 14.3 | 7.3 | 13.6 | 1.15 | 13.65 | 16.67 |
| Wheat | 33 | - | - | - | 27.6 | 31.9 | 41.2 | 8.34 | 22.26 | 28.57 |
| Lentil | 33 | - | - | - | 13.9 | 6.9 | 6.3 | 4.50 | 2.77 | 34.29 |
| Gram | 33 | - | - | - | 2.4 | 1.3 | 2.9 | 0.73 | 2.48 | 3.17 |
| Arhar | 50 | - | - | - | 11.7 | 23.6 | 21.6 | 3.26 | 6.70 | 25.00 |
| Mung | 33 | - | - | - | - | 21.2 | 19.0 | 7.03 | 0.00 | 25.00 |
| Urad | 33 | - | - | - | - | 6.9 | 5.1 | 2.96 | 10.01 | 20.00 |
| Pea | 33 | - | - | - | 17.4 | 9.4 | 10.2 | - | - | - |
| Horsegram | 33 | - | - | - | 23.7 | 7.6 | 8.8 | 0.0 | 0.0 | 80.0 |
| R/Mustard | 50 | - | - | - | 72.1 | 20.05 | 34.4 | 2.43 | 15.59 | 40.82 |
| Niger | 50 | - | - | - | - | - | 39.3 | 0.0 | 0.0 | 72.73 |

**Source:** Zonal Seed Review meetings, MOA-GOI (seednet.gov.in)

The productivity of different crops in Jharkhand agriculture suffered due to severe drought conditions during 2009-2015, except in 2011 in which normal rainfall was observed (Ministry of Agriculture, 2015).

Quality seed is most critical for enhancing agriculture production. In India in general and in Jharkhand in particular, farm saved seed (FSS) is the most prominent source of seed to raise crops year after year. Farmers are familiar with the seed they themselves grow and know that the variety is adapted to local conditions and preferences. In India more than 70 percent seed usage, particularly for food crops, is through FSS which leads to low SRRs. Use of FSS is the time-tested household tradition ingrained in the farming practices (Ayyappan and Kochhar 2010). In spite of the availability of quality seed of superior genotypes from the organized seed industry, the age-old tradition continues (Pattanaik, 2013). However, the yield potential cannot be realized if due care is not taken regarding varietal and physical purity, seed health and vigor. Seed viability and seed vigor are the two important traits that significantly influence crop performance. The use of quality seed solves this problem because the term 'quality' is maintained in the seed as per seed certification standards and the certified seed is supposed to have all the quality parameters as per norms.

## Impact of varietal and seed replacement rates (VRRs & SRRs) in Jharkhand

As per the National Seed Policy, 2002 emphasis was placed to increase the seed replacement rate (SRR) among different crops. The year wise progressive status of seed replacement rate among different field crops in Jharkhand is given in Table 4. Since an emphasis was made to increase SRRs in field crops, the state government adopted measures to strengthen seed chain management by establishing seed villages. In addition, through various central government central sector seed schemes and also through National Food Security Mission (NFSM), Rashtriya Krishi Vikas Yojana (RKVY), Bringing Green Revolution to Eastern India (BGREI), and Accelerated Pulses Production Program (A3P) for pulses etc., structural reforms took place. In order to increase varietal replacement rates (VRRs), under NFSM emphasis was placed upon only providing the seed of newly released varieties (<10 years old) for their adoption and diffusion among farming communities. The rice crop received the highest attention and the SRR has been achieved, because it excluded the area under hybrid rice which is quite significant (approximately ~25 percent). The case of maize is more or less the same. In wheat the SRR is increasing steadily, even though it has not reached the standard level of 33 percent. In pulses and oilseeds, the SRR is erratic except in chickpea (gram) where it is extremely low (<5 percent) and requires immediate attention, because chickpea is very important for food and nutritional security. For other crops, consistent and persistent efforts are needed to improve SRRs.

Now we can conclude that here seed is a critical input factor to increase production and productivity. The identified constraints to increasing VRRs and SRRs needs to be tackled on a priority basis with consistent and persistent efforts. The most challenging task is to correct the conversion ratio of seed. This can be improved through good agronomic practices. (GAPs).

## Plant Protection

After green revolution (1930-1960) we are in the post green revolution period. Now, the new dimensions of crop production, particularly the fertilizer responsive new hybrids, dwarf varieties and exotic high-yielding improved strains have raised new hopes and inspired new confidence. At present maximum people think that we have crossed traditional level of agriculture which is merely a slogan beyond the horizon of truth. But no doubt our irrigation, fertilizer and high-yielding seed consumption has increased its ratio of uses. However, there is a large number of factors that impose a serious limit on all human ingenuity, efforts and resources. Much of the crop losses (10 to 20 percent), if not all, inflicted by large variety of pests and weeds could be reduced by timely and adequate plant-protection measures. Plant-protection or chemical power has now been accepted as one of the major factors in increasing agricultural production, along with better seeds, fertilizer and irrigation. (Singh, 1976, P.220). Pests and diseases can seriously damage and even completely annihilate crops if adequate preventive and curative measures are not taken in time. Agronomists have discovered that the yield can be boosted substantially with proper spraying aerial as well as manual of the crops in particular.

At the time of its formation in 2000 much of Jharkhand state - aside from certain pockets - was almost organic by default. Use of pesticide by farmers has increased in recent years for two reasons. On the one hand pesticides have been promoted by government policies and programs, and by certain NGOs through projects. On the other hand, the lack of support provided to farmers (e.g., in Integrated Pesticide Management (IPM) or other agro-ecological techniques) has left them to rely solely on the private sector for advice and inputs. In the past 6 years, there has been a 6-fold increase in consumption of pesticide in Jharkhand, from 84 to 541 metric tons.

A 2017 study of 493 farming households located across the state showed that 75% use chemical pesticides.93 An analysis of average pesticide consumption (in kg per 1000 persons, using 2012 population census data) 94, shows that in Jharkhand, consumption has risen from 2.2 to 13.9 kg per 1000 persons over the past 6 years. During this period, it has overtaken per capita consumption in Bihar, which marginally increased from 6.8 to 8.5 kg/1000 persons over the same period. Positively, average consumption in Jharkhand is far below that of Punjab, which is relatively stable at 205 to 209 kg/1000 persons. Across many states, pesticide consumption is rising, i.e., in Jharkhand, Bihar, Punjab, Uttar Pradesh and Maharashtra. Some southern states show decreasing trends in overall consumption, e.g., Andhra Pradesh and Telangana (now separated) from 104 to 67 kg/1000 persons, and Karnataka from 29 to 20 kg/1000 persons. In Kerala, surprising given the initiatives taken by the government and civil society, consumption of pesticides has risen from 19 to 31 kg/1000 persons. On 9th December 2017 at a seminar at Birsa Agricultural University, Jharkhand's Agricultural Minister Shri Randhir Kumar Singh announced that Jharkhand will go fully organic by 2025. The department has also announced that organic agriculture will be promoted in several of Jharkhand's districts. The question of how the government envisages or conceptualizes a transition to organic is presently unclear. Jharkhand does not have an Agricultural Policy or an Organic Farming Policy unlike states like Sikkim, Uttarakhand, Karnataka, Kerala, MP, Bihar and Maharashtra that have announced their own organic farming policies. With the support of the state governments, in Karnataka and Andhra Pradesh Zero Budget Natural Farming (ZBNF) has picked up in a big way which might well explain these states' downward trend in consumption of chemical pesticide. Karnataka's 2017 Organic Farming Policy states that it aims to bring the net sown area under organic cultivation from 1% to 10% by 2022. Yet Karnataka's organic policy doesn't mention anything about the remaining 90% of net sown area in the state, neither does it express, for example, any desire to regulate the hazardous chemical pesticide trade.

## Hypothesis Testing

In the concerned synopsis, there were three hypotheses, two of them are related to this chapter. i.e., Farm Technology, the third one is related to the next chapter i.e., Cropping Intensity.

First hypothesis is - "Farmers of Jharkhand are using chemical fertilizers on the farms of all food crops regularly. Null hypothesis of this work is that the use of fertilizers over rice and maize (which are the larger crops) is same according to their ratio of hectare of area and production."

For the testing of my hypothesis, we prepared a simple questionnaire with six questions. Four questions are related to the first hypothesis and two questions for second hypothesis. Third hypothesis is related to secondary data.

Questionnaire were given to 100 medium class farmers of four villages i.e., Village Madhukarpur (Bokaro district),

Boreya (Ranchi district), Ormanjhi (Ranchi district) and village Panso (Gumla district). I have relatives in these villages who helped me carry out this activity. We distributed the questionnaires selecting 25 farmers from each village.

We processed the reply and very easily solved the hypothesis. Here our hypothesis is rejected and thus we can say that use of fertilizer over rice is more than maize. But effect on production could not be tested. For this one crop should be planted without fertilizer and farmers were not ready to do that. Hence comparison of use of fertilizers could not be tested.

Rice is most important crop in terms of both area and fertilizer use. Fertilizer use on irrigated paddy (155 Kg/hectare) is double than on rainfed paddy (77.6 Kg/ha). The average per-hectare use of fertilizer on paddy was 119.1 Kg (81.7 Kg/ha N, 24.3 Kg/ha P2O5 and 13.1 Kg/ha K2O).

Whereas maize is mostly grown under rainfed conditions. Per hectare fertilizer use is 60.2 Kg, which consists of 41.7 Kg/ha N, 14.7 Kg/ha P2O5 and 3.8 Kg/ha K2O. The rate of fertilizer use on irrigated maize is 92 Kg/ha (59.6 Kg/ha N, 27 Kg/ha P2O5 and 4.8 Kg/ha K2O).

My second hypothesis is related to the third chapter i.e., Farm Technology. Hypothesis is "State government tries to improve the quality of seeds and instruments of agriculture frequently. Null hypothesis for this work is that the changing ratio of improved seeds and instruments of agriculture is same."

The last two questions of the questionnaire are related to this hypothesis. These are - (i). What new equipment you are using for farming. And (ii). Name the seeds of the HYV you use in farm. When we analyze the results, we find that none of the farmers have any new equipment for farming. However, few replied that sometimes they use water pumps and tractors by hiring them on rent. But all have replied that they use HYV paddy, wheat and maize and also vegetables. Thus, my null hypothesis is rejected and we can thus conclude that maximum farmers (more than 80 percent) are using HYV seeds but new equipment for cultivation are being used by only around 5 percent of the farmers of four villages.

## References

Ayyappan S, & Kochhar, S. (2010); Indian Agriculture in the new millennium: Innovative strategies for productivity enhancement prospects and challenges. Indian Seed and Planning material, 3(1), P. 7002.

Chand, R (2007); The State of Indian agriculture and prospects for the future. In K. Chopra & C.H. Hanumantha Rao (Eds), Growth, Equity, Environment and Population: Economic and Sociological Perspectives, Sage Publication, New Delhi.

Eddowes, Maurice (1969); Crop Technology, London, Hutchinson Educational.

Ghosh, G.N. (2013); Sustainable agriculture: Strategies and approaches. In P.K. Shetty, S. Ayyappan and M.S. Swamynathan (Eds) climate change and sustainable food security. Bangalore; National Institute of Advanced Studies and New Delhi: Indian Council of Agricultural Research.

Ministry of Agriculture (2015); Proceedings of the workshop. In union Agriculture Minister (Chair), Agriculture scenario in Eastern India: Review and future prospects. Workshop held at NASC Complex, DPS Marg, Pusa Complex; New Delhi. Retrieved from bgrei- rkvy.nic.in/ minutes.

Ministry of Agriculture (2018); Final Report on Monitoring, Concurrent evaluation and impact assessment of Sub-Mission on Agricultural Mechanization.

Mishra, R.K. (2006) & V. Jairam; Agricultural development in Jharkhand: At a Glance through figures:2006. Agricultural Data Bank, A/13 Krishi Vihar, Post Boreya, Ranchi.

NITI Ayog (2015); Comment on the concept note on agriculture: Challenges and way forward Jharkhand. Retrieved from http://nti.gov.in.

Pattnaik, B. B (2013); Quality Seeds: Contribution of national seeds corporation. In A.K. Singh (Ed.) Golden Jubilee Souvenir 2012-13, New Delhi National Seeds Corporation Limited.

Simon Kuznets (1973); Population, Capital and Growth (Selected Essay), W.W. Norton & company, New York

Singh, H & Chand, R (2011); The Seeds Bill, 2011: Some Reflections. Economic and Political Weekly, XLVI (51)

Singh, Jasbir (1976); An agricultural Geography of Haryana, Vishal Publications, Kurukshetra, Haryana.

Singh, R. P & Singh, S. (2016); Optimizing seed replacement rates in Jharkhand: Present scenario, challenges and opportunities. Jharkhand Journal of Development and management studies, XISS, Ranchi, Vol. 14, No.-2 (June).

Singh, R.P. (2012); New Frontiers in seed sector in Jharkhand - Role of SAU. Seed Times, 5 (3).

Tiwari, Ram Kumar (2019); Jharkhand Ki Ruprekha, Bihar Hindi Granth Academy, Patna.

# 4

# Cropping Intensity

- Introduction
- Net Area Sown
- Total Cropped Area
- Shrinking in Net Area Sown and Improvement in Fallowing
- Intensification of Cultivation
    - Cropping Intensity during 2004-05
    - Cropping Intensity during 2016-17
    - Irrigation Intensity during 2004-05
    - Irrigation Intensity during 2016-17
- Correlation Consequence

## Introduction

Basic objective of this chapter lies in evaluating the intensity of cropping scenario in Jharkhand. Intensity of cropping is expressed as the ratio of gross cropped area to net sown area. In other words, we can say that the cropping intensity is closely related to the number of crops harvested per land unit per year. Thus, no doubt cropping intensity is highest in irrigated area and less in dry or non-irrigated area. Again, we can understand this concept in other way that where more than one crop is harvested per year, certainly the net sown area of that region will be high and ultimately the intensity of cropping will be high. So, the intensity of cropping is the study of 'net sown area' on one hand and 'total cropped area' on the other hand. Both are studied in the chapter of land use; thus, it is very clear that the study of intensity of cropping is indirectly the study of land use. Land use scenario of the state has been studied in preceding first chapter of this work in brief. Here we shall concentrate on only the net sown area and total cropped area of the state district wise. This study will make it clear that which district has more intensity and which district has less. It will be easy to understand through this study that which district has more capacity to increase the production of food grains to satisfy the increasing demand of population. "The problem of the widening gap between increasing human population and food production has been often emphasized, and the gravity of the situation in India is an open secret. It is not the problem of quantitative shortage alone, the qualitative aspect of the food available for ingestion is causing equally great anxiety to experts. (Singh, 1976, P.227).

## Net Area Sown

This represents the actual land sown with crops during the particular year. This is the actual arable land of any region. This area is representing the actual crop growing area which is in direct observation as green area of any region. In other words, this is the area where agricultural population (farmers) grow their:

1. Food Crops (Cereals, Pulses, Sugarcane, Spices, Fruits and Vegetables, Root crops and other food crops) and,
2. Non-Food Crops (Oil-Seeds, Fibers, Dyes & Tanning material, Drugs & Narcotics, Green Manuring crops, other non-food crops.

**Table 4.1**
**Net Area Sown: Jharkhand (In 000 hectares)**

| Year | Net Area Sown (Per cent to total area) | Cropping Intensity |
|---|---|---|
| 1955-56 | 1870<br>23.5% | 131 |
| 1999-2000 | 1809<br>22.7% | 114 |
| 2014-15 | 1385<br>17.3% | 112 |

| | | |
|---|---|---|
| 2016-17 | 1451<br>18.2% | 141 |

Temporal study of net area sown has depicted a decreasing tendency in Jharkhand. Net area sown has registered about 1870 thousand hectares land during 1955-56 which was 23.5 percent to the total geographical area of the state. During 1999-2000 it decreased to 1809 thousand hectares which was 22.7 % of the total area. Again, it was registered 1385 thousand hectares in 2014-15 and 1451 thousand hectares in 2016-17 which were 17.3 percent and 18.2 percent of the total area respectively.

Temporal study of net area sown has registered about 1870 thousand hectares land during 1955-56 which was 23.5 percent to the total geographical area of the state. During 1999-2000 it decreased to 1809 thousand hectares which was 22.7 percent. Again, it was registered 1385 thousand hectares in 1914-15 and 1451 thousand hectares during 2016-17 which were 17.3 percent and 18.2 percent of the total area respectively.

District wise study shows a highest area of net sown area in 2004-05 in Ranchi District (255.85 thousand hectares) followed by Gumla (161.75 thousand hectares), Dumka (119.10 thousand hectares), Hazaribagh (106 thousand hectares), Palamu (97.75 thousand hectare), East Singhbhum (83.28 thousand hectares) and so an. The share of net sown area was lowest in the district of Bokaro (15.51 thousand hectare).

Similarly, the study of the data of 2016-17 of net area sown reveal a highest coverage of the area in Ranchi district (199.05 thousand hectares) followed by Palamu (141.32 thousand hectares), Gumla (131.20 thousand hectares) and so on as mentioned in the table. But the lowest area registered in the district of Ramgarh (3.34 thousand hectares). In between the period of 12 years the area of net sown area has decreased to about311.21 thousand hectares.

Dr. R.C. Tiwari has mentioned in his book 'Krishi Bhoogol' (2016, P.89) that the net sown area has an average percentage to their total geographical area about 32 percent in the world,

40 percent in USA, 27.4 percent in USSR, 16.1 percent in Brazil and 6.1 percent in Canada. Similarly, India has about 45.82 percent (2009-10) and Jharkhand has an average of 18.2 percent only. He has also calculated that in India 8 persons are dependent on agricultural land; whereas in UK it is 7 persons, in Belgium 11 persons, in Germany 14 persons and in Japan 17 persons are dependent. When we calculate the ratio of Jharkhand, it is 18 persons per hectare of the net sown area.

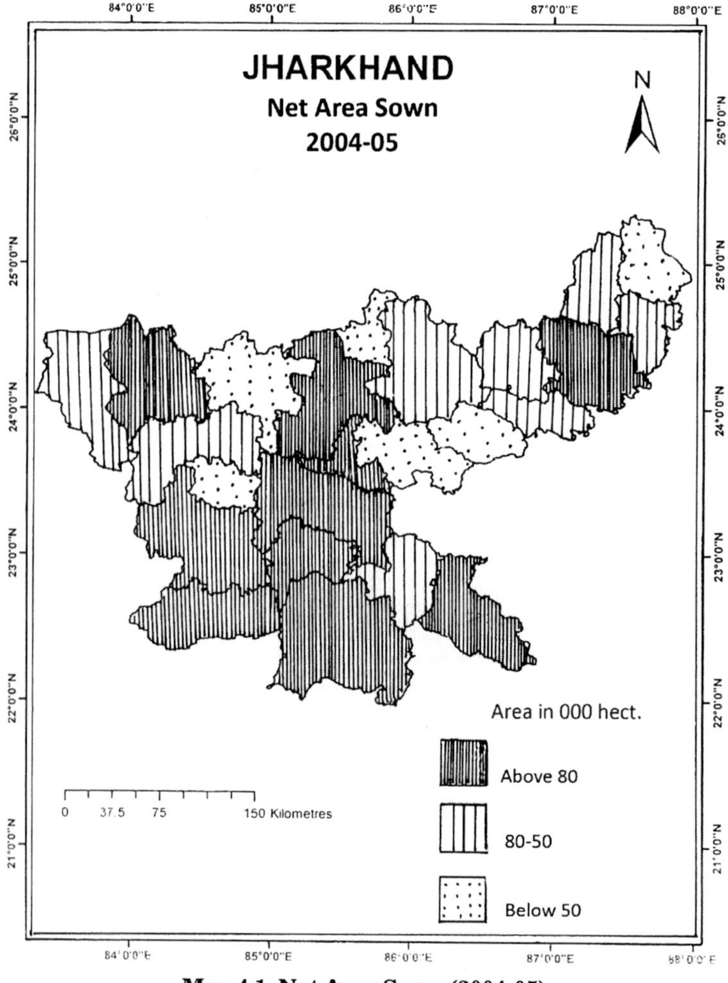


**Map 4.1: Net Area Sown (2004-05)**

Dr. V.N.P. Sinha (2008) has rightly observed that 'Jharkhand has 22.11 percent land but there is lack of uniformity in development of net area sown. Areas having homogeneous land in the form of flat plateaus, broad valleys or very gentle sloping land have higher percentage of land under cultivation than the areas having comparatively rough terrain, dense forests, sparse population etc. According to the data mentioned in the table, districts Ranchi and Godda

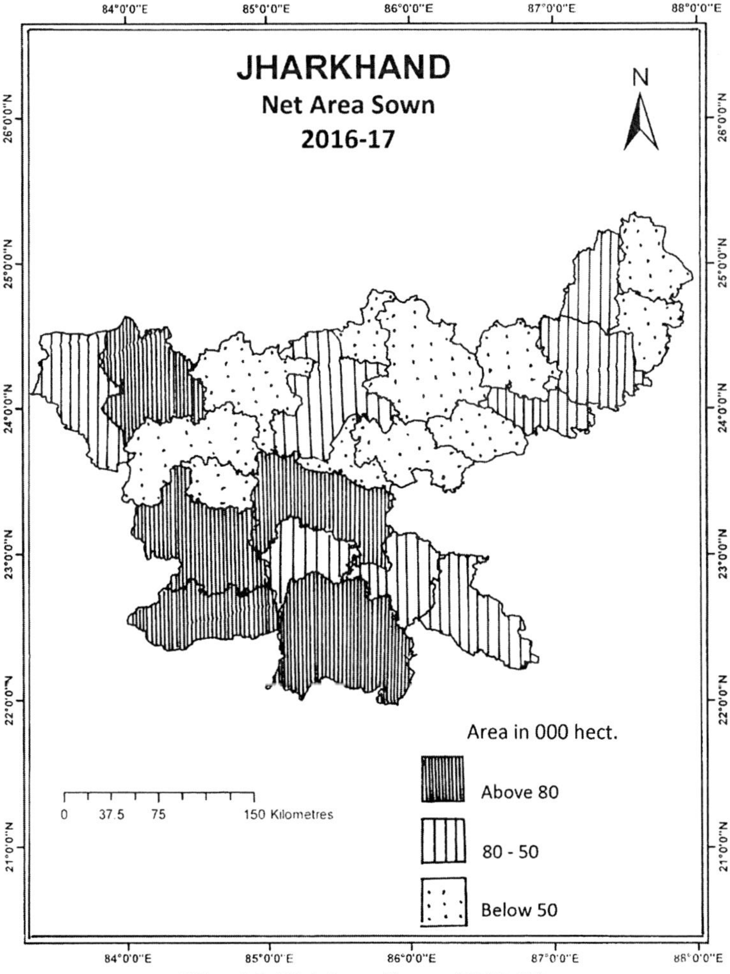


**Map 4.2: Net Area Sown (2016-17)**

have the same percentage of net sown area which is highest in the state because of their location over the gentle slope area. Maximum portion of the area of the districts is homogeneous land where agriculture has developed which is clear through the tables. It may be noted that where the net area sown is medium all such districts have suitable land for agricultural use. Districts having less net area sown have rough terrain, mines and forest etc.

When we calculate the change in net sown area district wise, only five districts have registered positive change or growth; they are Bokaro, where net sown area was 15.51 thousand hectares during 2004-05 and it became 29.38 thousand hectares during 2016-17. Garhwa district had 64.77 thousand hectares which grows to 66.43 thousand hectares. Palamu district had only 97.75 thousand hectares net sown area but it grew to 141.32 thousand hectares within this period. Sahebganj district has recorded 41.46 thousand hectares and Simdega has 81.67 thousand hectares of net sown area during 2004-05 census but both the districts have recorded 48.10 and 81.68 thousand hectares of net sown area during the census of 2017-18. Rest all 19 districts have recorded negative growth or a decrease in net sown area of Jharkhand. This is an overall decrease of 311.21 thousand hectares during the period of twelve years. This was a decrease of 17.66 percent.

In Jharkhand right from 1956 growth of net sown area is facing a fluctuating or irregular trend. Till date it has never touched its higher position which was in 1964-65 (2432 thousand hectares).

## Total Cropped Area

Total cropped area is also known as Gross cropped area. This represents the total area sown once and/or more than once in a particular year i.e., the area is counted as many times as there are sowing in a year. This total area is known as Total Cropped Area.

**Table 4.2**
**Jharkhand: Total Cropped Area (000 hectare)**

| Year | 1955-56 | 1999-2000 | 2014-15 | 2016-17 |
|---|---|---|---|---|
| Total Cropped Area (000 Hectare) | 2444<br>30.6 % | 2068<br>25.9% | 1554<br>19.4% | 2045<br>25.7% |

Jharkhand was at one time known for its 'forest land' but after arrival of tribal people they started to clear the forest to bring suitable land under the plough. Such land or farms (khet) are still known as 'Bhuinhari' land (Section 48, Chotanagpur tenancy act, 1908). Similarly, villages have been founded by them during their early settled time in Chhotanagpur and are known by 'Khuntkatti' right. "A raiyat having khuntkatti right means a raiyat in occupation of, or having any subsisting title to land reclaimed from jungle by the original founders of the village or their descendants in the male line, when such raiyat is a member of a family." (Section 7 of Chotanagpur tenancy act, 1908). Same was different types of land have been prepared for their agricultural purpose i.e., Korkar land, Bhutkheta land etc. In spite of all above facts, it has been mentioned in the literature of Sharatchandra Rai that before arrival of tribal people there were living a very developed community of race in Chotanagpur. No doubt, the agriculture was prevailing in this region. Nevertheless, period of 1951 to 1966 (after independence and abolition of zamindari), the period of planned development, saw much horizontal expansion of agriculture and the shrinking cultivable wasteland acted as a useful indicator of progress. It shrank to 4.4 percent in 2016-17 from 5.8 percent in 1955-56, while the cultivable wasteland increased to 49 percent in 2016-17 from 39 percent in 1955-56.

**Table 4.3**
**Cultivable Area in Jharkhand (Other fallow+ current fallow+ net sown area)**

| Year | 1955-56 | 1999-2000 | 2014-15 | 2016-17 |
|---|---|---|---|---|
| Cultivable Land (000 Hectare) | 3112<br>39% | 3907<br>49% | 3893<br>48.8% | 3913<br>49% |

Indeed, cultivable wasteland is still in existence at present about 4.4 percent as per the statistics of the state. This change shows the reduction of wasteland but this change is very slow. All efforts of the Agriculture department are in vain because the existence of this 4 percent has continued since last two decades. Statistics of total cropped area is neither increasing nor decreasing. It is fluctuating from 30.6 percent to 25.7 percent showing a decreased ratio. After two decades it has increased to the situation of 1990. This increase has been recorded due to an increase in double cropped area. It was only one percent during 1990 and now during 2017 it has grown to 7.45 percent. Details of the mechanics of land use show that now waste land is unable to convert into arable land, only the double cropped area can be increased. This depends upon the technical and chemical power enhancement of the state government.

Now the district wise study reveals a highest coverage of total cropped area in Ranchi district (272.65 thousand hectare) followed by Gumla district (169.32 thousand hectare), West Singhbhum (160.19 thousand hectare), East Singhbhum (146.88 thousand hectare), Dumka (125.15 thousand hectare), Palamu (118.60 thousand hectare), Hazaribagh (118.31 thousand hectare), Garhwa (89.11), Simdega (86.96), Godda (85.82), Giridih (84.19), Latehar (72.52), Deoghar (67.72), Saraikela (65.28), Pakur (63.05), Jamtara (58.16), Chatra (50.99), Sahebganj (48.96), Lohardaga (45.47), Dhanbad (40.00), Bokaro (25.94) and Koderma (21.54 thousand hectares) respectively during the census of 2004-05).

During the census of 2016-17 highest total cropped area was recorded in the district of Gumla (288.19 thousand hectares) followed by Ranchi (252.44), Palamu (188.09), Giridih (143.45), W. Singhbhum (117.55), Garhwa (91.38), Godda (85.89), Dumka (84.19), Simdega (84.73), Khunti (75.86), Hazaribagh (62.58), E. Singhbhum (62.37), Jamtara (61.93), Saraikela (59.39), Lohardaga (50.90), Sahebganj (48.61), Deoghar (48.23), Bokaro (41.19), Chatra (39.17), Ramgarh (21.60), Koderma (13.76) and Dhanbad (11.51 thousand hectares) respectively.

A study of changes in total cropped area represents a positive change among the ten districts of Jharkhand i.e., Bokaro, Garhwa, Giridih, Godda, Gumla, Jamtara, Khunti, Lohardaga, Palamu and Ramgarh. Rest districts have recorded a negative growth in total cropped area. That is why a total growth has been recorded in the state. During 2004-05 it was 2025.51 thousand hectares but it became 2044.97 thousand hectares during the census of 2016-17. There is a very little change of only 19.46 thousand hectares growth of total cropped area in Jharkhand within two decades but when we look into the past records of 1955-56 and 1964-65 decades it was recorded 30.6 and 34.4 percent respectively. Presently we have achieved only 25.7 percent. Now question is here that why the total cropped area has decreased its highest score.

An immediate cause of reduction of total cropped area is the fallow land. When the percentage of the total cropped area was higher during 1955- 56 and 1964-65 the percentage of fallow land was lesser. Fallow land has their two categories - other fallow land and current fallow land. Other fallow land was recorded 6.6 and 7.2 percent to the total geographical area during 1955-56 and 1964-65 respectively. Current fallow land recorded its share 8.7 and 7.5 percent during the census of 1955-56 and 1964-65 respectively. Now the Other fallow land grows to 12.8 percent and the current fallow land grows to 18 percent during 2016-17 census. I am looking at the answer of reducing total cropped area in increasing percentage of fallow land.

**Table 4.4**
**Comparison between Fallow Land and Total Cropped Area**

| Year | 1955-56 | 1964-65 | | 2014-15 | 2016-17 | |
|---|---|---|---|---|---|---|
| | | | Increase | | | Increase |
| Fallow Land | 15.5% | 14.7% | -5.4% | 31.3% | 30.8% | -1.6% |
| Total Cropped Area | 30.6% | 34.4% | +11% | 19.4% | 25.7% | 24% |

Fallow land was 15 percent which grows to 30 percent but at the same time total cropped area was 30 percent that reduced to 19.4 from 25.7 percent. Now again a question arises here that why fallow land is increasing decade by decade in this region. To answer this question, we can say that this category of land has been more pronounced in almost all districts of the state because of the hilly terrain. Abandoning of cultivable land is a common feature because of leaching of soil, lack of irrigation, economic condition of cultivator and uneconomic production etc.

## Shrinking in Net Area Sown and Rise in fallowing

The total cultivated area comprises the net area sown and the fallow land. The distributional patterns of the net area sown and cultivated area are identical. Which means the fallowing has no important status in land use and the major proportion in any cultivated area is that of the net area sown. The percentage strength of the net sown area to total area of Jharkhand has changed significantly during the past twelve years decreasing to 18.20 percent in 2016-17 from 22.11 percent in 2004-05. In case of the cultivated area the increase was to the extent of 0.93 percent, but in net area sown the magnitude of decrease was equivalent to 21 percent. The fallow land which was mostly current-fallow, was 6.8 percent in 1955-56 and increased to 12.8 percent in 2016-17. Hence the scope for increasing production by increasing the extent of fallowing is also very limited or surprising. 'The regional variations in the extent of the net area sown are linked with the same reasons that conditions the regional differences in the percentage strength of the cultivated area. Factors other than relief, soil, climate such as technical-organizational, institutional and socio-economic factors also contribute to regional contrasts in the distribution of the net area sown, chiefly by reinforcing natural environment' (Singh, 1976, P.232).

Map of net area sown for 2004-05 and 2016-17 shows that during 2004- 05 there were 8 districts in the category of high (over 80 percent) net area sown but there are only 5 districts in this high category in the map of 2016-17. Three districts i.e., Dumka, E. Singhbhum and Hazaribagh are absent from the map of 2016-17 because of the reduction in net area sown during this period. Similarly, the both maps depicted the changes in net area sown. During 2004-05 under the category of moderate (80-50 percent) net area sown there were 10 districts but in the map of 2016-17 there are only 9 districts. Under the category of low (below 50 percent) net area sown there were only 6 districts but in the map of 2016-17 there are 10 districts under this low category. Increasing the number of districts of low category proves the decreasing percentage of net area sown in the state.

When we compare both the maps of net area sown, following districts are continuously in the low category- Bokaro, Chatra, Dhanbad, Koderma, Lohardaga and Sahebganj. Some districts have changed their category from medium to low; those are - Deoghar, Giridih, Latehar and Pakur. In ordinary sense, the extent of fallowing is primarily determined either by inclement climatic conditions or by lack of irrigation facilities and migration of cultivators and agricultural laborers. Sometimes author has observed a less interest in cultivation due to occupational change of cultivator as there is not sufficient economic gain in farming. Here an instructive conclusion is to improve the net area sown within the cultivated area of the state that a radical modification in physical environment of the area should be initiated and expedited by the government chiefly through the provision of an additional agricultural water supply and improved dry farming techniques. Then only fallow land and other fallow land will disappear from the agricultural landscape of Jharkhand.

## Intensification of Cultivation

The extension of cultivation is not a matter of significance at present, except for achieving local gains. Intensity of cropping, extent of maturity and increase of the yield from the existing cultivated area are problems of paramount importance in the agricultural economy of India. These need a serious thought by planners. (Singh, 1974). As we have studied in the past pages that underutilization of land is no problem in Jharkhand because most of the topographically accessible area for cultivation is already under the plough (except 4.4 % of cultivable waste land), but the problems of under-use of the net area sown, under-productivity per unit area per unit time, and the risk of crop failures are taxing the rural society. It would be a useful gain to overcome these problems in the foreseeable future. Hence, it is imperative to investigate the degree of efficiency with which the net sown area is utilized.

Intensification of cultivation means the land use efficiency. It is defined as 'the extent to which the net area sown is cropped or resown'. The total cropped area as a percentage of the net area sown gives a measure of land use efficiency, which really means the intensity of cropping (Singh, 1976, P.235).

First of all, it should be clear that the cropping intensity has its two synonyms as used by different authors in their book i.e., Land use efficiency by Jasbir Singh as stated above and Agricultural intensity by R.R. Tripathy and G.S. Tyagi (1970). Secondly, the meaning of the degree of cropping intensity should be clear in mind. 'The intensity of cropping refers to the number of crops raised on a field during an agricultural year; for example, if one crop is grown on a field either as a Kharif or as a rabi crop in a year, the index of cropping is 100 percent and it can be termed as single-cropping; if two crops a year are procured as Kharif and Rabi crops, the intensity index will be 200 percent and such a case can be designated as double -cropping; if three crops a year are produced as kharif, rabi and zaid-rabi, the intensity of cropping will be 300 percent and it will be a case of multiple cropping (Singh, 1976, P235). Therefore, a higher index of cropping intensity indicates a higher efficiency and higher land-use, and lower the index, the lower the intensity of cropping and less utilized or under-utilized the net area sown.

When we calculate the cropping intensity of Jharkhand for different years in between 1955-56 to 2016-17 the percentage of intensity was recorded 131 during 1955-56, 113 during 1964-65, 107 during 1989-90, 114 during

1999-2000, 114 during 2014-15 and 141 in the year of 2016-17. All the intensities percentage are in the range of minimum 107 to maximum 141 in the period of 60 years. This reveals that Jharkhand is a mono crop area.

**Table 4.5**

**Jharkhand: Cropping Intensity 2004-05**

| Districts | Total Cropped Area (000 hectare) | Net Area Sown (000 hectare) | Cropping Intensity (%) |
|---|---|---|---|
| Bokaro | 25.94 | 15.51 | 167 |
| Chatra | 50.99 | 45.07 | 113 |
| Deoghar | 67.72 | 63.68 | 106 |
| Dhanbad | 40.00 | 34.60 | 116 |
| Dumka | 125.15 | 119.10 | 105 |
| E. Singhbhum | 146.88 | 83.28 | 176 |
| Garhwa | 89.11 | 64.77 | 138 |
| Giridih | 84.19 | 77.49 | 109 |
| Godda | 85.52 | 78.21 | 108 |
| Gumla | 169.32 | 161.75 | 105 |
| Hazaribagh | 118.31 | 106.14 | 111 |
| Jamtara | 58.16 | 55.34 | 105 |
| Khunti | - | - | - |
| Koderma | 21.54 | 17.04 | 126 |
| Latehar | 72.52 | 59.77 | 121 |
| Lohardaga | 45.47 | 39.64 | 115 |
| Pakur | 63.05 | 57.84 | 109 |
| Palamu | 118.60 | 97.75 | 121 |
| Ramgarh | - | - | - |
| Ranchi | 272.65 | 255.85 | 107 |
| Sahebganj | 48.96 | 41.46 | 118 |
| Saraikela | 65.28 | 60.42 | 108 |
| Simdega | 86.96 | 81.67 | 106 |
| W. Singhbhum | 160.19 | 145.99 | 110 |
| **Jharkhand** | **2025.51** | **1762.47** | **115** |

**Table 4.6**
**Jharkhand: Cropping Intensity 2016-17**

| Districts | Total Cropped Area (000 hectare) | Net Area Sown (000 hectare) | Cropping Intensity (%) |
|---|---|---|---|
| Bokaro | 41.19 | 29.38 | 140 |
| Chatra | 39.17 | 32.05 | 122 |
| Deoghar | 48.23 | 47.71 | 101 |
| Dhanbad | 11.51 | 8.27 | 139 |
| Dumka | 84.19 | 55.71 | 152 |
| E. Singhbhum | 62.37 | 61.09 | 102 |
| Garhwa | 91.38 | 66.43 | 138 |
| Giridih | 143.45 | 40.99 | 350 |
| Godda | 85.89 | 72.30 | 119 |
| Gumla | 288.19 | 131.20 | 220 |
| Hazaribagh | 62.58 | 52.25 | 120 |
| Jamtara | 61.93 | 51.42 | 120 |
| Khunti | 75.86 | 60.89 | 125 |
| Koderma | 13.76 | 13.43 | 102 |
| Latehar | 59.39 | 27.05 | 220 |
| Lohardaga | 50.90 | 36.87 | 138 |
| Pakur | 59.41 | 48.75 | 122 |
| Palamu | 180.09 | 141.32 | 127 |
| Ramgarh | 21.60 | 3.34 | 646 |
| Ranchi | 252.44 | 199.05 | 127 |
| Sahebganj | 48.61 | 48.10 | 101 |
| Saraikela | 59.72 | 55.82 | 107 |
| Simdega | 84.73 | 81.68 | 104 |
| W. Singhbhum | 117.55 | 86.02 | 137 |
| Jharkhand | 2044.97 | 1451.26 | 141 |

Now we shall examine the data of cropping intensity of the state for two census years i.e., 2004-05 and 2016-17 (Table 4.5 & 4.6) having the district wise analysis.

## Cropping Intensity During 2004-05

Average cropping intensity of the state was 115 percent during this census year. Only nine districts' intensity was more than the average of the state; they are Bokaro, Deoghar, E.

Singhbhum, Garhwa, Koderma, Latehar, Lohardaga, Palamu and Sahebganj, rest have less than the average of the state. It is very clear that 15 districts have the intensity below the average of the state. No doubt, maximum portion of the whole region have a less cropping intensity.

As it has been stated in preceding paragraph that if the cropping intensity is 100 for any region, it should be treated as mono-cropping area. Here not even one district has a cropping intensity of 200. We can say that those nine districts having higher intensity than the average are showing their growing tendency. As for example, E. Singhbhum have 176 intensity of cropping which shows that the district will touch the intensity of 200 very soon and the area will be double-cropped.

## Cropping Intensity During 2016-17

Average intensity of cropping during this census year rose to 141 percent. Only five districts are having more intensity than the average of the state. During previous census (2004-05) nine districts were above the average of the state which means a clear-cut decrease in the intensity of cropping but when we see the growth of intensity of different districts, we get highest intensity (646 percent) in Ramgarh district. It is an indicator of multiple cropping of the district. Similar case with the Giridih district where cropping intensity is more than 300. It is the indicator of multi- cropping for the district. Gumla (220) and Latehar (220) having the same intensity but indicate having more than 200 of intensity implies them as double-cropping districts. Rest all twenty districts are mono-cropping regions.

It is an agricultural truth that 'intensity of irrigation has evolutional command over the intensity of cropping and is the primary influencer'. It is, therefore, of interest to know how and to what extent this influencer influences the intensity of cropping in each region. Before examining the impact of the intensity of irrigation on the intensity of cropping it will be in order to construe the theoretical reasons for assuming an inter- relationship between the two variables, viz. independent (intensity of irrigation) and dependent (intensity of cropping). The theoretical reasoning is that in the case of time series data, with the increase in intensity of irrigation there should be corresponding increase in the intensity of cropping, and in the cross-sectional data, the higher the intensity of irrigation, the higher will be the intensity of cropping and more positive the coefficient of correlation. (Singh, 1976, P. 244-245).

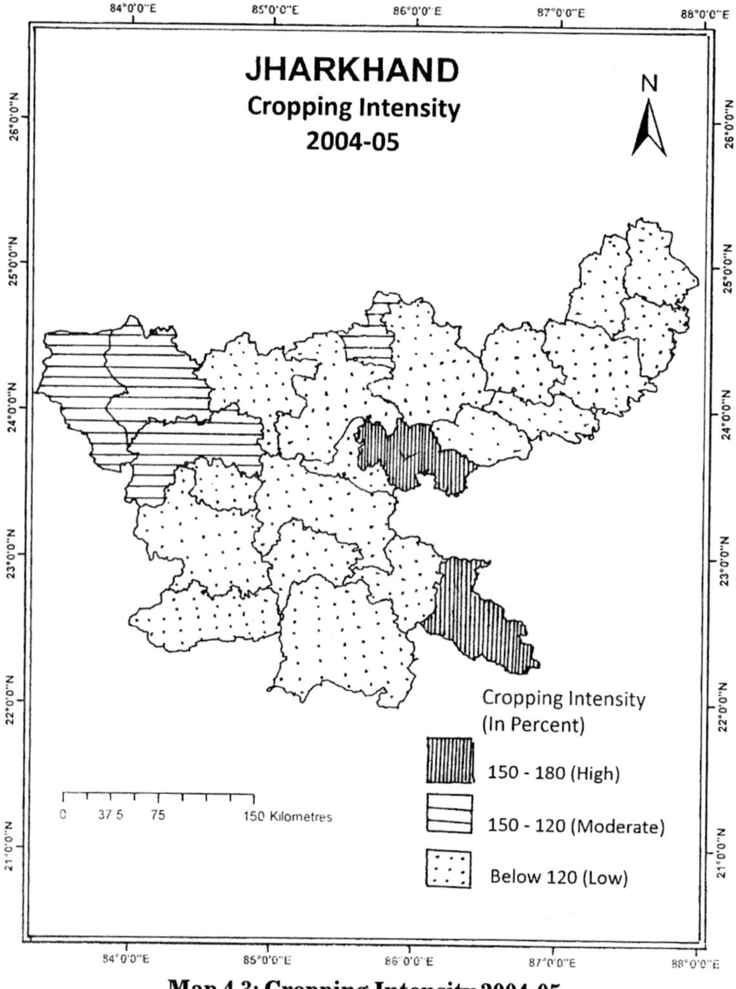


**Map 4.3: Cropping Intensity 2004-05**

Here at first, we shall discuss only irrigation intensity with the district wise scenario and then the correlation between cropping intensity and irrigation intensity will be discussed.

## Irrigation Intensity : 2004-05

Here we shall compare the irrigation intensity between 2004-05 and 2016-17 with district wise scenario but I believe it

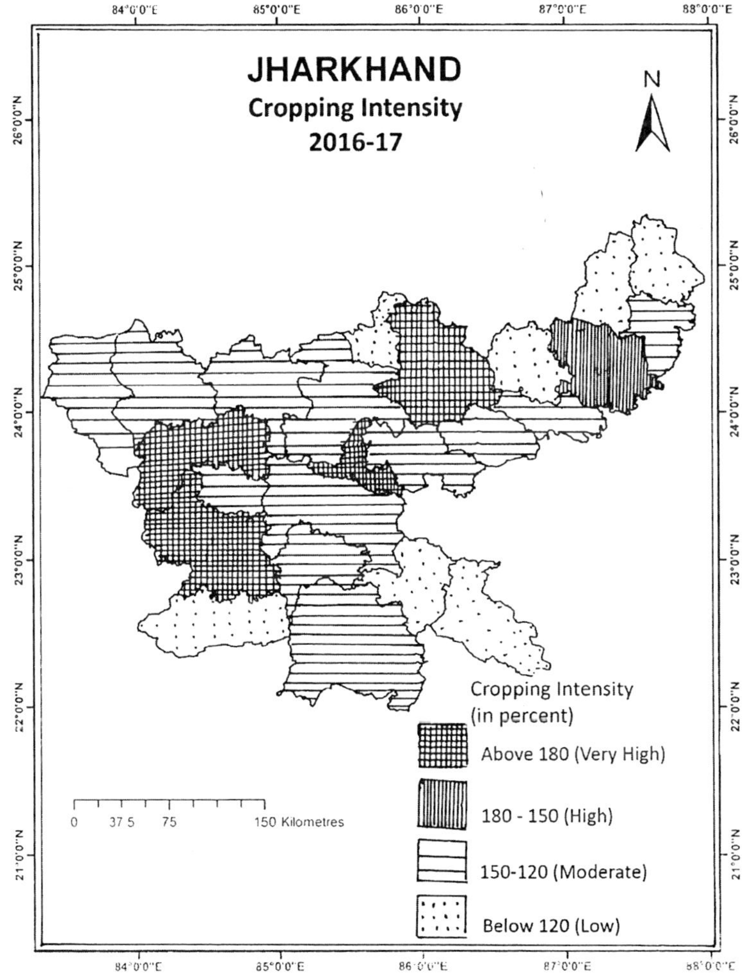


**Map 4.4: Cropping Intensity 2016-17**

will be better to see at first the irrigation intensity of the whole state from 1955-56 to 2016-17.

**Table 4.7**
**Irrigation Intensity of Jharkhand (000 hectare)**

| Districts | Total Irrigated Area (000 hectare) | Net Area Sown (000 hectare) | Irrigation Intensity (%) |
|---|---|---|---|
| 1955-56 | 187.0 | 1870 | 10.00 |
| 1964-65 | 223.7 | 2432 | 9.20 |
| 1985-86 | 138.0 | 2071 | 7.00 |
| 1986-87 | 147.0 | 2029 | 7.24 |
| 1987-88 | 124.0 | 1985 | 6.25 |
| 1988-89 | 146.0 | 2010 | 7.30 |
| 1989-90 | 152.0 | 2046 | 6.70 |
| 1995-96 | 177.4 | 1830 | 9.70 |
| 1997-98 | 157.0 | 1808 | 8.70 |
| 1999-2000 | 204.3 | 1808 | 11.30 |
| 2014-15 | 207.0 | 1385 | 14.95 |
| 2016-17 | 235.1 | 1451 | 16.20 |

**Source:** Jharkhand Ka Bhoogol, Ram Kumar Tiwari (Bihar Hindi Granth Academy, Patna)

**Table 4.8**
**Jharkhand: Intensity of Irrigation -2004-05 and 2016-17 (000 hectares)**

| District | Net Area Irrigated from all sources (2005-06) | Net Area Sown | Intensity of Irrigation (2004-05) | Net Area Irrigated from all sources (2016-17) | Net Area Sown | Intensity of Irrigation (2016-17) |
|---|---|---|---|---|---|---|
| Bokaro | 0.90 | 15.51 | 5.80 | 3.17 | 29.38 | 10.78 |
| Chatra | 2.40 | 45.07 | 5.33 | 7.16 | 32.05 | 22.34 |
| Deoghar | 0.42 | 63.68 | 0.66 | 8.99 | 47.71 | 18.84 |
| Dhanbad | 1.78 | 34.60 | 5.14 | 0.78 | 8.27 | 9.43 |
| Dumka | 16.29 | 119.10 | 13.68 | 6.57 | 55.71 | 11.79 |
| E. Singhbhum | 0.74 | 83.28 | 0.89 | 2.86 | 61.09 | 4.68 |
| Garhwa | 13.38 | 64.77 | 20.66 | 12.06 | 66.43 | 18.15 |
| Giridih | 0.15 | 77.49 | 0.19 | 33.42 | 40.99 | 81.53 |
| Godda | 7.43 | 78.21 | 9.50 | 14.84 | 72.30 | 20.52 |
| Gumla | 7.77 | 161.75 | 4.80 | 8.87 | 131.20 | 6.76 |
| Hazaribagh | 6.15 | 106.14 | 5.97 | 11.54 | 52.25 | 22.08 |

| | | | | | | |
|---|---|---|---|---|---|---|
| Jamtara | - | 55.34 | - | 3.75 | 51.42 | 7.29 |
| Khunti | - | - | - | 19.37 | 60.89 | 31.81 |
| Koderma | 0.03 | 17.04 | 0.18 | 3.00 | 13.43 | 22.33 |
| Latehar | 0.61 | 59.77 | 1.02 | 10.68 | 27.05 | 39.48 |
| Lohardaga | 1.83 | 39.64 | 4.61 | 11.88 | 36.87 | 32.22 |
| Pakur | 1.53 | 57.84 | 2.64 | 6.22 | 48.75 | 12.75 |
| Palamu | 5.93 | 97.75 | 6.06 | 35.38 | 141.32 | 25.03 |
| Ramgarh | - | - | - | 3.02 | 3.34 | 90.41 |
| Ranchi | 17.03 | 255.85 | 6.65 | 15.62 | 199.05 | 7.84 |
| Sahebganj | 0.86 | 41.46 | 2.07 | 3.61 | 48.10 | 7.50 |
| Saraikela | 4.64 | 60.42 | 7.67 | 6.51 | 55.82 | 11.66 |
| Simdega | 2.66 | 81.67 | 3.25 | 3.74 | 81.68 | 4.57 |
| W. Singhbhum | 12.53 | 145.99 | 8.58 | 2.05 | 86.02 | 2.38 |
| Jharkhand | 104.69 | 1762.47 | 5.93 | 235.12 | 1451.26 | 16.20 |

**Source:** Data from R.K. Mishra & Tiwari, Ram Kumar and calculated by author.

Before the separation of Jharkhand from Bihar, irrigation intensity was always below 10 percent but after creation of separate state intensity began to rise and now it is 16.20 percent as per the data from 2016-17. It is a notable fact that after the creation of Jharkhand as a new state, several irrigation projects have been started here i.e., Damodar Valley Project (1948) in which Tilaiya, Maithan, Aiyer, Konar, Panchet projects were under progress; Subarnarekha river project (1973), Koel-Karo project (1973), Mayurakshi Project, Kanhar project etc. were prepared. Similarly, several major and meso irrigation projects have been started i.e., North-Koel Project, Kanchi River Project, Bhairwa, Suru, Kansjor, Panchkhero, Upper Shankh, Satpotka, Nakti, Ramrekha, Jharjhara, Sonua, Surangi, Katri, Kesho, Basuki, Kashi, Salaya and Sukari. Through the major river projects 383.82 thousand hectares area was irrigated. There were 11 major irrigation projects completed till 2019. Through the meso irrigation projects about 33 thousand hectares area was irrigated. Both irrigation projects have potential to irrigate the area of 416.82 thousand hectares. In spite of all these efforts state has only 16.20 percent irrigation intensity.

Map of irrigation intensity clearly shows that intensity above 25 percent is nowhere in the state during 2004-05. Under the category of 10-30% irrigation intensity there were two districts i.e., Dumka (13.68) and Garhwa (20.66).

Under the category of below 10 intensity there were all the districts of state. During 2004-05 Jamtara, Khunti and Ramgarh were not created. Irrigation intensity was 5.93 percent as state average.

## Irrigation Intensity: 2016-17

During 2004-05 state average of irrigation intensity was only 5.93 but after a decade during 2016-17 state average has

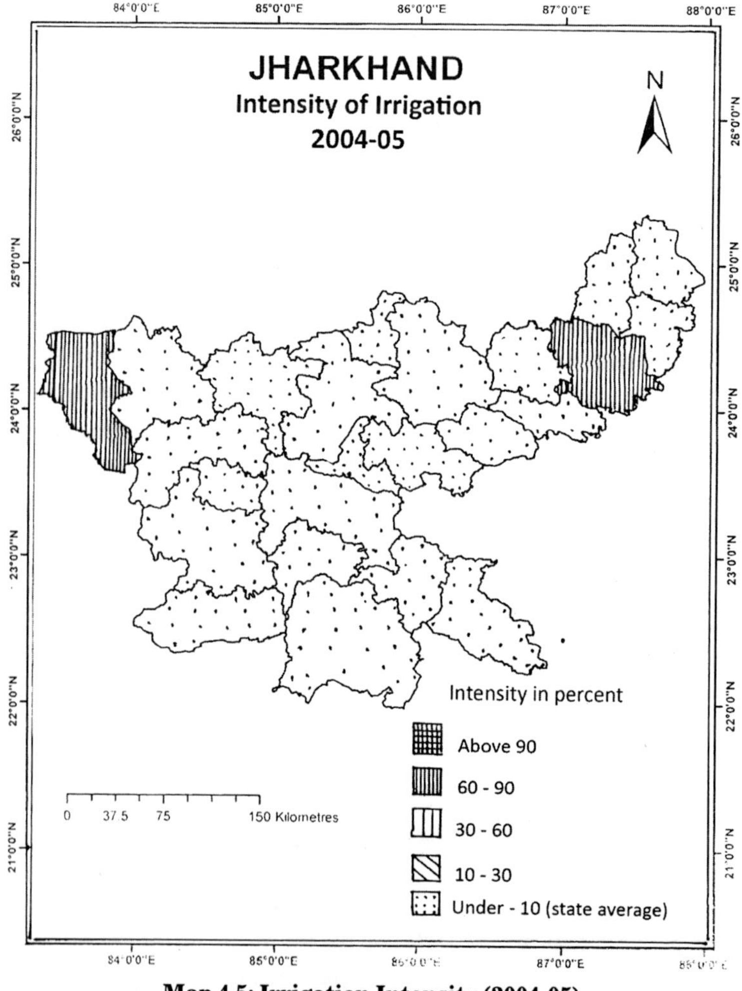


**Map 4.5: Irrigation Intensity (2004-05)**

grown up to 16.20 percent. During this period Ramgarh district has 90.41 percent irrigation intensity. In between 90 to 60 percent category of intensity only one Giridih district (81.53) was registered. Under the category of 30 to 60 percent, Khunti (31.81), Latehar (39.48), Lohardaga (32.22) districts has been recorded. Irrigation intensity of 10 to 30 percent was recorded in Deoghar (18.84), Dumka (11.79), Garhwa (18.15), Godda (20.52),

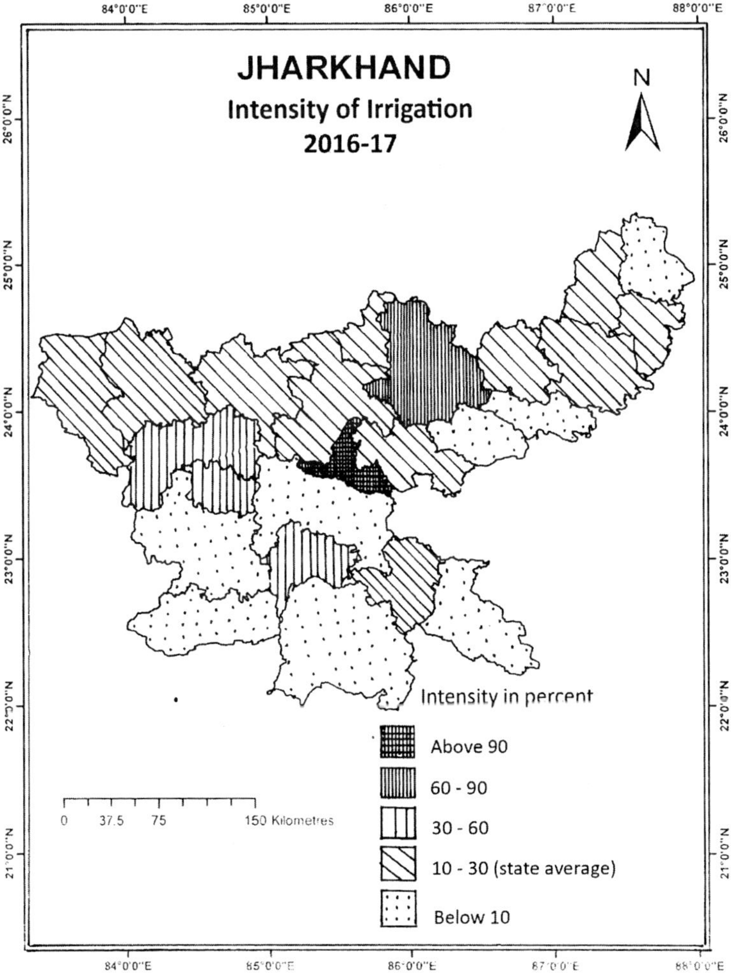


**Map 4.6: Irrigation Intensity (2016-17)**

Hazaribagh (22.08), Koderma (22.33), Pakur (12.75), Palamu (25.03) and Saraikela (11.66) districts. Under the category of below 10 percent Dhanbad (9.43), E. Singhbhum (4.68), Gumla (6.76), Jamtara (7.29), Ranchi (7.50), Simdega (4.57) and W. Singhbhum (2.38) have been recorded in the state.

## Correlation Consequence

Here the correlation between irrigation intensity and cropping intensity has been studied for the whole state and therefore we get a simple indication about the effect of irrigation intensity over cropping intensity. The result of correlation of the data of 2004-05 shows a very low correlation (+0.136) whereas data of 2016-17 shows a high correlation (+0.843) (See appendix for calculation). Hope it will be 0.9 and 1 in correlation in the coming decade. This clearly shows that during the period of creation of Jharkhand there were less irrigation facilities but later on it has been increasing gradually year by year. It would be better if a spatial variation of irrigation intensity and cropping intensity would have been studied in block level and then a district wise correlation study would portray the whole state. This would be a detailed spatial differential study to understand the iso-correlates of the state. But it would be a vast study and deviation from the present research problem. Such type of study has been suggested by Jasbir Singh (1976) in his book and he has done so for the mapping of the state of Haryana calculating more than 100 simple correlation coefficients and has got enlightening results.

Ultimately, we conclude the study regarding cropping intensity that if the irrigation intensity of the state will increase as because the irrigation is the sine qua non of agricultural development in Jharkhand. After becoming a separate state, the growing demand of housing by the exploding population and increasing urbanization and industrialization are creating direct competition for agricultural land. Not only this the demand for land, not only by humans, but also by animals which are the main mode of power agriculture and sources of milk, meat, wool etc. will increase in near future. Data shows that some areas have turned from single crop to double crop and some even multiple crops in the irrigated part. If cultivable waste land (mostly tanr land in Jharkhand) and fallow land would be provided irrigation facility no doubt the cropping intensity of the state will increase certainly.

## Hypothesis Testing

Third hypothesis testing is related to cropping intensity. It is as mentioned - "The cropping intensity of the nation and the state is same. It is assumed that highest intensity region lies within the triangle of DHANRAJ (Dhanbad, Ranchi, Jamshedpur)".

This hypothesis is based on secondary data. Through the secondary data we calculated the cropping intensity of the nation and the state. Null hypothesis is rejected. Now we can say that cropping intensity of India and Jharkhand is not the same. Values of both are different. Trend of country is decreasing or fluctuating while the trend of the state is increasing. India had cropping intensity in 2012-13 of about 138.8 which decreased to 136 in 2018-19; but the cropping intensity of Jharkhand has been recorded 117.9 in 2012-13 and 142.4 in 2018-19.

Our assumption that the highest intensity of crop is in the triangle of DHANRAJ fails too. District wise cropping intensity analysis proved that highest intensity during 2016-17 was recorded in Giridih district which is outside of the triangle.

## References

Akhouri, Rajshree & Achyut, Keshav (2017); Chotanagpur Tenancy Act, 1908. Malhotra Book Agency, Ranchi.

Mishra, R.K & Choudary, J.C (2011-12); Agricultural Resource database of Jharkhand. SAMETI, Department of Agri. and Cane development, Govt. of Jharkhand, Ranchi.

Mishra, R.K.; Agricultural Statistics of Jharkhand, Agricultural Data Bank, Birsa Agricultural University, Ranchi.

Singh, Jasbir (1974); The Green Revolution in India-How Green it is, Kurukshetra, Vishal Publications.

Singh, Jasbir (1976); An Agricultural Geography of Haryana, Vishal Publications, Kurukshetra University Campus, Haryana.

Sinha, V.N.P Singh, L.K.P & Manoj Kumar (2018); Jharkhand: Land and People, Rajesh Publications, New Delhi.

Tiwari, R.C. (2016); Krishi Bhugol, Prawalika Publications, Allahabad.

Tiwari, Ram Kumar (2019); Jharkhand Ka Bhugol, Bihar Hindi Granth Academy, Patna.

Tripathi, R.R. (1970); Changing Pattern of Agricultural Land Use in Upper Ganga-Gomati Doab, Unpublished PhD Thesis, Agra University.

# 5

## Cropping Pattern and Diversification

- Introduction
- Types of Cultivated Land
- Harvesting Calendar
- Areal Pattern of Food grains
  - Rice (Oryza Sativa)
  - Maize or Corn (Zea Mays)
  - Wheat (Triticum Asetivum)
  - Ragi or Marua (Elusine Coracana)
  - Jowar & Bajra (Sorghum Vulgara & Pennisetum Typhoides)
  - Pulses
  - Oil Seeds
  - Vegetables
- Patterns of Crop Diversification

### Introduction

Cropping pattern is a dynamic concept because it changes over space and time. It can be defined as the proportion of area under various crops at a point of time. In other words, it is a yearly sequence and spatial arrangement of sowing and fallow on a given area. In India, or therefore Jharkhand the cropping pattern is determined by Rainfall, Climate, Temperature, Soil Types and Technology. Every country has different cropping pattern depending on its geographical location, weather system and soil type etc. The crop statistics published by the government are used to denote the cropping patterns. The role of human in the cultivation of certain crops in a region is also quite important. Man, by his technological advancement, can ameliorate the physical limits. Those areas of the world where physical diversities are less, the cropping patterns are less diversified. For example, in the rainfall deficient area of Rajasthan, the farmers grow bajra, while in Brahmaputra Valley of Assam, Rice is the dominant crop. Similarly, Cotton is grown in Regur (black earth) soil of Maharashtra and Gujrat while the loamy soil of western U.P., Haryana and Punjab are ideally suited for wheat, rice and sugarcane crops. (Husain, 1996, P.218)

In Jharkhand there is a single physical factor that broadly determines the extension as well as intensity of land use and that is 'slope', as observed by Dr. Ayodhya Prasad (1973, P.58). In the same place he further adds 'as slope increases, the percentage of land under plough decreases.' Areas of higher elevation are the areas of steeper slopes. Percentage of cultivated area to the total area is, therefore, everywhere less than 50 in those parts of the plateau which lie above 1000 ft. and more than 50 in areas laying below 1000 ft. There are, however, five exceptions to this generalization. They are - The Ranchi plateau, Chaibasa plains, Lower South-Koel valley, Upper Shankh basin and Lower North-Koel Valley. The first four are above 1000 ft. but have more than 50 percent area under plough. The last one is below 1000 ft. but the cultivated area is less than 50 percent of the total. The Ranchi plateau is an uplifted peneplain with subdued relief. Chaibasa plain and Upper Shankh basin are just high-laying plains. The lower South-Koel valley is broad enough to permit the extension of cultivation on to the valley sides covered with a veneer of alluvium. The agricultural capability of the North-Koel valley is reduced by the presence of extensive sandy tracts.

## Types of Cultivated Land

In Jharkhand cultivated lands have been classified as 'Don' and 'Tanr' in local terminology. The surface of Jharkhand is, to a large extent, composed of long undulating ridges. Between these ridges' drainage runs off to join large streams. Between these undulations lie trough-like hollows which are surfaced with alluvial soil. The ridges, being made up of sterile gravels and stiff clay laying on a hard sub-soil, are very poor and locally known as 'Tanr'. Valley surface are known as 'Don'. "When the slopes are too steep for terracing or the soil too strong for cultivation, the bed of the stream is levelled up and made into one long narrow rice field". (D.G. Santhal Pargana, 1938, P198). On the basis of a rough calculation of productive equivalence each one of the two division is further divided into three classes- first, second and third. The first-class don or rice land is usually situated in the bed of a stream or immediately below an embankment. It retains moisture for the greater part of the year. Land of this class is invariably irrigated. "Best of this class is usually fed by perennial springs from which moisture oozes even in the hottest month of the year" (District Gazetteer, Santhal Pargana, 1938, P.198). The first- class don is everywhere the favorite habitat of slow ripening heavy 'agahani' rice. Second class don consists of the lower parts of the terraced slopes and is, thus confined to narrow strips between the first class and the third-class rice land. It receives considerable moisture through percolation from fields above through capillary action. The third-class rice land consists of the higher terraced field bounded by low embankments capable of detaining surface flow for a comparatively short period. This class of land is completely irrigated and retains surface moisture only a little longer than the tanr land.

Now the tanr land has been also divided into three categories like don land. First class tanr land is known as 'Bari land' or 'Gharbari' and includes only a small function of total cultivated area. Bari land is confined just near the settlement area. It is actually an extension of the homestead area. Bari land is usually fenced and is a multiple cropped area. Here cultivator always cultivates valuable cash crops like maize, vegetables, sugarcane etc. The second class tanr uplands are generally situated contiguous to the bari land. This land produces at least two crops- Bhadai followed by rabi crop. The third class tanr land includes all the uplands and slopes to which cultivation has been extended. Such lands are situated away from village site and are never manured.

In some villages it is sown every year and in others only once in a cycle of years. This class of land is invariably given to bhadai crops. (Prasad, 1973, P62).

## Harvesting Calendar

Agricultural harvesting calendar of Jharkhand is very much similar to its neighboring states i.e., Bihar, Bengal to some extent. There are four harvesting seasons - They are Bhadai (autumn or mid monsoon crops), Aghani (winter or end monsoon crops), Rabi (spring crops) and Garma (summer season crops).

Such crops which are quick maturing, hardy and less sensitive to the fertility, texture and depth of soil are bhadai crops. These crops are not fastidious about the perennial water supply and can withstand short rainless intervals without suffering from shunted growth. As a rule, none of the bhadai crops grown in Jharkhand require inundation. The important bhadai crops in Jharkhand are: Gora Rice, Maize, Marua, Gondli, Urid, Mung, Kodo, Til and Flaxes.

Aghani crops are of two types- Rice and Sarguja in Jharkhand. These crops take six months to be mature in the field. Rice is the dominant crop as aghani in the state. Both crops are grown in contrast land types- lowest land are given for rice but ridge tops higher slopes are occupied by sarguja. In Jharkhand rice forms the basis of economy in the entire rural area. It serves as reliable index of agricultural and general prosperity in the rural area.

Rabi crops in Jharkhand has a marginal importance but practiced in the whole state. The important crops of this season are gram, wheat, barley, pulses, mustard and tewa rice. This rice is also known as summer rice.

A fourth harvest consists of summer crops, called garma (i.e., of summer). This consists mostly of vegetables, millets and boro rice grown in moist riparian tracts or irrigated land near homesteads (Ahmad, 1965, P.131). As it is mentioned by Ivern (1969, P.55), summer paddy is known as 'tewa paddy' which has been also mentioned by Ayodhya Prasad (1973).

## Areal Pattern of Foodgrains

Food grains are important for human ingestion as grain and as straw for animals; hence it has an important role as food for both animal and human being. It is only under food grains that the land can produce enough to support the dense population base. The food grain crops are generally less demanding and less exacting in their soil and moisture requirements than cotton and sugar cane. Thus, the food grain crops tend to displace the fiber and other cash crops in major part of the state. It is said that cropping of some food pulses is unavoidable since these make a satisfactory rotation.

Figure 5.1 exhibits the geographic distribution of total food grains or the areal pattern of food grains in the state. The distribution corresponds with the spatial patterns of physical and non-physical variables. Areas of the dominant and major crops may be distinguished, within which certain other crops are characteristic and common. Though these reflect edaphic and rainfall differences, sociological and economic factors are also significant besides many other factors determining the choice and combination of crops. For example, rice is more suited to monsoon rain and loamy and clayey loamy soil of the central and south-eastern part of Jharkhand, and Wheat and maize as dominant crops in the western parts of the rain shadow area of Jharkhand. On the whole, food grain cultivation is intensive in major parts of the state, having more than 70 percent of the total harvested area. (Fig 5.1)

**Table 5.1**
**Jharkhand: Food Grains Hectarage (Cereals, Pulses and Oilseeds) 2015-16**

| District | Total Cropped Area (000 Hectares) | Area Under Food Grains (000 Hectares) | Percentage of total cropped area |
|---|---|---|---|
| Hazaribagh | 62.58 | 138.05 | 45.33 |
| Jamtara | 61.93 | 78.45 | 78.94 |
| Khunti | 75.86 | 108.08 | 70.18 |
| Koderma | 13.76 | 58.40 | 23.56 |

| | | | |
|---|---|---|---|
| Latehar | 59.39 | 129.67 | 45.80 |
| Lohardaga | 50.90 | 65.52 | 77.68 |
| Pakur | 59.41 | 101.85 | 58.33 |
| Palamu | 180.09 | 125.58 | 143.40 |
| Ramgarh | 21.60 | 70.29 | 30.72 |
| Ranchi | 252.44 | 227.13 | 111.14 |
| Sahebganj | 48.61 | 89.05 | 54.58 |
| Saraikela | 59.72 | 153.76 | 38.83 |
| Simdega | 84.73 | 171.61 | 49.37 |
| W. Singhbhum | 117.55 | 204.89 | 57.37 |
| Jharkhand | 2044.97 | 2707.49 | 75.53 |

**Source:** Agricultural Directorate, Govt. of Jharkhand

**Table 5.2**
**Food Grain Production Area in Jharkhand (000 hectares)**

| Year | Rice | Wheat | Barley | Maize | Marua | Total Food Grain Production Area |
|---|---|---|---|---|---|---|
| 1985-86 | 1529 | 73 | 15 | 137 | - | 1753 |
| 1986-87 | 1451 | 69 | 18 | 123 | - | 1661 |
| 1987-88 | 2172 | 67 | 18 | 122 | - | 2379 |
| 1988-89 | 1522 | 54 | 12 | 120 | - | 1708 |
| 1989-90 | 1532 | 67 | 13 | 121 | - | 1733 |
| 1990-91 | 1571 | 64 | 10 | 98 | - | 1743 |
| 1997-98 | 1444 | 63 | 11 | 122 | - | 1640 |
| 1998-99 | 1508 | 79 | - | 150 | - | 1737 |
| 1999-2000 | 1420 | 70 | 8 | 118 | 41 | 1957 |
| 2000-01 | 1464 | 66 | - | 157 | - | 1687 |
| 2001-02 | 1521 | 65 | - | 140 | - | 1726 |
| 2002-03 | 1383 | 68 | - | 157 | - | 1622 |
| 2003-04 | 1363 | 74 | - | 180 | - | 1639 |
| 2004-05 | 1276 | 64 | - | 191 | - | 1550 |
| 2015-16 | 1471 | 161 | - | 288 | 13 | 1933 |

**Table 5.3**
**Pulse Production Area in Jharkhand (000 hectares)**

| Year | Gram | Masur | Matar | Arhar | Urad | Moong | Kurthi | Others | Total Area |
|---|---|---|---|---|---|---|---|---|---|
| 1985-86 | 32 | 6 | 2 | 24 | - | - | - | - | 77 |
| 1986-87 | 35 | 6 | 2 | 26 | - | - | - | - | 84 |
| 1987-88 | 32 | 6 | 3 | 27 | - | - | - | - | 80 |
| 1988-89 | 23 | 3 | 2 | 23 | - | - | - | - | 62 |
| 1989-90 | - | - | - | - | - | - | - | - | - |
| 1997-98 | - | - | - | - | - | - | - | - | - |
| 1998-99 | 33.19 | 8.74 | 28.3 | 16.02 | 29.6 | 0.12 | 5.93 | - | 12.2 |
| 1999-2000 | 26.04 | 8.09 | 42.7 | 16.07 | 29.0 | - | 8.12 | 23 | 15.3 |
| 2000-01 | 15.04 | 3.96 | 19.3 | 17.29 | 20.5 | - | 8.82 | 19 | 103 |
| 2001-02 | 16.33 | 6.35 | 28.5 | 16.31 | 38.5 | 0.05 | 10.98 | 19 | 137 |
| 2002-03 | 34.56 | 15.41 | 19.02 | 56.23 | 56.93 | 9.65 | 13.45 | 40 | 245 |
| 2003-04 | 36.72 | 16.72 | 23.01 | 73.53 | 78.93 | 10.70 | 16.64 | 46 | 304 |
| 2004-05 | 33.17 | 15.47 | 19.93 | 88.64 | 77.80 | 8.59 | 12.37 | 35 | 290.92 |
| 2015-16 | 164.22 | 40.15 | 44.36 | 193.68 | 94.87 | 24.88 | - | 32.83 | 594.99 |

**Source:** Agricultural Directorate, Govt. of Jharkhand

**Table 5.4**
**Oil Seeds Production Area in Jharkhand (000 hectares)**

| Year | Mustard/ Rai | Tisi | Kusum | Sunflower | Groundnut | Til | Sarguja | Castor | Soyabean | Total Area |
|---|---|---|---|---|---|---|---|---|---|---|
| 1988-89 | 41.39 | 7.22 | 0.19 | 0.05 | 0.20 | 0.50 | 1.27 | - | - | 50.82 |
| 1999-2000 | 41.01 | 7.23 | 0.23 | 0.03 | 0.20 | 0.60 | 1.18 | - | - | 50.48 |
| 2000-01 | 21.83 | 5.96 | 0.08 | 0.04 | 0.21 | 0.49 | 0.88 | - | - | 24.49 |
| 2001-02 | 21.42 | 8.40 | 0.12 | 0.09 | 0.15 | 0.63 | 1.17 | - | - | 31.98 |
| 2002-03 | 47.96 | 13.62 | 0.63 | 0.68 | 15.81 | 7.15 | 5.91 | - | - | 91.76 |
| 2003-04 | 47.92 | 17.01 | 0.44 | 0.34 | 17.69 | 8.76 | 7.01 | - | - | 99.17 |
| 2004-05 | 44.91 | 16.05 | 0.28 | 0.29 | 18.34 | 9.57 | 9.43 | - | - | 98.85 |
| 2015-16 | 202.13 | 23.56 | 0.100 | 1.04 | 24.47 | 5.60 | 2.32 | 0.004 | 0.46 | 259.68 |

**Source:** Agricultural Directorate, Govt. of Jharkhand

**Table 5.5**
**Food Grains Production in Jharkhand (000 metric tons)**

| Year | Rice | Wheat | Barley | Marua | Maize | Millet | Total Production |
|---|---|---|---|---|---|---|---|
| 1985-86 | 1514 | 83 | - | - | 133 | - | 1738 |
| 1986-87 | 1351 | 89 | - | - | 105 | - | 1555 |
| 1987-88 | 1314 | 90 | - | - | 86 | - | 1501 |
| 1988-89 | 1466 | 59 | - | - | 94 | | 1624 |
| 1989-90 | 1537 | 76 | - | - | 152 | - | 1773 |
| 1990-91 | 1571 | 72 | - | - | 98 | - | 1751 |

| Year | Rice | Wheat | Barley | Marua | Maize | Millet | Total Production |
|---|---|---|---|---|---|---|---|
| 1997-98 | - | - | - | - | - | - | - |
| 1998-99 | 2556.53 | 139.90 | - | - | 248.40 | - | 2911.40 |
| 1999-2000 | 1755 | 83 | - | 38 | 187 | 4.7 | 2075 |
| 2000-01 | 2278.60 | 108.48 | - | - | 254.21 | - | 2641.29 |
| 2001-02 | 2732.58 | 115.36 | - | - | 209.18 | - | 3057.12 |
| 2002-03 | 2071.47 | 113.69 | 0.82 | 8.81 | 269.18 | 0.16 | 2464.13 |
| 2003-04 | 2143.43 | 128.14 | 0.64 | 13.47 | 296.19 | 0.26 | 2582.13 |
| 2004-05 | 1908.45 | 104.14 | 0.16 | 10.35 | 278.59 | 0.14 | 2301.83 |
| 2015-16 | 2471.37 | 274.06 | - | 7.78 | 375.47 | 0.10 | 3128.77 |

**Source:** Agricultural Directorate, Govt. of Jharkhand/ Bihar Through Figures, 1991, Patna

**Table 5.6**

**Pulses Production in Jharkhand (000 metric tons)**

| Year | Gram | Masur | Pea (Matar) | Arhar | Urad | Moong | Kurthi | Others | Total |
|---|---|---|---|---|---|---|---|---|---|
| 1985-86 | 16 | 4 | 0 | 15 | - | - | - | 10 | 45 |
| 1986-87 | 25 | 3 | 0 | 27 | - | - | - | 9 | 64 |
| 1987-88 | 21 | 4 | 1 | 32 | - | - | - | 8 | 66 |
| 1988-89 | 9 | 3 | 0 | 17 | - | - | - | 7 | 36 |
| 1989-90 | 16 | 3 | 1 | 25 | - | - | - | 8 | 53 |
| 1990-91 | - | - | - | - | - | - | - | - | - |
| 1997-98 | - | - | - | - | - | - | - | - | 117 |

| Year | Gram | Masur | Pea (Matar) | Arhar | Urad | Moong | Kurthi | Others | Total |
|---|---|---|---|---|---|---|---|---|---|
| 1998-99 | 27.30 | 5.57 | 2.38 | 13.46 | 2.16 | 0.06 | 2.71 | - | 52 |
| 1999-2000 | 24.47 | 6.02 | 4.10 | 13.51 | 1.35 | - | 3.44 | 15.6 | 68 |
| 2000-01 | 13.34 | 2.57 | 1.58 | 12.01 | 0.97 | - | 4.23 | 9.4 | 53 |
| 2001-02 | 14.99 | 4.13 | 2.22 | 14.92 | 2.56 | 0.03 | 6.28 | 13.1 | 165.82 |
| 2002-03 | 31.26 | 7.64 | 18.91 | 43.44 | 32.74 | 4.96 | 6.23 | 20.5 | 160.45 |
| 2003-04 | 37.52 | 11.51 | 25.13 | 52.01 | 42.16 | 6.03 | 7.31 | 26.5 | 208.17 |
| 2004-05 | 29.26 | 9.22 | 19.81 | 62.47 | 41.61 | 4.40 | 5.45 | 17.8 | 190.11 |
| 2015-16 | 172.88 | 32.03 | 52.21 | 173.56 | 72.06 | 18.35 | - | 26.06 | 547.15 |

**Source:** Agricultural Directorate, Govt. of Jharkhand/ Bihar Through Figures, 1991, Patna

**Table 5.7**

**Oil Seeds Production in Jharkhand (000 metric tons)**

| Year | Mustard/Rai | Tisi | Kusum | Sunflower | Groundnut | Til | Sarguja | Total |
|---|---|---|---|---|---|---|---|---|
| 1998-99 | 26.42 | 3.34 | 0.06 | 0.03 | 0.17 | 0.22 | 0.59 | 60.53 |
| 1999-2000 | 28.47 | 2.91 | 0.09 | 0.02 | 0.12 | 0.19 | 0.54 | 76.81 |
| 2000-01 | 15.26 | 2.50 | 0.03 | 0.02 | 0.19 | 0.16 | 0.45 | 44.91 |
| 2001-02 | 15.59 | 3.63 | 0.04 | 0.07 | 0.12 | 0.23 | 0.52 | 49.56 |
| 2002-03 | 25.06 | 6.06 | 0.30 | 0.41 | 17.65 | 2.79 | 1.33 | 53.67 |
| 2003-04 | 24.74 | 6.88 | 0.26 | 0.17 | 18.99 | 3.63 | 1.37 | 56.50 |
| 2004-05 | 25.37 | 6.12 | 0.13 | 0.17 | 16.70 | 3.60 | 0.89 | 52.98 |
| 2015-16 | 138.68 | 13.19 | 0.04 | 0.59 | 20.31 | 2.25 | 1.55 | 177.00 |

**Source:** Agricultural Directorate, Govt. of Jharkhand/ Bihar Through Figures, 1991, Patna

### Table 5.8
### District wise Food Crop Area in Jharkhand - 2015-16 (000 hectares)

| District | Rice | Wheat | Maize | Marua | Bajra |
|---|---|---|---|---|---|
| Bokaro | 32.177 | 1.153 | 7.646 | 0.100 | 0.000 |
| Chatra | 36.010 | 8.415 | 13.915 | 0.250 | 0.008 |
| Deoghar | 49.750 | 8.662 | 17.000 | 0.000 | 0.000 |
| Dhanbad | 37.325 | 0.905 | 2.193 | 0.051 | 0.046 |
| Dumka | 107.330 | 11.445 | 20.071 | 0.000 | 0.000 |
| E. Singhbhum | 46.369 | 3.900 | 11.327 | 1.100 | 0.000 |
| Garhwa | 39.379 | 17.500 | 25.665 | 0.000 | 0.000 |
| Giridih | 86.543 | 10.013 | 18.519 | 0.000 | 0.006 |
| Godda | 46.240 | 12.491 | 12.486 | 0.000 | 0.000 |
| Gumla | 173.145 | 7.716 | 6.037 | 1.481 | 0.000 |
| Hazaribagh | 77.285 | 8.650 | 11.105 | 1.572 | 0.010 |
| Jamtara | 49.00 | 6.661 | 9.900 | 0.000 | 0.000 |
| Khunti | 68.780 | 3.005 | 5.942 | 1.450 | 0.008 |
| Koderma | 16.081 | 6.749 | 8.673 | 0.615 | 0.016 |
| Latehar | 24.700 | 5.950 | 22.136 | 0.125 | 0.015 |
| Lohardaga | 34.232 | 2.018 | 6.150 | 0.000 | 0.000 |
| Pakur | 48.905 | 7.830 | 9.564 | 0.000 | 0.000 |
| Palamu | 35.024 | 9.462 | 26.224 | 0.000 | 0.000 |
| Ramgarh | 30.710 | 2.125 | 9.027 | 0.661 | 0.006 |
| Ranchi | 160.629 | 6.360 | 11.193 | 3.123 | 0.000 |
| Sahebganj | 42.359 | 8.193 | 11.243 | 0.000 | 0.024 |
| Saraikela | 99.000 | 1.600 | 5.913 | 0.000 | 0.040 |
| Simdega | 85.216 | 8.792 | 8.823 | 2.000 | 0.020 |
| W. Singhbhum | 162.690 | 1.515 | 7.275 | 0.325 | 0.017 |
| Jharkhand | 1471.683 | 161.110 | 288.047 | 12.853 | 0.2160 |

**Source:** Agricultural Directorate, Govt. of Jharkhand

### Table 5.9
### District wise Pulses Production Area in Jharkhand - 2015-16 (000 hectares)

| District | Arhar | Urad | Moong | Gram | Masur | Matar | Others |
|---|---|---|---|---|---|---|---|
| Bokaro | 2.953 | 0.470 | 0.128 | 0.855 | 0.000 | 0.663 | 0.999 |
| Chatra | 8.110 | 1.480 | 0.815 | 8.970 | 3.470 | 1.940 | 1.865 |
| Deoghar | 2.875 | 0.477 | 0.253 | 5.413 | 1.129 | 1.161 | 0.459 |
| Dhanbad | 1.125 | 0.195 | 0.358 | 2.091 | 0.000 | 0.000 | 0.946 |

| | | | | | | | |
|---|---|---|---|---|---|---|---|
| Dumka | 7.846 | 1.469 | 0.619 | 8.130 | 2.228 | 1.676 | 0.741 |
| E. Singhbhum | 3.840 | 3.840 | 2.386 | 4.885 | 1.568 | 1.270 | 2.446 |
| Garhwa | 28.550 | 6.694 | 1.200 | 12.000 | 4.000 | 4.500 | 2.32 |
| Giridih | 5.990 | 1.875 | 0.904 | 7.374 | 0.026 | 0.000 | 2.526 |
| Godda | 6.905 | 1.750 | 0.950 | 7.762 | 3.000 | 1.494 | 0.000 |
| Gumla | 9.415 | 5.557 | 0.000 | 6.577 | 2.120 | 2.380 | 0.000 |
| Hazaribagh | 7.164 | 1.766 | 0.820 | 8.254 | 0.746 | 8.803 | 1.465 |
| Jamtara | 1.195 | 0.000 | 0.250 | 5.714 | 0.076 | 0.987 | 0.000 |
| Khunti | 5.429 | 4.654 | 0.394 | 5.142 | 1.703 | 1.041 | 0.837 |
| Koderma | 5.515 | 0.820 | 0.134 | 6.767 | 2.157 | 0.942 | 0.298 |
| Latehar | 19.715 | 4.000 | 0.280 | 12.220 | 2.850 | 1.575 | 1.8 |
| Lohardaga | 5.800 | 2.670 | 0.049 | 3.585 | 0.360 | 2.025 | 2.054 |
| Pakur | 6.852 | 1.659 | 0.024 | 9.047 | 4.280 | 0.927 | 0.000 |
| Palamu | 19.884 | 6.926 | 0.728 | 8.921 | 0.923 | 2.268 | 1.108 |
| Ramgarh | 4.198 | 1.405 | 0.497 | 5.207 | 0.110 | 8.041 | 0.674 |
| Ranchi | 10.322 | 8.527 | 1.823 | 6.048 | 1.551 | 0.312 | 7.84 |
| Sahebganj | 6.311 | 1.870 | 0.040 | 5.198 | 3.415 | 1.577 | 0.000 |
| Saraikela | 9.500 | 7.500 | 6.600 | 9.830 | 2.441 | 0.000 | 0.000 |
| Simdega | 9.000 | 23.200 | 1.000 | 11.389 | 0.000 | 0.000 | 3.279 |
| W. Singhbhum | 5.636 | 6.072 | 4.605 | 2.845 | 2.002 | 0.786 | 2.004 |
| Jharkhand | 193.680 | 94.876 | 24.884 | 164.224 | 40.155 | 44.368 | 32.643 |

**Source:** Agricultural Directorate, Govt. of Jharkhand

### Table 5.11
### District wise Food Grain Production in Jharkhand - 2015-16 (000 metric tons)

| District | Rice | Wheat | Maize | Marua | Bajra |
|---|---|---|---|---|---|
| Bokaro | 33.528 | 2.075 | 12.215 | 0.017 | 0.010 |
| Chatra | 36.802 | 16.836 | 15.444 | 0.263 | 0.004 |
| Deoghar | 238.559 | 21.665 | 25.687 | 0.000 | 0.000 |
| Dhanbad | 41.629 | 2.263 | 6.247 | 0.041 | 0.030 |
| Dumka | 343.456 | 17.850 | 23.405 | 0.000 | 0.000 |
| E. Singhbhum | 81.037 | 7.895 | 11.964 | 0.470 | 0.000 |
| Garhwa | 28.352 | 18.900 | 17.211 | 0.000 | 0.000 |
| Giridih | 121.592 | 24.631 | 22.870 | 0.000 | 0.003 |
| Godda | 141.032 | 25.606 | 23.044 | 0.000 | 0.000 |
| Gumla | 181.110 | 13.156 | 8.844 | 1.448 | 0.000 |
| Hazaribagh | 106.421 | 16.003 | 12.762 | 1.140 | 0.008 |
| Jamtara | 122.500 | 6.300 | 8.09 | 0.000 | 0.000 |
| Khunti | 69.123 | 3.786 | 6.798 | 0.217 | 0.001 |
| Koderma | 27.418 | 12.148 | 14.313 | 0.338 | 0.006 |
| Latehar | 16.796 | 7.883 | 12.009 | 0.142 | 0.005 |

**Table 5.10**

**District wise Oil Seeds Production Area in Jharkhand - 2015-16 (000 hectares)**

| District | Ground nut | Castor | Til | Sun flower | Tisi | Mustard | Sarguja | Soy bean | Kusum |
|---|---|---|---|---|---|---|---|---|---|
| Bokaro | 0.307 | 0.000 | 0.000 | 0.000 | 0.524 | 2.264 | 0.128 | 0.000 | 0.000 |
| Chatra | 0.425 | 0.000 | 0.342 | 0.000 | 0.975 | 10.250 | 0.000 | 0.005 | 0.000 |
| Deoghar | 0.570 | 0.004 | 0.069 | 0.011 | 0.802 | 9.026 | 0.022 | 0.022 | 0.034 |
| Dhanbad | 0.043 | 0.000 | 0.182 | 0.000 | 0.182 | 8.016 | 0.000 | 0.000 | 0.000 |
| Dumka | 0.000 | 0.000 | 0.000 | 0.071 | 1.921 | 11.325 | 0.000 | 0.090 | 0.056 |
| E. Singhbhum | 0.091 | 0.000 | 0.124 | 0.000 | 0.600 | 6.581 | 0.095 | 0.015 | 0.000 |
| Garhwa | 2.350 | 0.000 | 1.535 | 0.000 | 0.100 | 17.000 | 0.000 | 0.000 | 0.000 |
| Giridih | 0.075 | 0.000 | 0.008 | 0.000 | 0.005 | 8.005 | 0.074 | 0.035 | 0.000 |
| Godda | 0.764 | 0.000 | 0.000 | 0.075 | 1.575 | 11.284 | 0.001 | 0.000 | 0.000 |
| Gumla | 4.750 | 0.000 | 0.000 | 0.000 | 1.110 | 8.085 | 1.050 | 0.000 | 0.000 |
| Hazaribagh | 1.330 | 0.000 | 0.000 | 0.000 | 0.304 | 8.676 | 0.106 | 0.000 | 0.000 |
| Jamtara | 0.000 | 0.000 | 0.000 | 0.000 | 0.000 | 4.675 | 0.000 | 0.000 | 0.000 |
| Khunti | 0.719 | 0.000 | 0.000 | 0.003 | 0.319 | 8.601 | 0.022 | 0.023 | 0.010 |
| Koderma | 0.439 | 0.000 | 0.029 | 0.002 | 0.948 | 8.160 | 0.036 | 0.026 | 0.000 |
| Latehar | 1.677 | 0.000 | 0.000 | 0.000 | 2.301 | 14.005 | 0.000 | 0.000 | 0.000 |
| Lohardaga | 2.000 | 0.000 | 0.000 | 0.000 | 0.360 | 4.115 | 0.102 | 0.000 | 0.000 |
| Pakur | 0.027 | 0.000 | 0.000 | 0.000 | 0.494 | 12.250 | 0.000 | 0.000 | 0.000 |
| Palamu | 0.968 | 0.000 | 2.449 | 0.003 | 1.058 | 9.608 | 0.004 | 0.003 | 0.000 |
| Ramgarh | 1.050 | 0.000 | 0.000 | 0.000 | 0.085 | 6.480 | 0.000 | 0.000 | 0.000 |
| Ranchi | 1.078 | 0.000 | 0.131 | 0.572 | 0.676 | 6.241 | 0.326 | 0.117 | 0.000 |
| Sahebganj | 0.000 | 0.000 | 0.000 | 0.300 | 0.300 | 8.228 | 0.000 | 0.000 | 0.000 |
| Saraikela | 0.090 | 0.000 | 0.000 | 0.000 | 2.400 | 8.810 | 0.000 | 0.400 | 0.000 |
| Simdega | 4.800 | 0.000 | 0.600 | 0.000 | 1.866 | 11.189 | 0.348 | 0.068 | 0.000 |
| W. Singhbhum | 0.110 | 0.000 | 0.137 | 0.006 | 4.655 | 4.206 | 0.006 | 0.000 | 0.000 |
| Jharkhand | 24.473 | 0.004 | 5.060 | 1.043 | 23.560 | 202.130 | 2.320 | 0.459 | 0.100 |

**Source:** Agricultural Directorate, Govt. of Jharkhand

| | | | | | |
|---|---|---|---|---|---|
| Lohardaga | 67.996 | 4.036 | 8.610 | 0.000 | 0.000 |
| Pakur | 116.785 | 12.528 | 16.347 | 0.000 | 0.000 |
| Palamu | 26.377 | 11.358 | 22.860 | 0.000 | 0.000 |
| Ramgarh | 16.860 | 3.910 | 10.945 | 0.496 | 0.004 |
| Ranchi | 183.117 | 12.126 | 12.670 | 1.337 | 0.000 |
| Sahebganj | 80.905 | 8.561 | 8.770 | 0.000 | 0.017 |
| Saraikela | 21.780 | 3.360 | 0.223 | 0.000 | 0.002 |
| Simdega | 255.648 | 19.782 | 22.613 | 1.600 | 0.013 |
| W. Singhbhum | 112.549 | 1.407 | 7.369 | 0.189 | 0.006 |
| Jharkhand | 2471.372 | 274.065 | 375.473 | 7.788 | 0.099 |

**Source:** Agricultural Directorate, Govt. of Jharkhand

## Table 5.12
## District wise Pulses Production in Jharkhand - 2015-16 (000 metric tons)

| District | Arhar | Urad | Moong | Gram | Masur | Matar | Others |
|---|---|---|---|---|---|---|---|
| Bokaro | 2.037 | 0.185 | 0.050 | 7.116 | 0.000 | 0.468 | 0.407 |
| Chatra | 8.678 | 1.518 | 0.611 | 9.419 | 3.054 | 2.813 | 1.17 |
| Deoghar | 2.013 | 0.935 | 0.152 | 7.848 | 1.072 | 2.089 | 0.235 |
| Dhanbad | 1.350 | 0.113 | 0.331 | 3.137 | 0.000 | 0.000 | 1.261 |
| Dumka | 8.630 | 1.689 | 0.665 | 6.504 | 3.560 | 1.508 | 0.552 |
| E. Singhbhum | 1.836 | 1.836 | 1.329 | 2.360 | 0.655 | 0.670 | 1.252 |
| Garhwa | 15.390 | 0.795 | 0.468 | 14.640 | 2.980 | 3.802 | 1.033 |
| Giridih | 6.037 | 0.903 | 0.461 | 7.853 | 0.011 | 0.000 | 1.633 |
| Godda | 6.559 | 1.350 | 0.902 | 8.227 | 2.580 | 2.136 | 0.000 |
| Gumla | 7.776 | 2.832 | 0.000 | 7.564 | 2.591 | 3.670 | 0.000 |
| Hazaribagh | 5.874 | 1.518 | 0.353 | 11.968 | 0.627 | 8.839 | 1.091 |
| Jamtara | 1.195 | 0.000 | 0.200 | 8.413 | 0.120 | 0.790 | 0.000 |
| Khunti | 5.147 | 3.276 | 0.201 | 6.206 | 0.523 | 0.604 | 0.201 |
| Koderma | 3.309 | 0.483 | 0.077 | 7.038 | 1.618 | 0.942 | 0.125 |
| Latehar | 30.065 | 0.625 | 0.227 | 7.625 | 1.578 | 7.418 | 3.600 |
| Lohardaga | 5.800 | 2.136 | 0.049 | 2.868 | 0.540 | 4.050 | 1.437 |
| Pakur | 5.139 | 3.480 | 0.015 | 4.290 | 3.210 | 0.738 | 0.000 |
| Palamu | 12.898 | 3.809 | 0.327 | 6.691 | 0.646 | 1.474 | 0.438 |
| Ramgarh | 3.484 | 0.625 | 0.211 | 8.592 | 0.080 | 8.072 | 0.327 |
| Ranchi | 9.030 | 3.410 | 0.684 | 6.653 | 1.215 | 0.328 | 7.289 |
| Sahebganj | 2.802 | 3.134 | 0.020 | 5.330 | 1.500 | 0.900 | 0.000 |
| Saraikela | 9.500 | 7.500 | 6.600 | 1.671 | 2.170 | 0.000 | 0.000 |
| Simdega | 13.500 | 25.250 | 0.900 | 18.507 | 0.000 | 0.000 | 3.518 |
| W. Singhbhum | 5.523 | 4.736 | 3.523 | 2.361 | 1.702 | 0.904 | 0.495 |
| Jharkhand | 173.563 | 72.062 | 18.356 | 172.881 | 32.032 | 52.215 | 26.064 |

**Source:** Agricultural Directorate, Govt. of Jharkhand

## Table 5.13
## District wise Oil Seeds Production in Jharkhand - 2015-16 (000 metric tons)

| District | Ground nut | Castor | Til | Sun flower | Tisi | Mustard | Sarguja | Soy bean | Kusum |
|---|---|---|---|---|---|---|---|---|---|
| Bokaro | 0.148 | 0.000 | 0.000 | 0.000 | 0.217 | 1.650 | 0.047 | 0.000 | 0.000 |
| Chatra | 0.404 | 0.000 | 0.109 | 0.000 | 0.750 | 5.633 | 0.000 | 0.004 | 0.000 |
| Deoghar | 0.428 | 0.001 | 0.024 | 0.006 | 0.521 | 8.123 | 0.311 | 0.000 | 0.022 |
| Dhanbad | 0.053 | 0.000 | 0.080 | 0.000 | 0.080 | 2.714 | 0.000 | 0.030 | 0.000 |
| Dumka | 0.000 | 0.000 | 0.000 | 0.056 | 0.768 | 7.130 | 0.000 | 0.000 | 0.021 |
| E. Singhbhum | 0.906 | 0.000 | 0.056 | 0.000 | 0.233 | 2.646 | 0.054 | 0.000 | 0.000 |
| Garhwa | 1.433 | 0.000 | 0.238 | 0.000 | 0.052 | 9.265 | 0.000 | 0.000 | 0.000 |
| Giridih | 0.047 | 0.000 | 0.023 | 0.000 | 0.023 | 6.564 | 0.022 | 0.003 | 0.000 |
| Godda | 0.764 | 0.000 | 0.000 | 0.056 | 0.708 | 4.334 | 0.001 | 0.000 | 0.000 |
| Gumla | 6.341 | 0.000 | 0.000 | 0.000 | 0.660 | 11.966 | 0.454 | 0.000 | 0.000 |
| Hazaribagh | 0.865 | 0.000 | 0.000 | 0.000 | 0.268 | 7.982 | 0.059 | 0.008 | 0.000 |
| Jamtara | 0.000 | 0.000 | 0.000 | 0.000 | 0.000 | 2.337 | 0.000 | 0.000 | 0.000 |
| Khunti | 0.582 | 0.000 | 0.000 | 0.001 | 0.122 | 8.569 | 0.011 | 0.001 | 0.001 |
| Koderma | 0.318 | 0.000 | 0.087 | 0.001 | 0.427 | 5.508 | 0.015 | 0.006 | 0.000 |
| Latehar | 1.434 | 0.000 | 0.000 | 0.000 | 1.205 | 12.212 | 0.000 | 0.005 | 0.000 |
| Lohardaga | 0.030 | 0.000 | 0.000 | 0.000 | 0.284 | 2.057 | 0.020 | 0.000 | 0.000 |
| Pakur | 0.021 | 0.000 | 0.000 | 0.000 | 0.174 | 7.524 | 0.000 | 0.000 | 0.000 |
| Palamu | 0.794 | 0.000 | 1.102 | 0.001 | 0.423 | 5.861 | 0.002 | 0.000 | 0.000 |
| Ramgarh | 0.693 | 0.000 | 0.000 | 0.000 | 0.071 | 5.508 | 0.000 | 0.054 | 0.000 |
| Ranchi | 1.105 | 0.000 | 0.063 | 0.358 | 0.385 | 4.369 | 0.196 | 0.000 | 0.000 |
| Sahebganj | 0.000 | 0.000 | 0.000 | 0.108 | 0.214 | 4.370 | 0.000 | 0.017 | 0.000 |
| Saraikela | 0.010 | 0.000 | 0.000 | 0.000 | 1.950 | 7.930 | 0.000 | 0.002 | 0.000 |
| Simdega | 3.840 | 0.000 | 0.420 | 0.000 | 1.147 | 6.993 | 0.348 | 0.013 | 0.000 |
| W. Singhbhum | 0.092 | 0.000 | 0.056 | 0.002 | 0.233 | 2.646 | 0.054 | 0.000 | 0.000 |
| Jharkhand | 20.308 | 0.001 | 2.258 | 0.589 | 13.197 | 138.684 | 1.550 | 0.099 | 0.044 |

**Source:** Agricultural Directorate, Govt. of Jharkhand

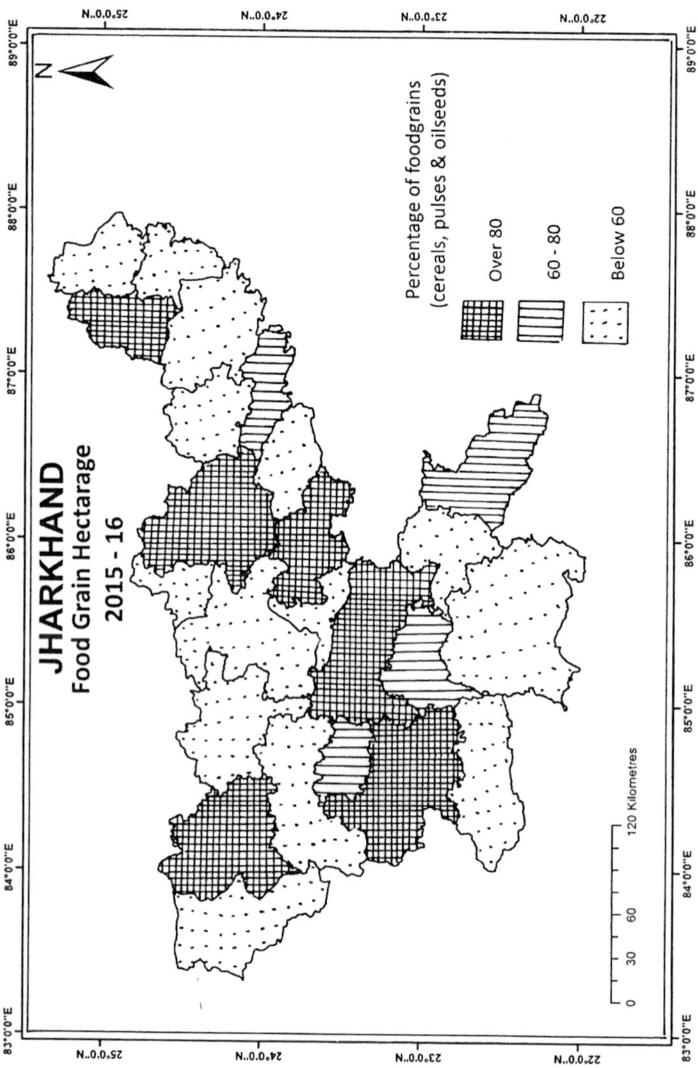


**Map 5.1: Foodgrain Hectarage (2015-16)**

Table 5.1 and Fig. 5.1 both have depicted that highest concentration of food grain hectarage (over 80 percent) in the district of Palamu (143.40 %), Ranchi (111.14 %), Giridih (107.10 %), Gumla (99.45%), Bokaro (81.79%) and Godda (81.99%). It is very much possible that the areas where food grains concentration is high, growing food grains has been main attraction for the farmers and the peasants. It is a relatively simple and streamlined system. Machinery and labor resources can be effectively integrated and deployed; chemical power can be used increasingly to solve problems of the practice of intensive food grain production.

On the other hand, where the percentage of total cropped area is less than 60 percent are Pakur (58.33 %), West Singhbhum (57.37 %), Garhwa (56.15 %), Sahebganj (54.58%), Simdega (49.37%), Deoghar (49.35%), Dumka (48.45%), Latehar (45.80%), Hazaribagh (45.33%), Chatra (40.28%), Saraikela (38.83%), Ramgarh (30.72%), Koderma (23.56%) and Dhanbad (21.44 %). Among these districts, lowest proportion of food grain area are concentrated as they do not reach even 50 percent share.

But there are only four districts in the state where food grain hectarage is in between 60 to 80 percent which is at the level of state average (75.53 %). These districts are Jamtara (78.94%), Lohardaga (77.68%), East Singhbhum (75.20 %) and Khunti (70.18%) respectively. Some of these districts are dominating vegetable growing i.e., Lohardaga, E. Singhbhum and Khunti. Here we can conclude that only 25 percent districts have highest food grain hectarage and 17 percent of the districts have moderate hectarage while 58 percent of districts have very poor hectarage in the state.

In the ensuing pages the food grain crops are discussed in order of importance in all Jharkhand cropping pattern i.e., Rice (paddy), Maize, Wheat and then Pulses and Oil Seeds respectively.

**Table 5.14**
**Percent to the total harvest area (2015-16)**

| District | Rice | Maize | Wheat | Pulses | Oil-Seeds |
|---|---|---|---|---|---|
| Bokaro | 64 | 2 | 2 | 12 | 20 |
| Chatra | 37 | 9 | 9 | 27 | 28 |
| Deoghar | 51 | 17 | 9 | 12 | 11 |
| Dhanbad | 70 | 4 | 2 | 9 | 18 |
| Dumka | 61 | 11 | 7 | 13 | 8 |
| E. Singhbhum | 56 | 14 | 5 | 24 | 1 |
| Garhwa | 24 | 16 | 11 | 36 | 13 |
| Giridih | 65 | 14 | 7 | 14 | 6 |
| Godda | 43 | 12 | 12 | 20 | 13 |
| Gumla | 75 | 3 | 3 | 11 | 8 |
| Hazaribagh | 56 | 6 | 6 | 21 | 11 |
| Jamtara | 62 | 8 | 8 | 10 | 12 |
| Khunti | 64 | 6 | 3 | 18 | 9 |
| Koderma | 28 | 15 | 12 | 28 | 17 |
| Latehar | 19 | 17 | 5 | 33 | 26 |
| Lohardaga | 52 | 10 | 3 | 25 | 10 |
| Pakur | 48 | 9 | 8 | 22 | 13 |
| Palamu | 28 | 21 | 8 | 32 | 11 |
| Ramgarh | 44 | 3 | 3 | 29 | 21 |
| Ranchi | 71 | 5 | 3 | 16 | 5 |
| Sahebganj | 48 | 13 | 9 | 21 | 9 |
| Saraikela | 64 | 4 | 1 | 23 | 8 |
| Simdega | 50 | 5 | 5 | 28 | 12 |
| W. Singhbhum | 79 | 4 | 1 | 12 | 4 |

### Rice (Oryza Sativa)

The rice enclosed in the hull is called paddy, it becomes rice only after the hull has been taken off. Rice is the vital staple food crop in India and therefore in Jharkhand too. Rice crop has primary importance as it supports millions in India. In the coastal regions of India rice has no competitor in its ability to support the dense population (Singh, 1976, P.268).

The total production of rice in Jharkhand is 2.609 million tons. Here, the majority of population is dependent on agriculture and the grains produced from these lands are sufficient only for a period of less than six months. The majority of cultivable land is covered by rice cultivation during wet season under rainfed conditions. (Singh, 2003, P.54). More than 46 percent (2002) of the total population in the state is below poverty line. (Tiwari, 2019, P.386). Maximum portion of this population is Agricultural Laborers in the cultivation of rice and other crops. Under this chapter we examine the trend in rice production and productivity and the causes of low rice productivity in the state during last 15 years on the paddy hectarage is limited by the rainfall and temperature. "The crop requires in most instances a relatively high atmospheric humidity, an average temperature of 7-degree F and a well distributed rainfall between 1250 mm and 1500 mm during the growing season within the area of production and in the catchment area of the streams and rivers furnishing water for irrigation. These very characteristics of rice climate exclude the cultivator of other food grains during the growing season of rice crop. To the climatic requirement must also be added an abundant supply of fresh water for irrigation where the rainfall is under 1250 mm. Rice thrives well on a variety of soils widely ranging from slightly acidic or slightly alkaline, particularly the heavy compact and acidic soils having the ability to hold the water over the surface for a considerable period, and providing a solid footing for power equipment where the crop is grown with the help of machinery. The heavy alluvial soils are more favorable than the lighter type, and there must be a sub-soil layer impervious to water underneath. Rice crop raised on clay-loams give higher yields, but higher yields are obtained from slightly acidic soils." (Singh, 1976, P.271)

In Jharkhand, the rainfall is received during the south-west monsoon period of June-October. The mean annual rainfall varies from 1000-1600 mm. From rainfed rice cultivation point of view, the rainfall is categorized from medium (1400-1600 mm), moderately medium (1200-1400 mm) and low (1000-1200 mm). In these categories, either intermittent or terminal drought limits the rice productivity.

The rainfall quantum has a relationship with the annual number of rainy days. A majority of districts of the state receive rainfall for 50 to 80 days. With intermittent dry spell during rainy season, the length of rainy season is always greater than the number of rainy days (Singh, 1999, P. 76). The crop growing season is always greater than the rainy season and dependent not only on rainy season but also on the water availability period and the soil characteristics like water holding capacity. The moisture availability periods in the state were estimated by Singh (1999, P.76) using long term rainfall and potential evapo-transpiration (PET). Depending on the balance between rainfall and PET, the moisture availability periods were divided into three sub-class: (a) Moist I, (b). Humid and (c) Moist II. A sub class when rainfall is lesser than the PET and more than ½ PET (PET > R > 1/2 PET) was considered as 'Moist period I or Moist period II', and when rainfall exceeded ½ PET, as 'Humid Period' sub class. 'Moist I' is the period at the beginning of rainy season when the PET is quite high. Most of the crop establishment operations are done in this period. The humid period is the moisture stress free period when the rainfall is always higher than PET. This period is the best period for the rice crops to enter into vegetative and reproductive stages. 'Moisture II' is the period at the end of the rainy season when, in spite of high crop canopy, the PET becomes low due to lowering of temperature. Most of the crop harvesting and processing operations are done in this period. (Singh, 2003, P.54-55)

Rice is grown in Jharkhand state is about 1471,68 thousand hectares. The actual share of the area and production by each district are given in table 5.2 to 5.12 in detail with all food grains. District wise percentage of area and production of rice is given in table 5.15.

**Table 5.15**
**Percentage of Rice Area and Production in Jharkhand (2015-16)**

| District | Rice Area (%) | Production (%) |
|---|---|---|
| Bokaro | 2.1 | 1.3 |

| | | |
|---|---|---|
| Chatra | 2.4 | 1.5 |
| Deoghar | 3.4 | 9.6 |
| Dhanbad | 2.5 | 1.6 |
| Dumka | 7.3 | 13.9 |
| East Singhbhum | 3.2 | 3.3 |
| Garhwa | 2.6 | 1.1 |
| Giridih | 5.9 | 4.9 |
| Godda | 3.1 | 5.7 |
| Gumla | 11.7 | 7.3 |
| Hazaribagh | 5.2 | 4.3 |
| Jamtara | 3.3 | 5.0 |
| Khunti | 4.7 | 2.8 |
| Koderma | 1.1 | 1.1 |
| Latehar | 1.6 | 0.7 |
| Lohardaga | 2.3 | 2.8 |
| Pakur | 3.3 | 4.7 |
| Palamu | 2.4 | 1.0 |
| Ramgarh | 2.0 | 0.7 |
| Ranchi | 10.9 | 7.4 |
| Sahebganj | 2.8 | 3.3 |
| Saraikela | 6.7 | 0.8 |
| Simdega | 5.8 | 10.3 |
| West Singhbhum | 11.0 | 4.5 |
| Jharkhand | 100.0 | 100.0 |

District Gumla has the maximum share of rice land (11.7 %) but maximum production has been recorded in Dumka district (13.9 %). In terms of percent share of rice area West Singhbhum is in second position having 11 percent of total area followed by Ranchi (1.9 %), Dumka (7.3 %), Saraikela (6.7 %), Simdega (5.8 %), Giridih (5.9 %), Hazaribagh (5.2 %), Khunti (4.7 %) and rest all other districts have less than 4 % share of rice land. Minimum share of rice land is with Koderma district (1.1 %). So far production share is concerned it is highest in Dumka (13.9 %) followed by Simdega (10.3 %), Deoghar (9.6 %), Ranchi (7.4 %), Gumla (7.3 %), Godda (5.7 %) and Jamtara (5.0 %) respectively; but all other districts have less than five percent share of the total production of rice in the state.

Share of rice area and production differ widely among the districts. Simdega has only 5.8 percent share of total area but share of production is 10.3 percent. On the other hand, Gumla has 11.7 percent of rice land but the production share is only 7.3 percent. This is due to the productivity difference.

Table 5.13 represents the hectarage of different food grains which has been calculated as percent to the total harvest area of state (2015-16). Accordingly, Fig 5.2 has depicted rice

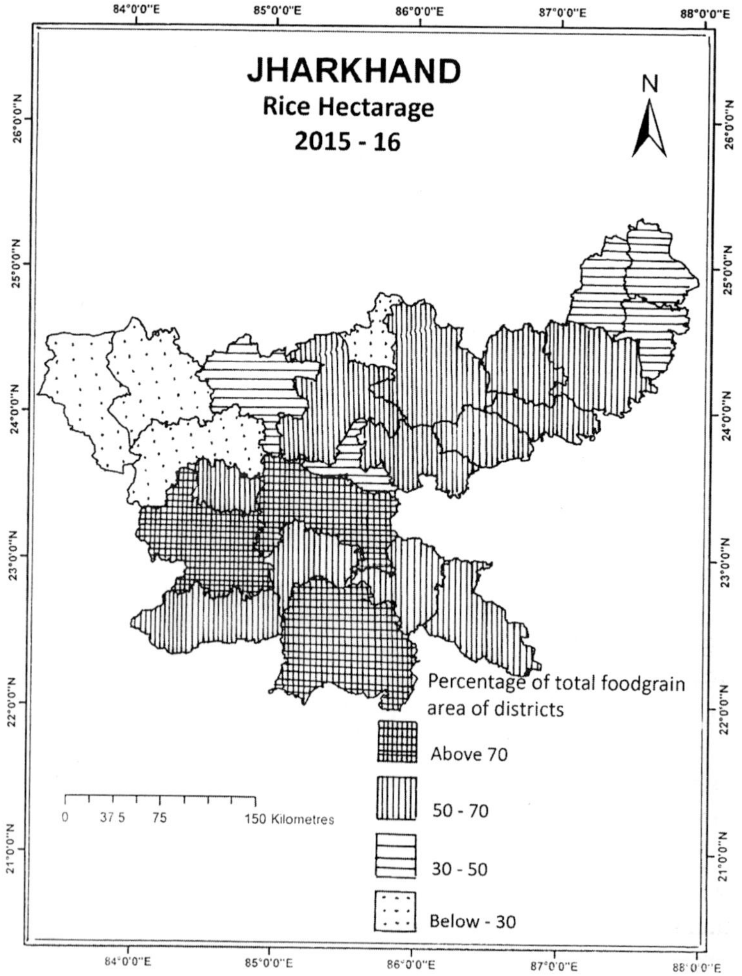


Map 5.2: Rice Hectarage (2015-16)

hectarage as percentage of total food grain area of the district. More than 70 percent area of the total food grain area has been covered by rice area. This area comes under West Singhbhum, Gumla and Ranchi district. Rice is much appreciated in this area and therefore among these areas particularly Gumla is known as the Rice-Bowl of Jharkhand. Cause behind this is only the favorable climate and suitable soil texture of the area. Paddy hectarage is limited only by rainfall and temperature i.e., climate. If you see the climate of Jharkhand (as stated in chapter-1), a separating line has been drawn which passes through the northern boundary of Gumla, Ranchi, Bokaro, Dhanbad and Jamtara by Koppen to depict the climatic map of Jharkhand and the northern part of this line has cwg climate and southern part has Aw climatic conditions. Climate of southern part is tropical rainy Savanna climate where as northern part has Humid Mesothermal Warm dry winter climate. Southern area where there is Aw climate is more suitable for rice cultivation and therefore cultivation of rice is much appreciated here.

Under the hectarage of 50 to 60 percent 12 districts have recorded their share i.e., Dhanbad (70%), Giridih (65%), Bokaro (64%), Khunti (64%), Saraikela (64%), Jamtara (62%), Dumka (61%), East Singhbhum and Hazaribagh (56%), Lohardaga (52%), Deoghar (51%), and Simdega (50%). This shows that maximum districts of the state are having more than 50 percent rice hectarage.

Hectarage of rice between 50 to 30 percent are recorded in Sahebganj (48%), Pakur (48%), Godda (43%), Ramgarh (44%), and Chatra (37%) respectively. Similarly, the hectarage below 30 percent are recorded in Koderma and Palamu (28%), Garhwa (24%) and Latehar (19%). All these districts are in the drought prone areas of Jharkhand and are situated in the western part of the state.

Now the changes in the area of rice production can be seen in table 5.2 where temporal change has been recorded. During 1985-86 there were 1529 thousand hectares involved in rice cultivation. It became 1522 thousand hectares during 1988-89 and 1571 thousand hectares in 1990-91. In 1999-2000 it recorded 1420 thousand hectares and 2004-05 recorded 1276 thousand hectares but again apart for these decreasing hectarages it has recorded 1471 thousand hectares during 2015-16. Similarly, the record of rice production is also fluctuating. During 1985-86 rice production of Jharkhand was 1514 thousand metric tons which decreased to 1314 thousand metric tons during 1987-88. But it grew again to 2556 thousand metric tons during 1998, 2732 during 2001-02. Again, it started to decline year by year till 2004-05 and was recorded 1908 thousand metric tons. Now in the year 2015-16 rice production in Jharkhand was recorded 2471.37 thousand metric tons. Thus, we may conclude that rice hectarage has recorded a decrease of 3.4 percent from 1986 to 2016 whereas production of rice has been increased to 63 percent during the time period of 30 years. Production has been increased just after being separate state as Jharkhand.

R.K. Singh (2003, P.58-60) has also observed rightly (scientist of rice research station, Hazaribagh) in his project report that "out of 12 districts in the state (during 1985-1999), 10 districts showed a declining trend in winter rice area. Average rice area decline is between -35.85 to -6.16 percent. Study of productivity has revealed a positive growth rate almost in all the districts of the state."

Methods of Rice Establishment has been mentioned in Annexure of the thesis in detail. At the same time the list of traditional rice cultivars has also been mentioned in the Annexure.

Directorate of Economics and Statistics has calculated that per capita net availability of rice per annum in India is 69 Kg. During 2018 census this is per head per day about 183 grams.

## Maize or Corn (Zea Mays)

Maize has emerged as the second largest crop in hectarage in Jharkhand. It is largest in production too though it is a less celebrated human food crop than rice and wheat. Corn is a coarse annual plant of the grass family Graminae. It originated in tropical South-America and Indians grew it on the west-coast slopes of Mexico about 7000 BC. Columbus introduced this American crop along with tobacco to the old world. Though tropical in origin, it is grown in over 100 countries extensively from equator to 50° North in Canada and Siberia and to 42° South in Chile and New Zealand. Many cultivars of maize varieties have been developed biotechnologically to adapt to widely different ecozones: 1. Lowland Tropics, 2. High Altitude Zones, 3. Warm Sub-

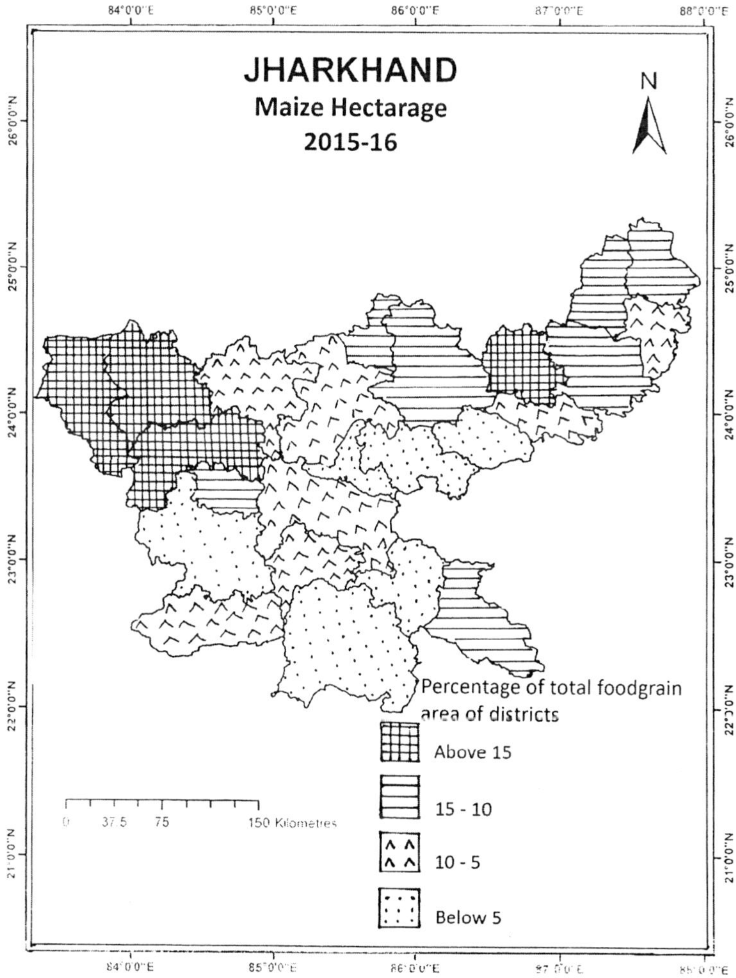


**Map 5.3: Maize Hectarage (2015-16)**

Tropics and 4. Warm Temperate latitudes and zones. On higher altitudes it does not ripen, but they grow it for green fodder. (Singh & Siddique, 2012, P.309).

Maize is a very hard crop but does not thrive well in fry areas unless irrigated. It prefers a frost-free growing season of 140-150 days, a mean temperature of 70°-80° F (20°-27° C), a mean night temperature of 58° F, and nightly rainfall, sunny days and high humidity. Annual rainfall of above 50 cm with fair distribution in growing season, particularly during pollination and grain-filling stages, and well retained soil moisture give high yields. Fertile deep, friable (crumbling), and well-drained soil with good tilt or suitable clustering (aggregates of soil particles into tiny clumps) and high moisture retaining capability are highly beneficial. Soils must have adequate porosity to permit aeration to provide oxygen to the roots as the soil water level varies. High fertilization increases the efficiency of water use and boosts high yields. (Singh & Siddique, 2012, P. 309-10).

**Table 5.16**

**Percentage of Maize Area and Production in Jharkhand (2015-16)**

| District | Maize Area (%) | Production (%) |
|---|---|---|
| Bokaro | 2.6 | 3.3 |
| Chatra | 4.8 | 4.1 |
| Deoghar | 5.9 | 6.8 |
| Dhanbad | 0.7 | 1.6 |
| Dumka | 6.7 | 6.2 |
| East Singhbhum | 4.0 | 3.2 |
| Garhwa | 8.9 | 4.6 |
| Giridih | 6.4 | 6.0 |
| Godda | 4.3 | 6.1 |
| Gumla | 2.0 | 2.4 |
| Hazaribagh | 3.8 | 3.4 |
| Jamtara | 3.1 | 2.1 |
| Khunti | 2.0 | 2.0 |
| Koderma | 3.0 | 3.8 |
| Latehar | 7.6 | 3.2 |

| | | |
|---|---|---|
| Lohardaga | 2.1 | 2.3 |
| Pakur | 3.3 | 5.1 |
| Palamu | 9.1 | 6.0 |
| Ramgarh | 3.1 | 2.9 |
| Ranchi | 3.9 | 3.4 |
| Sahebganj | 3.9 | 2.3 |
| Saraikela | 2.0 | 0.06 |
| Simdega | 3.0 | 6.0 |
| West Singhbhum | 2.5 | 2.0 |
| Jharkhand | 100.0 | 100.0 |

Maize is grown in Jharkhand state in about 288 thousand hectares. The actual share of area and production by each district is given in table 5.15 in detail with all food grains. Palamu district has the maximum share of maize land (9.1 %) but maximum production has been recorded in Deoghar district (6.8%). In terms of percent share of maize area Garhwa district has second position having 8.9 percent area of total maize producing area of district followed by Latehar (7.6%), Dumka (6.7%), Giridih (6.4 %), Deoghar (5.9%), Chatra (4.8%), Godda (4.3%) and East Singhbhum district with 4 percent respectively. Rest all 15 districts are having less than 4 percent share of land in maize cultivation. A simple calculation shows that Palamu division or only three districts i.e., Palamu, Garhwa and Latehar have more than 25 percent area covered in Maize cultivation. So, no doubt, Palamu division is the largest cultivator of maize in the Jharkhand state. But these three districts have only 14 percent of the total production. This is due to the low productivity in the area. Relatively high productivity in Jharkhand of maize crop is in Santhal Pargana division where among the six districts about 27.2 percent of area has been covered by Maize but there is 28.6 percent of total Maize production.

In relation to the total food grain area of the district share of maize crop is shown in table 5.13 where the hectarage of all food grain has been calculated. Fig 5.3 represents the Maize hectarage in the state which shows the total food grain area of the district. Palamu, Deoghar, Latehar and Garhwa have recorded the highest hectarage of Maize crop. These districts have more than 15 percent area to the total food grain area of the district. 15 to 10 percent area has been recorded by Koderma, East Singhbhum, Giridih, Sahebganj, Godda, Dumka and Lohardaga. Under the area of 5 to 10 percent in Maize crop Chatra, Pakur, Jamtara, Khunti, Hazaribagh, Ranchi and Simdega districts have recorded their share. West Singhbhum, Saraikela, Dhanbad, Ramgarh, Gumla and Bokaro districts have depicted lowest share of area in maize crop. High quality of Maize which are high yielding and disease resistant varieties are - Suwan Composite- I, HQPM-1, Birsa Makka-I, Birsa Vikas Makka-2, Priya, Ganga White-2, Ganga White-5, Vinay (Composite Makka) and Shaktimaan-1 (hybrid), Ambar Pop-Corn, Priya Sweet Corn and as baby corn - BVM-2, HM-4 and VL baby corn-1 are the best commercial maize crop varieties.

When we study the changes in area and production of Maize in state; tables 5.2 and 5.4 reveals a temporal change. During 1985-86 there were 137 thousand hectares of land under the cultivation of maize. It changed to 121 thousand hectares during 1989-90. This was a decrease of 13 percent during this census. Again in 1997-98 land under maize cultivation was 122 thousand hectares. It grew to 150 thousand hectares during 1998-99 and in the immediate year 1999 it decreased to 118 thousand hectares. This was the period of separation of Jharkhand from Bihar as a separate state. Since then, the area under maize cultivation has been growing continuously in the state. During 2000-01 there were 157 thousand hectares under maize cultivation and now after 15 years (2015-16) it has grown to 288 thousand hectares. It is a growth of about 83 percent in the maize area in a span of 15 years.

When the growth in production is calculated we get a fluctuating scenario right from 1985 to 2015. During the period of 30 years, growth of maize production was never stable. But an overall growth is seen across the time period. Total maize production was only 135 thousand tons during 1985 but it was recorded 375 thousand tons during 2015-16. This is a growth of about 177 percent. A gradual increase in production and area are the matter of interest, high yield seeds and introduction of pesticides and fertilizers.

## Wheat (Triticum Aestivum)

Wheat is the third crop of the state in respect of hectarage and production. Wheat is the leading crop among all the 'Rabi' crops. Wheat is generally a dry zone crop occurring mostly in the areas under 40 inch or 1016 millimeters of annual rainfall. Among the important requirements of wheat are nitrogen, sufficient moisture and a cool growing period characterized by a falling temperature in the beginning which cools the seed bed, and a rapidly rising temperature towards harvest. Nitrogen is generally produced by the decay of organic matter during fallowing and moisture is supplied by irrigation or winter rain or capillary attraction from rivers in newer alluvium.

Dr. R.C. Tiwari (2010, P.232) has mentioned in his book 'Bharat ka Bhugol' on map of wheat producing area depicted as the percentage of wheat to the total cropped area where Jharkhand has no place in the list of wheat producing area as Jharkhand has recorded below 1 percent area in wheat cultivation in respect to the total cropped area. On the other hand, Dr. Enayat Ahmed (1965, P.138) in his book 'Bihar: Physical, Economic, Regional Geography' has mentioned that the wheat belt of Jharkhand includes the canal irrigated plain districts. Dr. Ayodhya Prasad (1973, P. 67) has mentioned in his book 'Chotanagpur: Geography of Rural Settlement' that cultivation of wheat is restricted to a narrow strip of land in the north of Chotanagpur. But among the 33 different important crops of Jharkhand, wheat has third position in area and production both.

**Table 5.17**
**Percentage of Wheat Area and Production in Jharkhand (2015-16)**

| District | Wheat Area (%) | Production (%) |
|---|---|---|
| Bokaro | 0.7 | 0.7 |
| Chatra | 5.2 | 6.1 |
| Deoghar | 5.4 | 7.9 |

| | | |
|---|---|---|
| Dhanbad | 0.6 | 0.8 |
| Dumka | 7.1 | 6.5 |
| East Singhbhum | 2.4 | 2.9 |
| Garhwa | 10.9 | 6.9 |
| Giridih | 6.2 | 9.0 |
| Godda | 7.8 | 9.3 |
| Gumla | 4.8 | 4.8 |
| Hazaribagh | 5.4 | 5.8 |
| Jamtara | 4.1 | 2.3 |
| Khunti | 1.9 | 1.4 |
| Koderma | 4.2 | 4.4 |
| Latehar | 3.7 | 2.8 |
| Lohardaga | 1.2 | 1.5 |
| Pakur | 4.9 | 4.6 |
| Palamu | 5.9 | 4.1 |
| Ramgarh | 1.3 | 1.4 |
| Ranchi | 3.9 | 4.4 |
| Sahebganj | 5.0 | 3.1 |
| Saraikela | 1.0 | 1.2 |
| Simdega | 5.5 | 7.2 |
| West Singhbhum | 0.9 | 0.5 |
| Jharkhand | 100.0 | 100.0 |

Jharkhand cultivates wheat in about 161 thousand hectares and produces 274 thousand metric tons per annum. This is 7.1 percent to the total crop production of the state. The percent share of area and production by each district are given in table 5.16. District Garhwa has the maximum share of wheat land (10.9%) followed by Godda (7.8%), Dumka (7.1%), Giridih (6.2%), Palamu (5.9%), Simdega (5.5%), Hazaribagh (5.4%), Deoghar (5.4%), Chatra (5.2%), Sahebganj (5.0%), Pakur (4.9%), Gumla (4.8%), Koderma (4.2%) and Jamtara (4.1%) and all other districts have below four percent share of area in wheat cultivation. Such districts are about ten in number which is about the half area of the state.

When we study the wheat production ratio of the state, Godda has the highest percent (9.3) share of wheat production. Giridih district has second position with the 9.0 percent share while Deoghar is in third (7.9 percent) and

Simdega is in fourth (7.2 percent) positions followed by Garhwa (6.9 percent), Dumka (6.5 percent), Chatra (6.1 percent), Hazaribagh (5.8 percent), Gumla (4.8 percent), Pakur (4.6 percent), Koderma and Ranchi (4.4 percent) and Palamu (4.1 percent) and all other district have below 4 percent share of wheat production. These districts are eleven in number mean about half of the state.

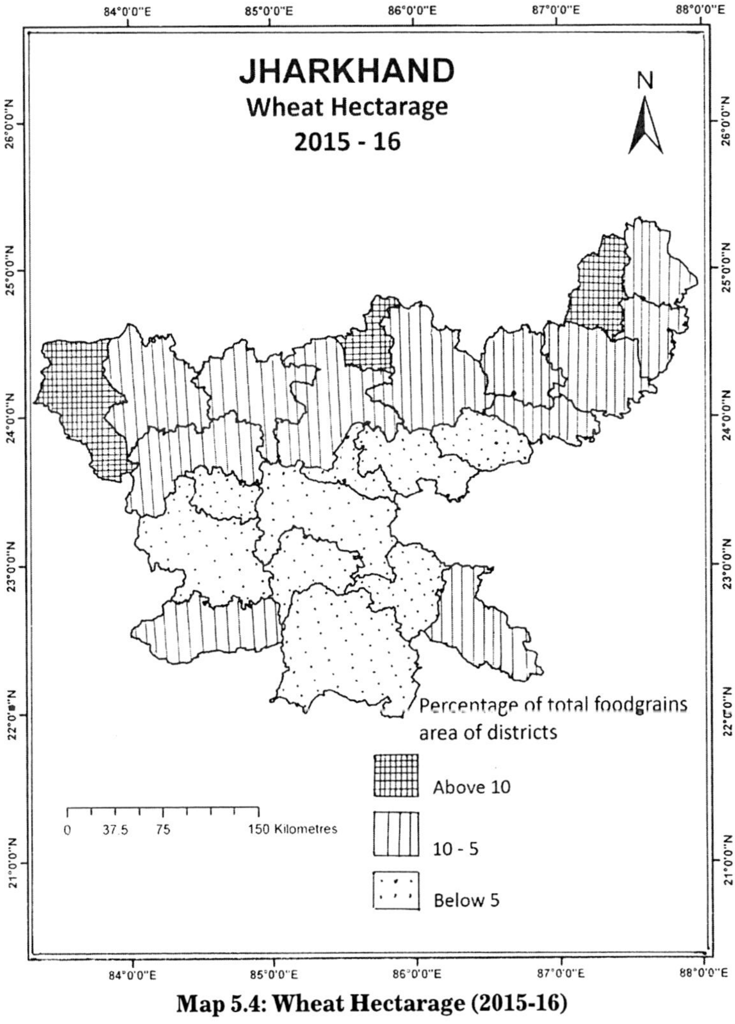


Map 5.4: Wheat Hectarage (2015-16)

Now the map of wheat hectarage (which is district wise percent) depicted the highest percentage in hectarage in Godda, Koderma and Garhwa district i.e., above 10 percent of the total food grain area. (fig 5.4). Under the percentage between 5 to 10 there are about 50 percent districts out of total 24 of the state in this category. These are - Deoghar, Sahebganj, Chatra, Giridih, Jamtara, Pakur, Palamu, Dumka, Hazaribagh, Simdega, Latehar and East Singhbhum. These districts are in the moderate category. This shows that the wheat area has moderate to low hectarage in the state. Nine districts are under the category of below 5 percent hectarage. Hence, we can say that though wheat cultivation is present in the state but it needs to enhance its area and production both.

Here we can conclude that Wheat is the third important crop in the state occupying 5.77 percent of the cropped area under food grains. The present yield of wheat in the state is 1.70 tons per hectare in 2015-16 as against 3 tons per hectare in the country. In Jharkhand, the scientists of Birsa Agricultural University have done a great job attracting the farmers to cultivate wheat through a targeted program SAMETI (State Agricultural Management & Extension Training Institute). The target is to grow wheat at the present national level of 3.3 tons per hectare by 2025. This will however require extension of irrigated area and adoption of HYV technology by the farmers.

## Ragi or Marua (Eleusine Coracana)

Ragi is locally known as Marua and Finger Millet in English. It is a very important crop which is substantially grown in India and Africa as a staple food for a certain population (Devi, 2014, P.6). Many epidemiological researchers have shown that routine consumption of Ragi whole grain and its derived products lowers the risk of type II diabetes, cardiovascular diseases, gastrointestinal cancer and a number of other disorders (Nishar Akhtar, 2020, P. 467-476). Ragi is consumed in Jharkhand in different food preparations such a porridge, chapati, cake, cookies, kurha and sweetmeat. In addition to these boiled and malted germinated grains are used as a food supplement to infants and pregnant women as it does not form acid and is easy to digest. Ragi is considered to be one of the most nutritious, highly digestible and least allergic food grains. Apart from this Ragi has the ability to grow on least fertile soils without the application of chemical fertilizers, thus making itself suitable for cultivation in arid and semi-arid areas. It is also considered as a drought tolerant crop due to higher water use efficiency with enhanced fixation of CO2 and minimal leaf area. Moreover, it also utilizes soil nitrogen very efficiently and is resistant to various storage pests. There is also increased recognition of ragi now a days as a better food crop among consumers which present an opportunity for the farming community to derive economic benefit from ragi production. All these characteristics of ragi make it a potential candidate as a substitution crop in the changing weather conditions of Jharkhand without hampering the economic returns and nutritive component. (Nishar, 2020, P. 467-476).

Dr. Ayodhya Prasad (1973, P.65) has mentioned in his book that Marua like Maize is extensively grown, though it fails to secure even a third position in the crop list. Being "essentially a crop of undulating, well- drained soil... that is too shallow or too poor for rice or too steep for terracing", marua is most extensively grown in the central part of the region where it commands more than 6% of the cropped area. It attains some importance in the marginal area also where it covers less than 20% of the total cropped area.

Now in Jharkhand, according to the data of 2015-16 ragi crop is cultivated in about 12.85 thousand hectares area where 7.78 thousand metric tons ragi has been produced. When we calculate the percentage of the area and production; about 46 percent area has been covered by ragi crop area; whereas production is only 20 percent of the total crop production.

**Table 5.18**
**Percentage of Ragi Area and Production in Jharkhand (2015-16)**

| District | Ragi Area (%) | Production (%) |
|---|---|---|
| Bokaro | 0.8 | 1.3 |

| | | |
|---|---|---|
| Chatra | 1.9 | 3.3 |
| Deoghar | - | - |
| Dhanbad | 0.4 | 0.5 |
| Dumka | - | - |
| East Singhbhum | 8.6 | 6.0 |
| Garhwa | - | - |
| Giridih | - | - |
| Godda | - | - |
| Gumla | 11.5 | 18.5 |
| Hazaribagh | 12.2 | 14.6 |
| Jamtara | - | - |
| Khunti | 11.3 | 2.7 |
| Koderma | 4.7 | 4.2 |
| Latehar | 0.9 | 1.8 |
| Lohardaga | - | - |
| Pakur | - | - |
| Palamu | - | - |
| Ramgarh | 5.1 | 6.3 |
| Ranchi | 24.3 | 17.0 |
| Sahebganj | - | - |
| Saraikela | - | - |
| Simdega | 15.6 | 20.5 |
| West Singhbhum | 2.5 | 2.3 |
| Jharkhand | 100.0 | 100.0 |

Table 5.16 has depicted district wise distribution of Ragi in percentage of area and production. At first, we shall study the distribution of area under ragi in state. Ranchi district has recorded highest (24.3 percent) share of the total ragi area of the state. This is very important to say that only one district covers around one fourth of the total area of the state. Simdega has second position (15.6 %) in area coverage in ragi cultivation. Hazaribagh, Gumla and Khunti have third, fourth and fifth position in the state with 12.2%, 11.5% and 11.3% share respectively. East Singhbhum takes sixth position in the state with 8.6% area while Ramgarh district has seventh position in the state with 5.1 percent area under cultivation. Koderma and West Singhbhum take eighth and ninth position with 4.7 and 2.5 percent respectively. Latehar,

Chatra and Dhanbad have lowest coverage of ragi in the state with 0.9, 0.8 and 0.4 percent share respectively.

Production of Ragi is highest (20.5, 18.5,17.0 and 14.6 percent) in Simdega, Gumla, Ranchi and Hazaribagh districts respectively. These four districts have about 70 percent of total production of the state. On the other hand, West Singhbhum, Khunti, Ramgarh, Latehar, Koderma, E. Singhbhum, Dhanbad, Chatra and Bokaro have altogether 30 percent share of production for the state.

Study shows that Santhal Pargana is totally absent in Ragi cultivation. Similarly, except Latehar entire Palamu division is absent from Ragi cultivation. Here climate is the only factor responsible for this. For Ragi cultivation, Ranchi and Hazaribagh plateau, Koderma plateau, Chaibasa plains (or Dharwarian country) and Shankh and South-Koel Basin provide suitable climatic and soil conditions in Jharkhand. Eastern part of the state due to high moisture content and western part of state due to highly dry conditions, both areas are not suitable for Ragi cultivation. Middle portion of Jharkhand from North to South is a suitable area. Nevertheless, among all ragi producing districts, the climate, soil and physiography of Chatra, Bokaro, Ramgarh and Gumla are most productive ones.

Study shows that during 2015-16 only 0.46 percent of the total grain growing area in under ragi cultivation. "During the normal period 0.70 percent covers in Ragi cultivation. Productivity is static during the period of last 8 to 10 years. It is almost 6.5 tons per hectare to 7.4 tons per hectare. Ragi, Gundli and Kodo are the first crops to mature in Kharif, thus provide precious food in the lean period in Jharkhand. The crop is handy to farmers because of the earliness, providing grain as well as fodder. The production of this crop can be doubled, provided both - improved seeds and optimum manuring" (Mishra, 2011-12, P. 102).

## Other Food Grains (Jowar and Bajra)

### Jowar

Jowar is the local name. In English it is 'Sorghum' and

botanically known as 'Sorghum Vulgare'. The word Sorghum is derived from the Latin word 'Sorgo' meaning 'rising above'. It is also known as 'giant or great millet', being the largest of the food grains called millets both in terms of plant and grain. The sweet sorghum is grown primarily for grain which may be used as food and food by livestock respectively (Singh, Jasbir, 1976, P.278).

Jowar yields well in areas of high rainfall but it is not popular because it stores badly and, in such areas, maize can be grown equally well. This competition limits the distribution of the Jowar in the area. The Jowars, however, unlike corn (maize), seem to have the ability to become practically dormant during periods of severe droughts and to renew growth, without much apparent injury, when conditions become more favorable (Wolfe and Kipps, 1970, P.312). Therefore, the adaptation of jowar, especially the grain type, to dry climates where annual rainfall is less than 750 mm is well known. The plant produces a very efficient rooting system which develops quickly and extensively taps large areas of the surface layers of soils (Cobley, 1962, P.156).

In Jharkhand jowar is grown in those area where climatic conditions are too dry for maize and irrigated rice. It is totally absent in Santhal Pargana area, but somehow grown in the north-western and south-western parts of the state. Sometimes it has been grown all over the state except four districts i.e., Dhanbad, Bokaro, Lohardaga and Simdega.

District wise production area of Jowar has recorded highest (2922 hectares) during 2000 in Bokaro districts, followed by Garhwa (443 hectares), Latehar (114 hectares), Palamu (118 hectares) and all other districts have either below 100 or nil area in Jowar cultivation. So far production is concerned, in year 2000 it was highest in Bokaro (1616 metric ton), Garhwa (408 metric tons) and Palamu (100 metric tons) respectively. Others have below 100 metric tons or nil.

During 2004-05 highest jowar cultivating area was (0.077 thousand hectares) Sahebganj district followed by West Singhbhum (0.062 thousand hectares), Giridih (0.035

thousand hectares), Chatra (0.027 thousand hect.), Koderma (0.020 thousand hect.), Godda (0.019 thousand hect.) and Chatra (0.027 thousand hect.) etc. Similarly, the production was highest in Sahebganj (0.029 thousand metric tons) followed by West Singhbhum (0.028 thousand metric tons), Giridih (0.021), Ranchi (0.013). Rest other districts have less than 10 thousand metric tons.

**Table 5.19**
**Jowar area and production in Jharkhand**

| Year | Area (000 hectares) | Production (000 metric tons) | Yield Growth Rate (%) |
|---|---|---|---|
| 1999-2000 | 8.80 | 5.52 | 94.4 |
| 2004-05 | 0.36 | 0.18 | 76.6 |
| 2009-10 | 0.20 | 0.10 | 75.2 |

**Source:** Data Base of Jharkhand, R.K. Mishra

As we see the above table 5.17 where area and production of Jowar has been illustrated for three census periods of one decade. It is very much clear that the Jowar cultivation area is reducing year by year. It was 8.8 thousand hectares during 1999-2000 which reduced to 0.36 thousand hectares during 2004-05 and became 0.20 thousand hectares during 2009-10. It is a surprising change or decrease of 4300 percent in Jowar crop area. On the other hand, production of jowar has also reduced from 5.52 thousand metric tons in 1999-2000 to 0.18 thousand metric tons in 2004-05 and 0.10 thousand metric tons in 2009-10. It is a decrease of 54.20 percent which is more than the areal decrease. Yield growth rate of jowar was 94.4 percent during 1999-2000 that decreased to 76.6 percent in 2004-05 and 75.2 percent in 2009-10.

### Bajra

Bajra is a millet crop known an Bajra locally but it has other popular names i.e., pearl millet, bulrush millet and spiked millet. Its botanical name is 'Pennisetum typhoides' which belongs to the family Gramineae. It is a good source of energy, promotes heart health, helps in reducing weight, helps in digestion disorders, may help in preventing cancer and controls blood sugar levels. In spite of all these benefits bajra is considered to be the poor man's staple nourishment. Though it gives a good yield of a high protein heat-giving grain that stores well, suffering less from biological hazards than all other food grains, unfortunately it is generally unpopular as a food and has little market. The advantage of this crop is that it succeeds in warm area having low rainfall (under 500 mm) because it is a quick growing crop. It does best under conditions of intermittent light showers and benefits from bright sunshine, low humidity and considerable diurnal range of temperature during the growth period. It is a heat and drought tolerant crop. Heavy rains are disastrous during inflorescence because the pollen is washed away and many of the female flowers are not fertilized, thereby depressing the yield per unit area. Therefore, the crop is very fastidious about its climatic requirements. Consequently, specialization in bajra growing occurs in a climate that does not favor the growing of other crops during kharif season. Its concentration is a reflection of the distribution of rain, the length of rainy season and soil. It grows best on shallow, well drained, light loamy or sandy soils rather lacking in humus: generally, comes up very well in poorer type of sandy soils requiring between 300 mm and 400 mm of rainfall (Singh, 1976, P.268).

Bajra is grown on 1.80 percent (2009-10 average) of the harvested area of the state. It was 0.2% during 1999-2000. Hence, a growth of 88 percent in the area of Bajra cultivation. Production of Bajra has recorded 7.78 percent of the total food grain production in the state. District wise area and production shows an irregular cultivation of Bajra. As per census 2004-05 there is no production of Bajra at all in six districts of the state i.e., Simdega, Lohardaga, East Singhbhum, Palamu, Latehar, Dhanbad and Bokaro as per the data given by Agricultural directorate of Ranchi. District Gumla has recorded highest (0.045 thousand hectares) in Bajra cultivation followed by Gumla (0.045), Hazaribagh (0.035), Giridih (0.035), West Singhbhum (0.029), Chatra (0.025), Kodarma (0.022), Godda (0.021), Pakur (0.020),

Sahebganj (0.018), Ranchi (0.014) and rest districts have below 10 thousand hectares share in the cultivation area.

District wise production of the crop Bajra has been recorded highest in Gumla (0.019 thousand metric tons) followed by Giridih (0.018), West Singhbhum (0.013), Godda (0.012), Chatra (0.011), Ranchi (0.011) and all other districts have production less than 0.009 thousand metric tons. These are Koderma, Pakur, Sahebganj, Garhwa, Saraikela, Dumka, Jamtara and Deoghar.

**Pulses**

There are many varieties of pulses in Jharkhand. These are - Gram (Chana), Lentil (Masur), Pea (Matar), Pigeon Pea or Red Gram (Arhar), Black Gram or Mash (Urad), Green Gram (Moong), Horse Gram (Kurthi) rtc.

In respect of area of cultivation among all pulses, gram was dominating crop in between 1985-90 but in between 2000-2005 Urad was dominating crop in area and during 2015-16 Arhar has a highest coverage area of 193.68 thousand hectares among all pulses in Jharkhand. Gram is in second position (164.22 thousand hectares) and Urad is in third rank (9.87 thousand hectares). No doubt, these three pulses are most popular among the people of Jharkhand. Matar has fourth rank with 44.36 thousand hectares of coverage. Masur comes fifth, Mung and Kurthi take sixth and seventh rank in terms of coverage area of cultivation. Jharkhand has 594.99 thousand hectares area engaged in the cultivation of pulses which is about 21 percent of the total food grain area (2788 thousand hectares) of the state (2015-16) (Tiwari, 2019, P. 211-212)

Production of pulses have grown from 45 thousand metric tons in 1985- 86 to 547.15 thousand metric tons in 2015-16. During these 30 years pulse production has grown to about 1115 times in the state. Production rate has changed just after separation from Bihar. During 2000-01 the production of pulses was only 53 thousand metric tons and suddenly from next year (2001-02) it became more than 165 thousand tons. During 2004- 05 it has grown to 190.11 thousand metric tons and in 2015-16 it was recorded 547.15 metric tons. Among all above mentioned pulses highest production was recorded by Arhar (173.56 thousand metric tons) followed by Gram (172.88), Urad (72.66), Matar (52.21), Masur (32.03), Kurthi (26.06) and Mung (5.45 thousand metric tons). Thus, we can say that Arhar, Gram and Urad are three important and major pulse crops in Jharkhand.

Now the district wise pulse distribution as area and production can be studied through the tables of food grain distribution in the state. Analysis shows that during 2015-16 about 594.83 thousand hectares area of cultivation has been recorded under Pulses among the 24 districts of Jharkhand. This year highest pulse cultivation area (59.21 thousand hectares) was recorded in Garhwa district. Garhwa has a dominant cultivation of Arhar, Gram and Urad, though all other varieties of pulses are also being cultivated in the district. Second position in Pulse cultivation was recorded by Simdega district with a cultivated area of 47.86 thousand hectares under pulses. Latehar comes next with a cultivated area of 42.44 thousand hectares. Palamu and Ranchi stand at fourth and fifth place with 40.75 and 36.42 thousand hectares of area under cultivation of Pulses respectively. Next in rank are Saraikela (35.87), Hazaribagh (29.01), Chatra (26.65), Hazaribagh (29.01), Gumla (26.04), West Singhbhum (23.95), Dumka (22.70), Godda (21.86), East Singhbhum (20.73), Ramgarh (20.13), Khunti (19.20), Giridih (18.60), Sahebganj (18.41), Koderma (16.63), Lohardaga (16.53), Deoghar (11.76), Jamtara (8.22), Bokaro (6.06) and Dhanbad (4.74) respectively. Consequently, we may say that 25 percent of the total pulse area have been covered by only three districts (Garhwa, Simdega Latehar). These are the major pulse cultivating districts of the state.

The distribution of pulse production is very similar to distribution of area. Total production in the state was 547.08 thousand metric tons which is equal to 14 percent of the total food grain production (3853 thousand metric tons) of the state. Here we should compare the area and production to feel the ratio of area and production. Total area under pulses in the state is 21 percent to the total area where as pulse production is only 14 percent of the total production. This ration shows an under growth of pulse production in the state.

District wise production depicted highest production of pulses (61.67 thousand metric tons) in Simdega district. The second position is taken by Latehar district (51.13) and third

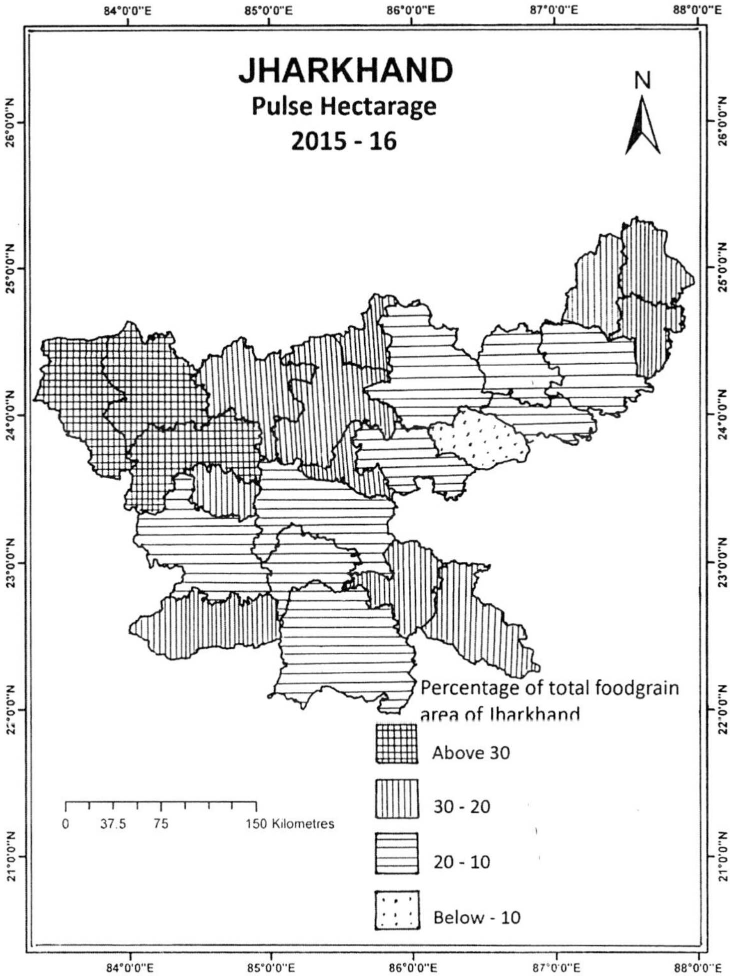


**Map 5.5: Pulse Hectarage (2015-16)**

rank goes to Garhwa district with 39.10 thousand metric tons. Only these three districts account for around 27 percent of the total production of the entire state. In fourth position we have Hazaribagh (30.27 thousand metric tons) followed by Ranchi (28.60), Saraikela, Chatra (27.26), Palamu (26.28), Gumla (24.42), Dumka (23.10), Godda (21.75), Ramgarh (21.39), West Singhbhum (19.24), Giridih (16.89), Pakur (16.87), Lohardaga (16.87), Khunti (16.15), Deoghar (14.34), Koderma (13.54), Sahebganj (13.68), Jamtara (10.71) and Bokaro (10.26) respectively.

At the end of the discussion of pulses an important fact about pulse cultivation is required to explain here. It is a matter of common understanding that cultivation of pulses is required to increase the soil fertility. Pulses are leguminous crops and it is an established scientific fact that they increase the amount of nitrogen in the soil. This fact was recognized in the middle of the 19th century. Till then it was not known exactly what factor is actually responsible for improving the soil fertility, until the scientists discovered that leguminous plants were capable of utilizing the nitrogen in the air and transforming it into substances assimilable by the cultivable crops. It was discovered that it was the bacteria in the root nodules of the leguminous plants which played an important role in this transformation. (Indian Council of Agricultural Research, 1970, P. 10-11)

**Oil Seeds**

Jharkhand is fortunate enough to devote a considerable proportion of the land to oilseed crops, which greatly strengthen the farmer's economy and improve the soil fecundity and productivity. Oil seeds have long been planted for local consumption, and older days the oil pressed from oilseeds were put to a variety of use, such as, lighting, cooking, industrial use etc. Today vegetable oils are increasingly used in lubricants, soap manufacturing, varnish and paint industry. The residue after the extraction of oil, which is also known as oilcake, serves as a food for cattle or fertilizer for soil.

The importance of oilseeds in the subsistence agricultural

economy of Jharkhand has a very little and unfortunately little hectarage just next to pulses. Not only do seeds supply an essential requirement in the nutrition of the farm economy, but they also are an important source of plant food. The effect of oilcake as a fertilizer lasts longer and fulfils role similar to that of pulses in fixing atmospheric nitrogen in soil and improving its fertility. Demand for oil is increasing

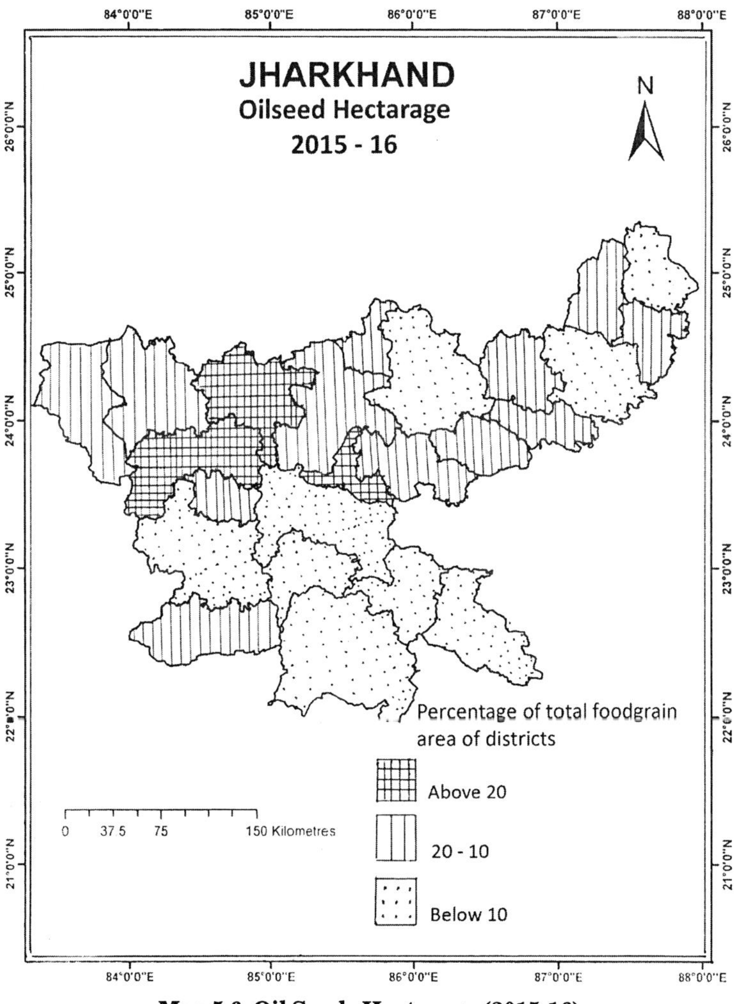


**Map 5.6: Oil Seeds Hectarage (2015-16)**

more rapidly than its supply. The primary solution is to increase production. It should be remembered that in India no expansion of the area under oil seeds at the cost of food grains can possibly be appreciated because of increasing population. The present problems pertinent to oilseeds production are: first, how to increase the area under oil seed crops without encroaching on the food grain fields and second, how to increase the yield per unit area and check the diseases and pests. (Singh, 1976, P.293)

Oilseed crops form a minor item if all-Jharkhand statistics are considered. Jharkhand has only 259.68 thousand hectares under oilseed crop which is only 9.31 percent to the total food grain area of Jharkhand during 2015-16. Area under oilseed was only 50 thousand hectares during 1988-89 that decreased to 24.49 thousand hectares till 2000. After separation from Bihar, it began to grow and gradually reached to 99 thousand hectares during 2004-05. Very surprisingly it became 259.68 thousand hectares during 2015-16. In spite of increasing in area it has only third position after food crops and pulses in the state.

Similarly, production of oilseeds has decreased from 60.53 thousand metric tons to 52.98 thousand metric tons in between 1999 to 2005. But it has grown up to 177 thousand metric tons during 2015-16.

The principal oilseed crops in Jharkhand area Rai or Toria or Indian rapeseed (Brassica Campestris) and Sarson (Brassica Campestris) are known as rape and mustard collectively. These are Rabi crops. Further Tisi (Linseed), Kusum (lac tree), Surajmukhi (Sunflower), Moongfali (Groundnut/Peanut), Til (Sesamum), Sarguja and Soybean etc. are the main oilseeds cropped in Jharkhand.

Total cropping area of oilseeds in Jharkhand has been distributed all over the state having a highest area (20.98 thousand hectares) in Garhwa district followed by Simdega (18.87), Latehar (17.98), Gumla (14.99), Palmu (14.09), Godda (13.69), Dumka (13.46), Ranchi (13.14), pakur (12.77), Chatra (11.99), Saraikela (11.34), Hazaribagh (10.41) and other districts having less than 10 thousand hectares.

Production of oilseeds in state is 181.07 thousand metric tons which is only 4.6 percent of the total food grain production of the state. Gumla district has recorded a highest production (19.42 thousand metric tons) in the state followed by Latehar (14.85), Simdega (12.76), Garhwa (10.98), Saraikela (9.89), Deoghar (9.43), Hazaribagh (9.18), Palamu (8.18), Pakur (7.71), Dumka (7.97), Chatra (6.90), Giridih (6.68), Koderma (6.36) and Ranchi (6.00) districts respectively. Rest all other districts have less than 6 thousand metric tons of oilseeds production in the state during 2015-16.

It would be worthy to discuss here that too much moisture is injurious to rabi oilseeds and loamy sand is suited for their growth, but alluvium is preferred. Alluvium cannot be spared for oil crops because of the importance of wheat and gram as rabi food grain crops in a nation where population pressure is increasing at a faster rate than ever before and hence in Jharkhand too.

**Vegetables**

A general scenario of vegetable growing in country is first in potato cultivation, other vegetables are in second position, Tomato and Onion take the third and fourth rank while Brinjal takes fifth rank. Further Cabbage, Cauliflower, Ladyfinger and peas take subsequent ranks. In Jharkhand Other Vegetables are in the first rank, Potato in second while Cabbage is in third rank statewide. Apart from these Ladyfinger, Pea, Onion, Brinjal and Tomato hold their subsequent ranks respectively.

Comparative year wise area and production of vegetable crops in Jharkhand can be studied in the following table 5.18.

**Table 5.20**
**Comparative Year Wise Area & Production of Vegetables in Jharkhand**

| Year | Area (000 hectares) | Production (000 metric tons) |
|---|---|---|
| 2000-01 | 149.80 | 2109.50 |
| 2001-02 | 158.50 | 1736.30 |
| 2002-03 | 118.20 | 1300.10 |

| | | |
|---|---|---|
| 2003-04 | 110.60 | 1197.26 |
| 2004-05 | 223.60 | 3394.90 |
| 2005-06 | 224.20 | 3401.30 |
| 2006-07 | 223.60 | 3394.90 |
| 2007-08 | 238.90 | 3639.70 |
| 2008-09 | - | - |
| 2009-10 | 229.63 | 3727.01 |

**Source:** Economic Survey of Jharkhand 2008-09

Area of vegetable growing was 149.80 thousand hectares during 2000-01 which increased to 229.63 thousand hectares during 2009-10. This is about 8 percent of total food grain area. When we compare the vegetable cultivating area of the state to the other neighboring states, Jharkhand falls short to all of them. For example, during the 2007-08 census Bihar has 823 thousand hectares, Chhattisgarh has 292 thousand hectares, West Bengal has 1313.20 thousand hectares and Orissa has 660 thousand hectares as compared to 238.90 thousand hectares of Jharkhand involved in vegetable cultivation. Undoubtedly, Jharkhand is under-represented and has a lot of scope for improvement in this area.

For detail concept or for the study of areal inequality of vegetable cultivation area of the state, district wise study depicted highest areal coverage (36121 hectares) in Ranchi district, followed by Hazaribagh (21501 hectares), East Singhbhum (16866), Giridih (16771), Dhanbad (14813), Dumka (14262), Palamu (11133), West Singhbhum (9986). All other districts have less than 10 thousand hectares of area in vegetable growing.

Dr. P.K. Gupta, Director of National Horticultural Research and Development foundation has rightly observed that the overall growth rate of Agriculture in the state is around 9% which had a contribution of 14 percent to the GSDP in 2014-15. About 516 Seed Villages have been established through 24 Krishi Vigyan Kendra (KVK) in year 2016 and 9 KVK have been planned during 2017. The state is now self-sufficient in paddy seed production. It produces about 2669.80 thousand tons of Paddy, thousand tons of wheat, 384.93 thousand tons of Maize, 495.13 thousand tons of pulses and 183.43 thousand tons of oilseeds. Farmers in the state cultivate many varieties of fruits and vegetables, both for domestic consumption and export. Many vegetables including Cauliflower area cultivated round the year in the state. Some high-value vegetables like French-Beans, Capsicum, Chilies and Carrot are also produced in large quantities. In fruits the state has largest area under Mango followed by Jackfruit, Banana, Guava and Lemon. Production-wise Mango and Jackfruit are the highest produced fruits followed by Papaya, Guava, Lemon and Litchi in the state. As per Ministry of Agriculture and Farmer welfare (2016-17), the state produces approximately 1.13% of the total fruits of the country. Now a days Jharkhand is also exporting vegetables nationally and Internationally as well.

## Patterns of Crop Diversification

"Diversification in cropping pattern means raising a variety of crops involving intensity of competition among field crops for arable land: The keener the competition the higher the magnitude of diversification. In the case of mono-culture or crop specialization-The ultra-simple cropping system-the competition for cropped land occupancy is weak. The magnitude of crop diversification mirrors the impact of physical, socio- economic and techno-organizational influents but stronger are the physical environs. A study of the spatial patterns of crop diversification and specialization is of vital importance in understanding the contemporary competition amongst crops for area, scope for rotation and effect on productivity. In general, it appears that keen competition, intensive and judicious crop rotation for the maintenance of soil fecundity and higher profits are associated with mixed cropping system rather than with crop specialization, but the choice of a cropping system in dependent primarily on physical and economic considerations. On the other hand, crop specialization or a monoculture system leads to progressive reduction in yield or to the yield falling below a profitable level". (Singh, 1976, P.309)

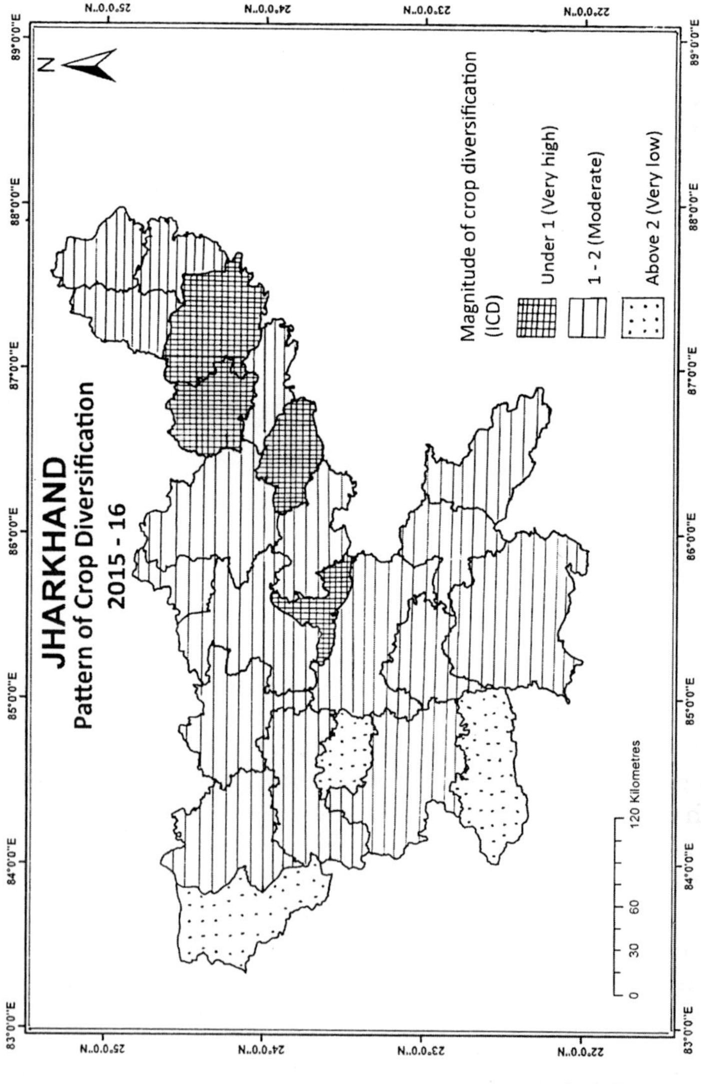


**Map 5.7: Pattern of Crop Diversification (2015-16)**

Agricultural diversification is now almost a normal feature of stable agriculture and progressive farm management in most of the extensive agricultural parts of the world. In fact, this has been made possible by modern irrigation and the use of fertilizers, high yielding varieties, pesticides and mechanization technologies. Besides, there are cultivators to take to crop, livestock or enterprise diversification. First, vagaries of weather compel the cultivators to sow a number of crops or practice a variety of enterprises on their operational holdings so as to get some return under adverse conditions of weather. Second, rural life and the orthodox farm practices force them to obtain most of their domestic requirements from their holdings. (Singh, Dhillon, 1984, P.220)

For complete comprehension of the geography of crops, livestock or agricultural enterprises of a region, the interpretation of their diversification is essential. In view of the importance of these attributes of agriculture, many agricultural geographers have attempted to measure and map them. Nevertheless, the degree of diversification may be measured by relating the number of agricultural elements in an 'established element combination' in a regional unit to the total percentage occupied by those elements in the same unit. Normally, it is assumed that if the number of crops grown in an enumeration unit is large, say about ten, each occupying only 10 percent of the cropped area, it would mean that the cropped area is uniformly distributed among all the ten crops. In such a case, crop diversification is of a very high degree and may be taken as an extreme case of crop diversification. In contrast, if a particular crop occupies 100 percent of the cropped area, then there is no diversification at all. Here crop specialization or monoculture is the rule rather than an exception. It may be inferred that lower the value of the index, higher is the degree of crop diversification and vice versa. In most parts of India, the number of crops grown varies from two to ten or twelve. In case, as noted earlier, there are ten crops grown in a regional unit each occupying more or less a uniform share of land, crop diversification will be of a very high magnitude. This can be explained thus: suppose, in the regional unit 'A' there are six crops occupying 30, 20, 18, 12, 8 and 6 percent of the cropped area respectively, and in the regional unit 'B' three crops cover 67, 20 and 10 percent of the cropped area. The degree of crop diversification in the former case is higher than in the latter. Bhatia (1965) has evolved a simple formula by taking into account the cropped area to make an objective measurement of crop diversification. The formula is:

$$\text{Index of crop diversification} = \frac{\text{Percentage of total harvested area under 'n' crops}}{\text{Number of 'n' crops}}$$

For the study of crop diversification in Jharkhand, Table 5.19 is prepared through the data mentioned in the formula. In the below table total number of crops and total harvested area of that crop has been calculated and ultimately Index of crop diversification (Icd) was inferred. As it has been suggested that lower the Index more the diversification and higher the Index, more the specialization. The result of the cropping Index in the table shows that Dumka, Deoghar, Dhanbad and Ramgarh districts have lowest diversification index which implies higher diversification of crops in these districts. On the other hand, highest index has been recorded for the districts of Lohardaga, Garhwa and Simdega. In these districts the diversification is lowest which means these districts are into more of monoculture. Rest all districts are moderately diversified. No doubt, maximum districts in Jharkhand are neither monocultural nor more diversified.

## Table 5.21
## Crop Diversification in Jharkhand (2015-16)

| District | Area of District | Food Grain (No. and % of area) | Pulses (No. and % of area) | Oil Seeds (No. and % of area) | Cash Crops (No. and % of area) | Fruits (No. and % of area) | Vegetables (No. and % of area) | Total No. of Crops | % of total cropped area | Icd |
|---|---|---|---|---|---|---|---|---|---|---|
| Dumka | 379030 | 5/6.88 | 9/2.2161 | 8/12.694 | 5/2.562 | 9/9.355 | 15/14.400 | 51 | 48.09 | 0.94 |
| Deoghar | 248150 | 5/3.71 | 9/9.295 | 9/7.450 | 3/0.892 | 7/4.337 | 15/10.345 | 48 | 36.01 | 0.75 |
| Godda | 231840 | 5/3.69 | 9/27.845 | 7/16.147 | 4/2.641 | 7/2.182 | 14/4.287 | 46 | 56.77 | 1.23 |
| Pakur | 181700 | 7/3.29 | 9/23.679 | 6/12.160 | 1/1.208 | 8/9.384 | 15/5.526 | 41 | 55.22 | 1.34 |
| Sahebganj | 201750 | 6/3.32 | 6/23.205 | 5/14.689 | 1/2.05 | 9/9.475 | 14/4.691 | 41 | 57.41 | 1.40 |
| Jamtara | 179170 | 5/3.06 | 6/14.55 | 2/3.962 | 1/3.241 | 6/1.736 | 15/11.902 | 35 | 38.44 | 1.09 |
| Hazaribagh | 604630 | 8/5.14 | 9/19.288 | 6/7.766 | 3/2.561 | 9/5.076 | 15/12.860 | 50 | 52.67 | 1.05 |
| Koderma | 130200 | 7/1.59 | 9/18.495 | 6/11.534 | 3/4.215 | 6/1.638 | 15/18.959 | 46 | 56.40 | 1.22 |
| Chatra | 375520 | 8/2.90 | 9/28.595 | 6/16.045 | 1/2.382 | 6/1.590 | 14/5.190 | 44 | 57.14 | 1.29 |
| Giridih | 493220 | 7/5.57 | 8/23.211 | 8/8.544 | 1/3.421 | 8/1.567 | 14/5.740 | 46 | 48.04 | 1.04 |
| Bokaro | 288970 | 6/1.94 | 5/27.112 | 2/0.388 | 2/0.116 | 6/1.474 | 15/7.492 | 36 | 38.51 | 1.06 |
| Dhanbad | 204160 | 8/1.51 | 7/6.022 | 2/5.466 | 1/0.690 | 7/2.793 | 14/4.809 | 39 | 21.27 | 0.54 |
| Ranchi | 622420 | 6/7.87 | 7/30.243 | 3/6.910 | 2/5.301 | 9/9.000 | 15/5.680 | 42 | 65.00 | 1.55 |
| Lohardaga | 153620 | 5/2.33 | 5/17.883 | 4/13.449 | 2/1.256 | 9/6.156 | 15/60.481 | 40 | 101.53 | 2.53 |
| Gumla | 538920 | 6/9.44 | 7/29.319 | 4/15.746 | 2/8.204 | 8/2.331 | 15/14.198 | 42 | 79.21 | 1.88 |
| Palamu | 524690 | 5/2.29 | 8/45.005 | 8/14.848 | 2/9.242 | 7/1.905 | 15/13.68 | 45 | 86.95 | 1.93 |
| Garhwa | 428820 | 5/2.12 | 8/47.885 | 4/13.546 | 2/8.250 | 6/1.205 | 15/19.720 | 40 | 80.71 | 2.01 |
| Simdega | 371630 | 8/4.82 | 9/52.360 | 8/18.745 | 2/9.200 | 7/2.325 | 15/7.961 | 39 | 90.58 | 2.32 |

| District | Area of District | Food Grain (No. and % of area) | Pulses (No. and % of area) | Oil Seeds (No. and % of area) | Cash Crops (No. and % of area) | Fruits (No. and % of area) | Vegetables (No. and % of area) | Total No. of Crops | % of total cropped area | Icd |
|---|---|---|---|---|---|---|---|---|---|---|
| Latehar | 319190 | 8/2.46 | 7/46.788 | 9/18.773 | 2/8.260 | 5/1.810 | 14/3.315 | 45 | 78.68 | 1.74 |
| East Singhbhum | 556690 | 6/5.36 | 8/21.311 | 8/9.184 | 1/5.510 | 6/1.256 | 15/9.112 | 44 | 51.72 | 1.17 |
| West Singhbhum | 562700 | 8/9.15 | 8/24.080 | 9/15.688 | 2/5.230 | 6/1.245 | 15/6.324 | 48 | 61.70 | 1.28 |
| Saraikela | 237230 | 6/5.53 | 7/35.635 | 4/3.313 | 1/1.540 | 6/1.404 | 15/6.620 | 39 | 54.03 | 1.38 |
| Ramgarh | 136008 | 7/1.96 | 9/9.414 | 4/6.295 | 1/2.864 | 9/6.183 | 15/12.370 | 46 | 39.07 | 0.85 |
| Khunti | 175120 | 8/3.77 | 9/24.269 | 9/14.209 | 1/1.165 | 7/1.493 | 15/7.550 | 49 | 52.43 | 1.07 |
| TOTAL | | 155/99.7 | 187/607.70 | 141/267.55 | 47/92.45 | 173/86.91 | 354/273.15 | 1057 | 1427.46 | 1.35 |

## References

Ahmad, Enayat (1965); Bihar: Physical, Economic, Regional Geography, Ranchi University, Ranchi.

Cobley, LS (1962); An Introduction to Botany of Tropical crops, Longmans, London.

Devi, PB, Vijaybharti, R. Satyabhama, S. Malleshi, NG and Priyadarshini, VB (2014); Health Benefits of Finger Millets, Polyphenols and dietary fiber: A review. Journal of food science and technology, 51 (6) 1021-1040.

Gupta, PK (2017-18); Value Chain Study of Tomato of Ranchi, Jharkhand, National Horticultural Research & Development Foundation (NHRDF).

Husain, Majid (1996); Systematic Agricultural Geography, Rawat Publication, Jaipur/New Delhi.

Indian Council of Agricultural Research, New Delhi (1970); Pulse crops of India, PP. 10-11.

Ivern, S.J.F. (1969); Chotanagpur Survey, Indian Social Institute, New Delhi.

Mishra, R.K (2011-12); Agricultural Resource Database of Jharkhand, SAMETI, Dept of Agricultural & Cane Development, Govt of Jharkhand.

Nishar Akhtar, Anwesha Dey, Gufran Ahmad, Firoz Ahmad, Md. Mahtab Rashid (2020); Potential and Economics of Ragi cultivation under changing weather condition of Ranchi, Jharkhand, International Journal of Current Microbiology, Applied Science, 9, 2002.

O'Malley, LSS (1938); Bihar District Gazetteers, Santhal Pargana (Revised by S.C. Mukherjee), Patna.

Prasad, Ayodhya (1973); Chotanagpur: Geography of Rural Settlement, Ranchi University, Ranchi.

Singh, Jasbir (1976); Agricultural Geography of Haryana, Vishal Publications, Kurukshetra University Campus Haryana.

Singh, Jasbir and Dhillon, S.S (1984); Agricultural Geography, Tata Mc Graw Hill Publishing company limited, New Delhi.

Singh, KN & AR Siddique (2012); Economic Geography, Prayag Pustak Bhawan, Allahabad.

Singh, R.K. (2003); Socioeconomic Dynamics of Rice Production System in Jharkhand State, National Agricultural Technology Project, Indian Council of Agri Research, Central Rainfed Upland Rice Research Station (CRRI), Hazaribagh, Jharkhand.

Singh, V.P., Shastri, R.K, ASRAS, Baghel, SS and Choudhry, JL (1999); Rice growing environment in East India: An agroclimatic analysis. International Rice Research Institute, Philippines and Indira Gandhi Agricultural University, Raipur, Chhattisgarh.

Tiwari, R.C. (2010); Bharat ka Bhugol, Prayag Pustak Bhawan, Allahabad.

Tiwari, Ram Kumar (2019); Jharkhand Ki Ruprekha, Bihar Hindi Granth Academy, Patna.

Wolfe, TK and Kipps, MS (1959); Production of crops, Mc Graw Hill Book Company, INC, New York.

# 6

# Agricultural Regions

- Introduction
- Methods of Agricultural Regionalization
- Regions by Crop Concentration
- Regions by Crop Combination
- Regions by Cropping Pattern
- Previous Work

## Introduction

Regionalization is the process of dividing an area into territorial units of uniformities and is the result of a set of processes. A necessity to divide the state of Jharkhand into regions or one can say the regionalization of the state inevitably arises at some stage or other in almost every geographical investigation.

Geographical reality is a complex situation of closely interwoven and diversely interlinked phenomena. Regions may be delimited, when analyzing these phenomena, based on separate features or a set of them. This regionalization elaborated by scientists conform to the existing reality as a taxonomic coordinate - and therefore we can speak about macro, meso and micro regions.

Regionalization owes its status as an important logical element of Geography due to its present development level as a fundamental system of sciences of spatial processes and states occurring in geosphere. However, we have to distinguish between the two approaches to regions while investigating the natural features and criteria of nature and while investigating the socio-production-phenomena (Pokshishevskiy, 1968, P- XXV).

The need of agricultural regionalization can arise at any stage of Geographical investigation of agriculture. It is an important logical fundamental element of explaining spatial processes and variations. Agricultural regionalization is not simply an operation of dividing the area or region into a number of territorial units but is also a method of understanding the agricultural patterns, designed in such a way so as to gain mastery over the processes under study. In addition, it represents the actual method of Geographic perception of the entire agro-spatial diversity of a region designed to identify territorial production units or complexes, and these will then become regional units in agricultural planning with which a complete transformation of agriculture of a region may be accomplished. Regionalization is important specially for large countries like India for the purpose of formulating both short-term and long-term agricultural plans for different units. Broadly, these are two types of regionalization - 1. General purpose regionalization and 2. Special purpose regionalization. The former deals with the delimitation of territorial production complexes, such as economic regions, whereas the latter is associated with the demarcation of some selective resources, such as agricultural regions. Agricultural regions should be delimited for the following two distinct purposes - 1. Academic and 2. Utilitarian.

A purely academic approach leads to further research based on geographical thinking. While the utilitarian approach serves the needs of planning and agricultural administration. Agricultural regionalization is a tool for developing the national economy of a country or state on a regional basis, so planned as to enable each part of the country or state to march at the same pace towards economic prosperity (Singh & Dhilon, 1984, P.209).

Regional differences in growth performance have always attracted considerable attention in academic writing as well as in popular imagination. The fact that different regions and countries have followed different growth trajectories is a stylized fact in development economics. What is perhaps more serious is the consideration that such diverse trajectories have resulted in widening the gap between regions in terms of economic and social opportunity of their people. Hence, we need to look closely at how regional differences shape 'regional inequalities' and result in growth processes that are polarizing, exploitative and undesirable. While some regions have seen substantial improvements in living standards, backwardness is still concentrated in a few regions. Hence agricultural regionalization is essential because it is intimately connected with the eradication of agricultural backwardness by rational allocation of new enterprises. This is so because agricultural development of a region is not an automatic process, rather it is created and stimulated. Besides, agricultural regionalization exhibits the territorial differentiation of agriculture which in turn depends on the structure of agricultural enterprises. The basic purpose of planning, therefore, should be to strive for the optimum development of agricultural resources of different regions for which they are best suited (Singh & Dhillon, 1984, P.210).

## Methods of Agricultural Regionalization

As mentioned by Majid Husain (1996, P.214) the main characteristics of agricultural regions are (i). They have location, (ii). They have transitional boundaries, (iii). They may be either formal or functional and (iv). They may be hierarchically arranged. Since the boundaries of agricultural region are transitional and not sharply dividing lines, their precise delimitation is a difficult task. The main techniques used by geographers for delimitation of agricultural regions are:

(i) Empirical Technique
(ii) Single element technique,
(iii) Multi-element (statistical) technique,
(iv) Quantitative-cum-qualitative technique, and
(v) Normative technique.

**Empirical Technique:** It is largely based on experience of the farmers and observed facts. Baker was the first geographer who adopted this technique and demarcated the agricultural belts of USA. The cotton belt, the corn belt and the wheat belt were demarcated on the basis of observation data. This technique was adopted by Jonsson, Jones Taylor, Valkenburg and Cressey to demarcate different agricultural regions in different parts of the world.

**Single Element Technique:** This technique considers one single element of agriculture to demarcate a region. In this technique the relative position of different agricultural enterprises is taken into consideration. The demarcation of Rice, Wheat and Maize region of any area on the basis of first ranking crops is an illustration of this technique. The main weakness of this technique is that it conceals the position and importance of other crops grown in the region. This is overgeneralized method. One can understand the demerit through this example that when we use this method and prepare a map of first ranking crops, we will observe that we have demarcated Punjab as a wheat region and Western UP as Sugarcane region which conceals the fact that Rice is grown prominently in these areas along with some other crops too.

**Multi-Element or Statistical Method:** This is an improved method than the previously discussed techniques. In statistical technique combination of the closely associated features are taken into consideration. The crop combination and livestock regions as demarcated by Weaver, Doi and Coppock are the examples of this methodology. The main advantage of this technique is that it is free from biasness and does not conceal the agricultural phenomena which are significant. In the absence of reliable data this technique may not depict the ground reality of agricultural landscape of a region.

**Quantitative-cum-Qualitative Methods:** For the demarcation of an agricultural region, when the physical, social and economic factors are taken into consideration, such a technique is known as quantitative-cum-qualitative technique of agricultural regionalization. This technique has been applied by Baker (1926), Whitsley (1936) and Carol (1952). The 14 main factors which should be taken into consideration for the demarcation of agricultural regions include six physical - Relief, Climate, Water, Soil, Sub-soil and Natural

Vegetation; two cultural - cultural vegetation and cultural structures; and six functional - Rural population, cultural and technological stage, farming operation, organization for providing the rural population with economic and cultural goods and commerce.

As discussed at the outset, regionalization is the process of dividing an area into territorial units of uniformities and is the result of a set of processes. The main advantage of agricultural regionalization lies in the fact that it helps in the formulation of agricultural plans which may go a long way in reducing the regional disparities and inequalities.

The agricultural regionalization may be done with the following methods (Husain, 1996, P.216 and Singh & Dhillon, 1984, P.210):

1. Cropping Pattern
2. Crop Combination, Crop Concentration and Crop diversification.
3. Regional Imbalances in Levels of Agricultural Productivity
4. Degree of Commercialization
5. Pattern of crop rotation or crop-sequence analysis.

Here we can work out regionalization of Jharkhand on the basis of above- mentioned methods. However, considering all the methods will be very elaborated and exhaustive which is not the purpose of this chapter; hence we are discussing only three methods for the agricultural demarcation of Jharkhand. These methods are:

1. Crop Concentration
2. Crop Combination and
3. Cropping Pattern (through relative yield and relative spread index)

It would be worth mentioning here that these methods are chosen here as they have not been discussed anywhere else in this thesis. Some methods which have already been discussed in previous chapters such as cropping intensity and crop diversification are not considered here to discard repetition.

## Regions by Crop Concentration

From the concentration of a specific element of agriculture we usually make out that there is a concentration of a recognizable degree of a particular element in an area (Singh & Dhillon, 1984, P.216). But Majid Husain has mentioned that 'crop concentration means the variations in the density of any crop in an area or region at a given point of time (Husain, 1996, P.220). The concentration of crop in an area largely depends on its terrain, temperature, moisture and pedagogical conditions. Each crop has a maximum, minimum and optimum temperature. It has a tendency to have high concentration in the areas of ideal agroclimatic conditions and the density declines as the geographical conditions become less conducive. It is because of the suitability of agroclimatic conditions that maize and gram have high concentration in the western rain-shadow area of Palamu, Garhwa and Latehar while rice dominates in Gumla, West-Singhbhum and Ranchi. Delineation of crop concentration region helps in ascertaining the areas where a particular crop grows well even with the help of minimum inputs, and thus has great significance for agricultural development and planning.

A number of statistical techniques have been evolved and applied for the demarcation of crop concentration regions. The percentage share of a crop in the total cropped area and the determination of relative density with the help of location quotient are some of the techniques that are frequently used for the demarcation of crop concentration regions. The Department of Agriculture, Govt. of India has adopted the following technique for the determination of crop concentration at the local, regional and national level.

SS Bhatia (1965, P. 3-4) has mentioned the below formula in his article to determine the crop concentration index for India.

$$\text{Crop Concentration of 'x' crop} = \frac{\text{Area of 'x' crop in the component areal unit}}{\text{Area of all crops in the component areal unit}} \div \frac{\text{Area of 'x' crop in the entire region}}{\text{Area of all crops in the entire region}}$$

With the help of above formula, the crop concentration index has been calculated for the districts of Jharkhand. Table 6.1 shows the crop concentration of Rice in Jharkhand for the year 2017-18.

**Table 6.1**
**Crop Concentration of Rice in Jharkhand (2017-18)**

| District | Area of Rice in District (000 hectares) | Area of all crops in District (000 hectares) | Crop Concentration Index (Rice) |
|---|---|---|---|
| Bokaro | 32 | 37.910 | 1.00 |
| Chatra | 36 | 56.820 | 0.73 |
| Deoghar | 51 | 72.710 | 0.81 |
| Dhanbad | 42 | 29.650 | 1.64 |
| Dumka | 110 | 134.66 | 0.94 |
| E. Singhbhum | 100 | 105.00 | 1.10 |
| Garhwa | 53 | 41.67 | 1.47 |
| Giridih | 82 | 109.12 | 0.87 |
| Godda | 38 | 72.42 | 0.61 |
| Gumla | 180 | 184.45 | 1.13 |
| Hazaribagh | 84 | 100.60 | 0.97 |
| Jamtara | 47 | 60.03 | 0.91 |
| Khunti | 76 | 73.95 | 1.19 |
| Koderma | 16 | 31.13 | 0.59 |
| Latehar | 25 | 48.31 | 0.60 |
| Lohardaga | 45 | 63.49 | 0.82 |
| Pakur | 49 | 64.42 | 0.88 |
| Palamu | 46 | 44.83 | 1.19 |
| Ramgarh | 33 | 38.45 | 0.99 |
| Ranchi | 169 | 154.26 | 1.27 |
| Sahebganj | 40 | 65.00 | 0.71 |
| Saraikela | 97 | 108.30 | 1.04 |
| Simdega | 85 | 94.49 | 1.04 |
| West Singhbhum | 168 | 179.25 | 1.08 |
| Jharkhand | 1704 | 1971.06 | 1.19 |

By applying the above technique, index value was calculated with a simple calculation like location quotient. This formula is exactly location quotient and nothing more. The value has been arranged either in ascending or descending order. If higher the value of crop concentration index, higher is the level of interest in the production of that

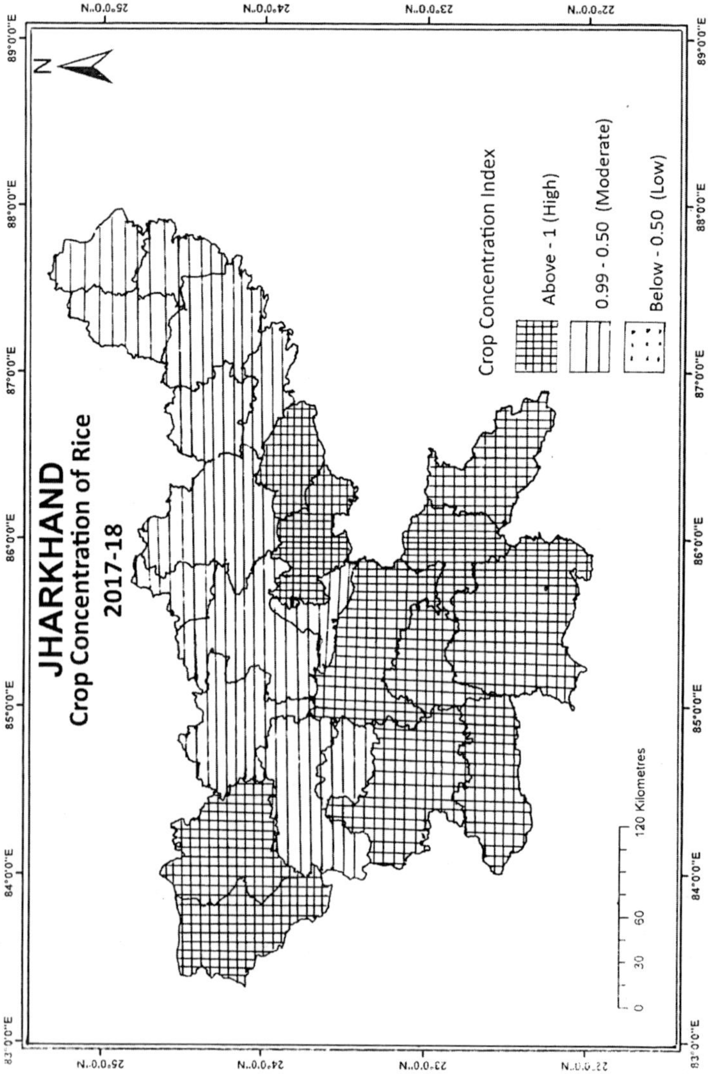


**Map 6.1: Crop Concentration of Rice (2017-18)**

crop. Result shows that highest concentration of crop (Rice) is in Ranchi, Gumla, Simdega, East Singhbhum, West Singhbhum, Saraikela, Palamu, Garhwa, Dhanbad, Bokaro and Khunti. All these districts cover 67 percent area of the total rice cropping area in the state. These districts have cropping index of more than 1.00. In second category where the index value is in between 0.50 and 0.99 is called moderate value, these districts are Lohardaga, Latehar, Hazaribagh, Chatra, Koderma, Giridih, Dumka, Jamtara, Deoghar, Godda, Sahebganj, Pakur and Ramgarh. In lowest category where the Index value is less than 0.50, we do not see any districts in the state. So, there are only two categories based on crop concentration in Jharkhand.

The main advantage of the crop concentration lies in the fact that it enables the geographers and planners to understand the area of specialization of different crops grown in a region at a given point of time.

Temporal study of crop concentration of Rice has shifting tendency in Jharkhand. For this study we calculated the Crop concentration Index of Rice for years 2004 and 2010. Both calculations gave a different result.

Here is a comparison between the results for both the years.

**Crop Concentration during 2004-05**

| | |
|---|---|
| Index Value > 1 | Ranchi, Gumla, Simdega, W. Singhbhum, Saraikela, Dhanbad, Bokaro, Dumka, Jamtara, Godda |
| Index Value = 0.50-0.99 | Lohardaga, E. Singhbhum, Hazaribagh, Giridih, Deoghar, Sahebganj, Pakur |
| Index Value < 0.50 | Palamu, Latehar, Garhwa, Chatra, Koderma |

**Crop Concentration during 2009-10**

| | |
|---|---|
| Index Value > 1 | Ranchi, Gumla, Simdega, E. Singhbhum, W. Singhbhum, Saraikela, Dumka, Jamtara, Godda |

| | |
|---|---|
| Index Value = 0.50-0.99 | Simdega, Hazaribagh, Chatra, Koderma, Giridih, Dhanbad, Bokaro, Deoghar, Sahebganj, Pakur |
| Index Value < 0.50 | Palamu, Latehar, Garhwa, Lohardaga |

Above three calculations give a detailed picture of crop concentration of rice spanning over thirteen years (2004-17). The data shows that Ranchi, Gumla, West Singhbhum and Saraikela are the regular high rice yielding area. These four districts are always under the high concentration area. Shifting tendency indicate the interest of cultivator is increasing towards rice cultivation in the state.

Similarly, crop concentration for second crop of the state - 'Maize' has been calculated in the table 6.2. Results has been depicted in Map 6.2. Highest Maize crop concentration has been recorded in Latehar, Garhwa and Palamu. These districts are in the whole Palamu Division. Climatically this is a suitable area for Maize cultivation along with other Kharif crops. About 13 districts (more than half of the state) recorded moderate maize crop concentration which is more than the average (1.0) of the state. These districts have sufficient maize producing area and cultivators have keen interest in this crop. Low cropping concentration has been recorded in Dhanbad, Gumla, Hazaribagh, Khunti, Ranchi, Saraikela, Simdega and West Singhbhum. These districts have least interest in Maize crop, inclined towards either Rice or Wheat.

Third main food crop of the state is Wheat by production and the hectarage both. Crop concentration Index (Wheat) has been recorded highest in Koderma, Garhwa and Godda. These three districts are located in three different direction in the state - North, West and East respectively. This shows that there is no particular area of wheat cultivation. It depends upon just favorable irrigation facilities which are provided by the government which resulted in cultivators show interest in cultivating wheat. Upon keen observation I intend to say that these three districts are the three growth centers

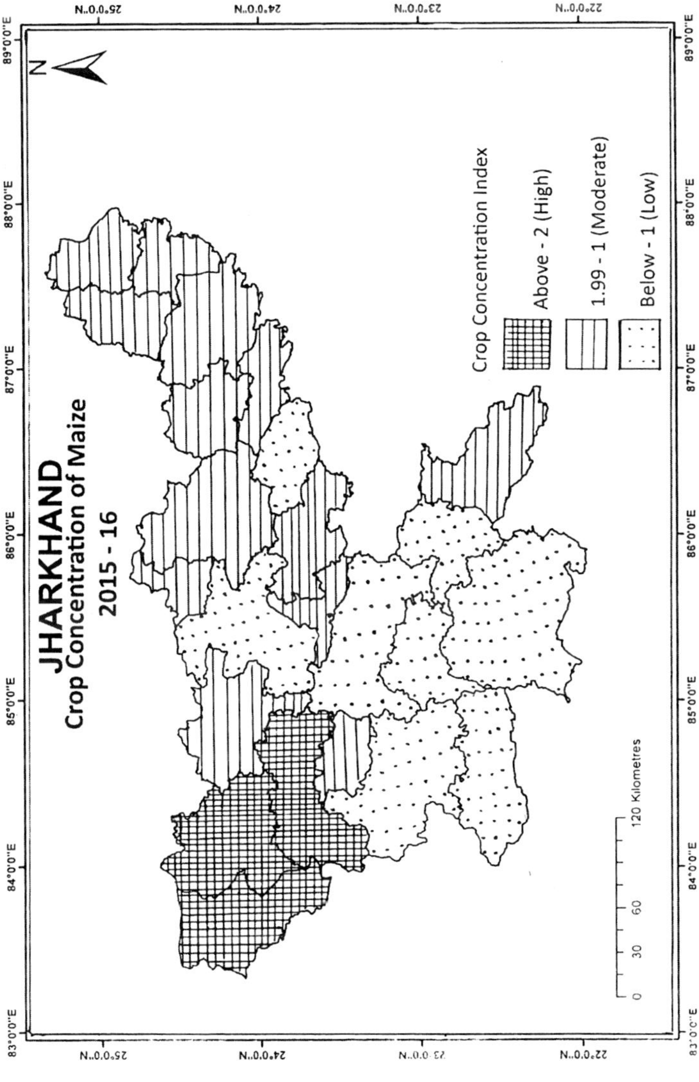


**Map 6.2: Crop Concentration of Maize (2015-16)**

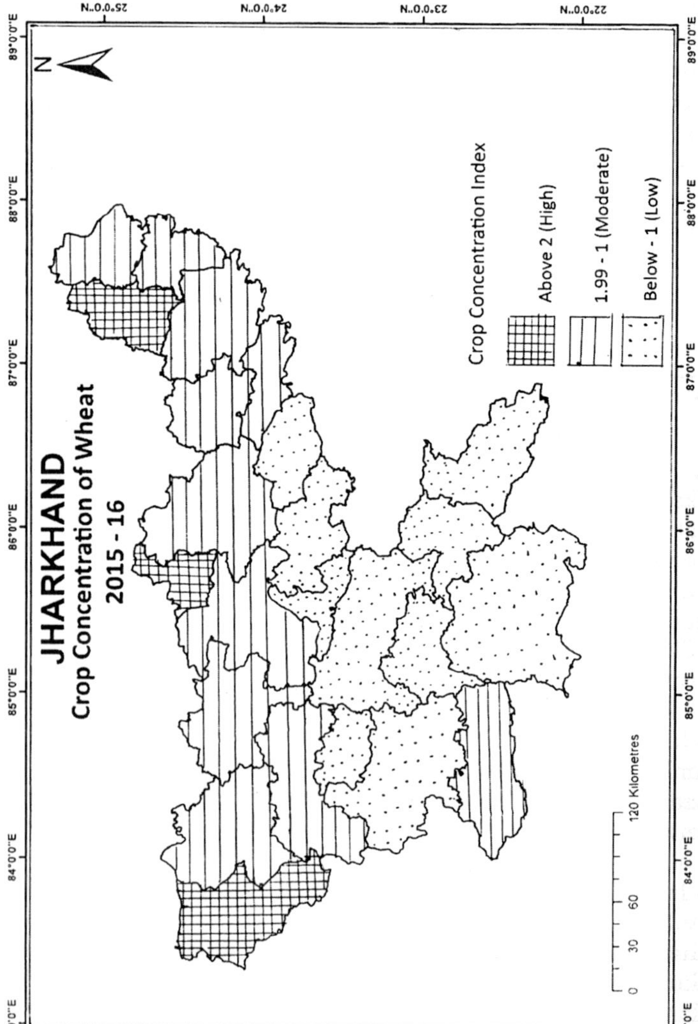


**Map 6.3: Crop Concentration of Wheat (2015-16)**

for wheat cultivation. These districts and have concentration value of more than 2 and certain districts around them have concentration value of 1. In other words, area with Index value 2 is surrounded by areas of Index value 1. It means Index value 2 is the growth center as wheat growing area.For example, Garhwa (2) is surrounded by Palamu (1.72) and Latehar (1.44); Koderma (2) is surrounded by Chatra (1.84), Hazaribagh (1.12) and Giridih (1.12) and Godda is surrounded by Sahebganj (1.70), Pakur (1.51) and Dumka (1.06). No Doubt, if irrigation, high yield seeds, technical equipment facilities will be provided to the cultivators of these areas, it will grow as wheat basket of Jharkhand in near future.

**Table 6.2**
**Crop Concentration of Maize in Jharkhand (2015-16)**

| District | Area of Rice in District (000 hectares) | Area of all crops in District (000 hectares) | Crop Concentration Index (Maize) |
|---|---|---|---|
| Bokaro | 7.65 | 41.07 | 1.33 |
| Chatra | 13.92 | 58.59 | 1.70 |
| Deoghar | 17.00 | 75.41 | 1.61 |
| Dhanbad | 2.19 | 40.52 | 0.39 |
| Dumka | 20.07 | 138.84 | 1.03 |
| E. Singhbhum | 11.33 | 62.69 | 1.29 |
| Garhwa | 25.66 | 82.54 | 2.22 |
| Giridih | 18.52 | 115.08 | 1.15 |
| Godda | 12.49 | 71.21 | 1.25 |
| Gumla | 6.04 | 188.37 | 0.23 |
| Hazaribagh | 11.11 | 98.62 | 0.80 |
| Jamtara | 9.90 | 65.56 | 1.08 |
| Khunti | 5.94 | 79.18 | 0.54 |
| Koderma | 8.67 | 32.13 | 1.93 |
| Latehar | 22.14 | 52.92 | 2.99 |
| Lohardaga | 6.15 | 42.40 | 1.04 |
| Pakur | 9.56 | 66.29 | 1.03 |
| Palamu | 26.22 | 70.71 | 2.65 |
| Ramgarh | 9.03 | 42.52 | 1.52 |
| Ranchi | 11.19 | 181.30 | 0.44 |
| Sahebganj | 11.24 | 61.81 | 1.30 |
| Saraikela | 5.91 | 106.55 | 0.40 |

| | | | |
|---|---|---|---|
| Simdega | 8.84 | 104.87 | 0.60 |
| West Singhbhum | 7.28 | 171.82 | 0.30 |
| Jharkhand | 288.05 | 2051.00 | 1.00 |

**Table 6.3**
**Crop Concentration of Wheat in Jharkhand (2015-16)**

| District | Area of Rice in District (000 hectares) | Area of all crops in District (000 hectares) | Crop Concentration Index (Wheat) |
|---|---|---|---|
| Bokaro | 1.15 | 41.07 | 0.35 |
| Chatra | 8.42 | 58.59 | 1.84 |
| Deoghar | 8.66 | 75.41 | 1.47 |
| Dhanbad | 0.91 | 40.52 | 0.29 |
| Dumka | 11.45 | 138.84 | 1.06 |
| E. Singhbhum | 3.90 | 62.69 | 0.80 |
| Garhwa | 17.50 | 82.54 | 2.72 |
| Giridih | 10.01 | 115.08 | 1.12 |
| Godda | 12.49 | 71.21 | 2.24 |
| Gumla | 7.12 | 188.37 | 0.48 |
| Hazaribagh | 8.65 | 98.62 | 1.12 |
| Jamtara | 6.66 | 65.56 | 1.30 |
| Khunti | 3.00 | 79.18 | 0.49 |
| Koderma | 6.45 | 32.13 | 2.57 |
| Latehar | 5.95 | 52.92 | 1.44 |
| Lohardaga | 2.02 | 42.40 | 0.61 |
| Pakur | 7.83 | 66.29 | 1.51 |
| Palamu | 9.46 | 70.71 | 1.72 |
| Ramgarh | 2.13 | 42.52 | 0.64 |
| Ranchi | 6.36 | 181.30 | 0.45 |
| Sahebganj | 8.19 | 61.81 | 1.70 |
| Saraikela | 1.60 | 106.55 | 0.19 |
| Simdega | 8.79 | 104.55 | 0.19 |
| West Singhbhum | 1.52 | 171.82 | 0.11 |
| Jharkhand | 161.11 | 2051.00 | 1.00 |

## Regions by Crop Combination

The crop combination is a method of analysis used for the delineation of agricultural regions. The foundation of the method is the realization that single-crop regions are rare and that in any one area farmers usually grow a variety of crops of different degrees of importance. Crop combination

Table 6.4
**Ranking of Main Crops and their percentage for Crop Combination**

| District | Net Sown Area (000 hect.) | Rice (hect.) | Wheat (hect.) | Maize (hect) | Pulses (hect.) | Rice (%) | Pulses (%) | Wheat (%) | Maize (%) |
|---|---|---|---|---|---|---|---|---|---|
| Ranchi | 256 | 169 | 12 | 11 | 29 | 66 | 11 | 06 | 04 |
| Khunti | 276 | 76 | 06 | 07 | 16 | 28 | 05 | 02 | 03 |
| Gumla | 162 | 180 | 07 | 07 | 25 | 111 | 15 | 02 | 04 |
| Simdega | 82 | 85 | 09 | 08 | 59 | 104 | 72 | 11 | 10 |
| Lohardaga | 60 | 45 | 03 | 06 | 10 | 75 | 16 | 05 | 10 |
| Garhwa | 65 | 53 | 19 | 21 | 21 | 81 | 32 | 29 | 32 |
| Palamu | 98 | 46 | 11 | 27 | 47 | 46 | 47 | 11 | 27 |
| Latehar | 60 | 25 | 08 | 19 | 32 | 41 | 53 | 13 | 31 |
| E. Singhbhum | 83 | 100 | 06 | 12 | 26 | 120 | 31 | 07 | 14 |
| W. Singhbhum | 146 | 168 | 01 | 07 | 29 | 161 | 48 | 07 | 14 |
| Saraikela | 60 | 97 | 04 | 07 | 29 | 161 | 48 | 07 | 12 |
| Hazaribagh | 106 | 84 | 12 | 11 | 26 | 79 | 24 | 11 | 10 |
| Ramgarh | 21 | 33 | 04 | 09 | 12 | 157 | 57 | 19 | 42 |
| Chatra | 45 | 36 | 11 | 15 | 18 | 80 | 40 | 24 | 33 |
| Koderma | 17 | 16 | 07 | 08 | 10 | 94 | 58 | 41 | 47 |
| Giridih | 78 | 82 | 10 | 21 | 13 | 105 | 16 | 13 | 26 |
| Dhanbad | 35 | 42 | 03 | 02 | 05 | 120 | 14 | 09 | 06 |
| Bokaro | 16 | 32 | 03 | 09 | 14 | 200 | 14 | 09 | 06 |
| Dumka | 119 | 110 | 12 | 21 | 20 | 92 | 17 | 10 | 18 |

| | | | | | | | | | |
|---|---|---|---|---|---|---|---|---|---|
| Deoghar | 64 | 51 | 09 | 15 | 10 | 79 | 16 | 14 | 23 |
| Jamtara | 55 | 47 | 06 | 12 | 05 | 85 | 09 | 11 | 21 |
| Godda | 78 | 38 | 12 | 12 | 18 | 48 | 23 | 15 | 15 |
| Sahebganj | 42 | 40 | 12 | 10 | 16 | 95 | 38 | 29 | 23 |
| Pakur | 58 | 49 | 07 | 10 | 12 | 84 | 21 | 12 | 17 |

**Source:** Directorate of Agriculture, Jharkhand, Ranchi

**Table 6.5**
**Calculation of Crop Combination-Jharkhand**

| Number of crops in the combination scheme | Percent of Hypothetical Agricultural Area | Percent of Actual Cropping Area | Difference (δ) $\delta^2$ | Square of Difference | Total of Square |
|---|---|---|---|---|---|
| **Ranchi District** | | | | | |
| One Crop Combination | 100 | 66 | 34 | 1156 | 1156 |
| Two Crop Combination | 50.00 | 66 | -16 | 256 | 1777 |
| | 50.00 | 11 | 39 | 1521 | |
| Three Crop Combination | 33.33 | 66 | -32.67 | 1067.32 | |
| | 33.33 | 11 | 22.33 | 489.62 | 2434.28 |
| | 33.33 | 04 | 29.33 | 877.34 | |

| | | | | | |
|---|---|---|---|---|---|
| Four Crop Combination | 25.00 | 66 | 41 | 1681 | 2759 |
| | 25.00 | 11 | 14 | 196 | |
| | 25.00 | 04 | 21 | 441 | |
| | 25.00 | 04 | 21 | 441 | |
| **Khunti District** | | | | | |
| One Crop Combination | 100 | 28 | 72 | 5184 | 5184 |
| Two Crop Combination | 50.00 | 28 | 22 | 484 | 2509 |
| | 50.00 | 05 | 45 | 2025 | |
| Three Crop Combination | 33.33 | 28 | 5.33 | 28.40 | 1812.36 |
| | 33.33 | 05 | 28.33 | 802.58 | |
| | 33.33 | 02 | 31.33 | 981.56 | |
| Four Crop Combination | 25.00 | 28 | 03 | 09 | 1422 |
| | 25.00 | 05 | 20 | 400 | |
| | 25.00 | 02 | 23 | 529 | |
| | 25.00 | 03 | 32 | 484 | |
| **Gumla District** | | | | | |
| One Crop Combination | 100 | 111 | -11 | 121 | 121 |
| Two Crop Combination | 50.00 | 111 | 61 | 3721 | 4946 |
| | 50.00 | 15 | 35 | 1225 | |

| | | | | | |
|---|---|---|---|---|---|
| Three Crop Combination | 33.33 | 111 | -77.67 | 6032.62 | |
| | 33.33 | 15 | -18.33 | 335.98 | 7350.16 |
| | 33.33 | 02 | 31.33 | 981.56 | |
| Four Crop Combination | 25.00 | 111 | -86 | 7396 | |
| | 25.00 | 15 | 10 | 100 | 8466 |
| | 25.00 | 02 | 23 | 529 | |
| | 25.00 | 04 | 21 | 441 | |
| **Simdega District** | | | | | |
| One Crop Combination | 100 | 104 | -4 | 16 | 16 |
| Two Crop Combination | 50.00 | 104 | -54 | 2916 | 3400 |
| | 50.00 | 72 | -22 | 484 | |
| Three Crop Combination | 33.33 | 104 | -70.67 | 4994.24 | |
| | 33.33 | 72 | -38.67 | 1495.36 | 6641.62 |
| | 33.33 | 11 | 12.33 | 152.02 | |
| Four Crop Combination | 25.00 | 104 | 79 | 521 | |
| | 25.00 | 72 | 47 | 2209 | 2946 |
| | 25.00 | 11 | 14 | 96 | |
| | 25.00 | 10 | 15 | 225 | |
| **Lohardaga District** | | | | | |
| One Crop Combination | 100 | 75 | 25 | 625 | 625 |

| Two Crop Combination | 50.00 | 75 | 25 | 625 | 1781 |
|---|---|---|---|---|---|
| | 50.00 | 16 | 34 | 1156 | |
| Three Crop Combination | 33.33 | 75 | -4.67 | 21.8 | |
| | 33.33 | 16 | 17.33 | 300 | 1124.30 |
| | 33.33 | 05 | 28.33 | 802.50 | |
| Four Crop Combination | 25.00 | 75 | -50 | 2500 | |
| | 25.00 | 16 | 9 | 81 | 3206 |
| | 25.00 | 05 | 20 | 400 | |
| | 25.00 | 10 | 15 | 225 | |
| **Garhwa District** | | | | | |
| One Crop Combination | 100 | 81 | 19 | 361 | 361 |
| Two Crop Combination | 50.00 | 81 | -31 | 961 | 1285 |
| | 50.00 | 32 | 18 | 324 | |
| Three Crop Combination | 33.33 | 81 | -47.67 | 6527.57 | |
| | 33.33 | 32 | 1.33 | 18.74 | 6548.07 |
| | 33.33 | 29 | 4.33 | 18.74 | |
| Four Crop Combination | 25.00 | 81 | -56 | 6536 | |
| | 25.00 | 32 | 49 | 2401 | 11594 |
| | 25.00 | 29 | 16 | 256 | |
| | 25.00 | 32 | 49 | 2401 | |

| | | | | | |
|---|---|---|---|---|---|
| **Palamu District** | | | | | |
| One Crop Combination | 100 | 46 | 54 | 2916 | 2916 |
| Two Crop Combination | 50.00 | 46 | 04 | 16 | 25 |
| | 50.00 | 47 | 03 | 09 | |
| Three Crop Combination | 33.33 | 46 | -12.67 | 160.52 | |
| | 33.33 | 47 | -13.67 | 186.86 | 846 |
| | 33.33 | 11 | 22.23 | 498.62 | |
| Four Crop Combination | 25.00 | 46 | -21 | 441 | |
| | 25.00 | 47 | -22 | 484 | 948 |
| | 25.00 | 11 | 14 | 196 | |
| | 25.00 | 27 | -2 | 04 | |
| **Latehar District** | | | | | |
| One Crop Combination | 100 | 41 | 59 | 3481 | 3481 |
| Two Crop Combination | 50.00 | 41 | 09 | 81 | 90 |
| | 50.00 | 53 | -3 | 09 | |
| Three Crop Combination | 33.33 | 41 | -7.67 | 58.82 | |
| | 33.33 | 53 | -19.67 | 386.9 | 859.02 |
| | 33.33 | 13 | 20.33 | 413.30 | |

| | | | | | |
|---|---|---|---|---|---|
| Four Crop Combination | 25.00 | 41 | -16 | 256 | |
| | 25.00 | 53 | -31 | 961 | 1397 |
| | 25.00 | 13 | 12 | 144 | |
| | 25.00 | 31 | -6 | 36 | |
| | **East Singhbhum District** | | | | |
| One Crop Combination | 100 | 120 | -20 | 400 | 400 |
| Two Crop Combination | 50.00 | 120 | -70 | 4900 | 5216 |
| | 50.00 | 31 | 19 | 361 | |
| Three Crop Combination | 33.33 | 120 | -86.67 | 7511.68 | |
| | 33.33 | 31 | 2.33 | 5.42 | 8210.38 |
| | 33.33 | 07 | 26.33 | 693.28 | |
| Four Crop Combination | 25.00 | 120 | -95 | 9025 | |
| | 25.00 | 31 | 6 | 36 | 9506 |
| | 25.00 | 7 | 18 | 324 | |
| | 25.00 | 14 | 11 | 121 | |
| | **West Singhbhum District** | | | | |
| One Crop Combination | 10 | 115 | -15 | 225 | 225 |
| Two Crop Combination | 50.00 | 115 | -65 | 4225 | 2509 |
| | 50.00 | 13 | 37 | 1369 | |

| | | | | | |
|---|---|---|---|---|---|
| Three Crop Combination | 33.33 | 115 | -81.67 | 6669.98 | |
| | 33.33 | 13 | 10.33 | 106.70 | 7523.60 |
| | 33.33 | 6 | 27.33 | 746.92 | |
| Four Crop Combination | 25.00 | 115 | 90 | 8100 | |
| | 25.00 | 13 | 12 | 144 | 9005 |
| | 25.00 | 6 | 19 | 361 | |
| | 25.00 | 5 | 20 | 400 | |
| **Saraikela District** | | | | | |
| One Crop Combination | 100 | 161 | -61 | 3721 | 3721 |
| Two Crop Combination | 50.00 | 161 | -111 | 12321 | 12325 |
| | 50.00 | 48 | 02 | 04 | |
| Three Crop Combination | 33.33 | 161 | -127.67 | 16229 | |
| | 33.33 | 48 | -14.67 | 215.20 | 17207.46 |
| | 33.33 | 07 | 26.33 | 693.26 | |
| Four Crop Combination | 25.00 | 161 | -136 | 18496 | |
| | 25.00 | 48 | -23 | 529 | 19518 |
| | 25.00 | 07 | 18 | 324 | |
| | 25.00 | 12 | 13 | 169 | |
| **Hazaribagh District** | | | | | |
| One Crop Combination | 100 | 79 | 21 | 441 | 441 |

| | | | | | |
|---|---|---|---|---|---|
| Two Crop Combination | 50.00 | 79 | -29 | 841 | 1517 |
| | 50.00 | 24 | 26 | 676 | |
| Three Crop Combination | 33.33 | 79 | -45.67 | 2085.74 | |
| | 33.33 | 24 | -9.33 | 87.04 | 2670.66 |
| | 33.33 | 11 | 22.33 | 498.62 | |
| Four Crop Combination | 25.00 | 79 | -54 | 2916 | |
| | 25.00 | 24 | 01 | 01 | 3388 |
| | 25.00 | 11 | 14 | 196 | |
| | 25.00 | 10 | 15 | 225 | |
| **Ramgarh District** | | | | | |
| One Crop Combination | 100 | 157 | -57 | 3249 | 3249 |
| Two Crop Combination | 50.00 | 157 | -107 | 11449 | 11498 |
| | 50.00 | 57 | -7 | 49 | |
| Three Crop Combination | 33.33 | 157 | -123.67 | 15294.26 | |
| | 33.33 | 57 | -23.67 | 560.26 | 16059.86 |
| | 33.33 | 19 | 14.33 | 205.34 | |
| Four Crop Combination | 25.00 | 157 | -132 | 17424 | |
| | 25.00 | 57 | -32 | 1024 | 18773 |
| | 25.00 | 19 | 6 | 36 | |
| | 25.00 | 42 | -17 | 289 | |

| | | | | | |
|---|---|---|---|---|---|
| | | **Chatra District** | | | |
| One Crop Combination | 100 | 80 | 20 | 400 | 400 |
| Two Crop Combination | 50.00 | 80 | -30 | 900 | 1000 |
| | 50.00 | 40 | 10 | 100 | |
| Three Crop Combination | 33.33 | 80 | -46.67 | 2178 | |
| | 33.33 | 40 | -6.67 | 44.4 | 2309.4 |
| | 33.33 | 24 | 9.33 | 87.04 | |
| Four Crop Combination | 25.00 | 80 | -55 | 3025 | |
| | 25.00 | 40 | -15 | 225 | 3315 |
| | 25.00 | 24 | 1 | 1 | |
| | 25.00 | 33 | 8 | 64 | |
| | | **Dumka District** | | | |
| One Crop Combination | 100 | 92 | 8 | 64 | 64 |
| Two Crop Combination | 50.00 | 92 | -41 | 1764 | 2853 |
| | 50.00 | 17 | 33 | 1089 | |
| Three Crop Combination | 33.33 | 92 | -58.67 | 3442.16 | |
| | 33.33 | 17 | 16.33 | 266.66 | 4122.12 |
| | 33.33 | 10 | 20.33 | 413.30 | |

| | | | | | |
|---|---|---|---|---|---|
| Four Crop Combination | 25.00 | 92 | -67 | 4489 | |
| | 25.00 | 17 | 8 | 64 | 4827 |
| | 25.00 | 10 | 15 | 225 | |
| | 25.00 | 18 | 7 | 49 | |
| **Deoghar District** | | | | | |
| One Crop Combination | 100 | 79 | 21 | 441 | 441 |
| Two Crop Combination | 50.00 | 79 | -29 | 841 | 1997 |
| | 50.00 | 16 | 34 | 1156 | |
| Three Crop Combination | 33.33 | 79 | -45.67 | 2085 | |
| | 33.33 | 16 | 17.33 | 300.32 | 2758 |
| | 33.33 | 14 | 19.3 | 373.64 | |
| Four Crop Combination | 25.00 | 79 | -54 | 2916 | |
| | 25.00 | 16 | 9 | 81 | 3122 |
| | 25.00 | 14 | 11 | 121 | |
| | 25.00 | 23 | 2 | 4 | |
| **Jamtara District** | | | | | |
| One Crop Combination | 100 | 85 | 15 | 225 | 225 |
| Two Crop Combination | 50.00 | 85 | -35 | 1225 | 2906 |
| | 50.00 | 9 | 41 | 1681 | |

| | | | | | |
|---|---|---|---|---|---|
| Three Crop Combination | 33.33 | 85 | -51.67 | 2669.78 | |
| | 33.33 | 9 | 24.33 | 591.94 | 3760.34 |
| | 33.33 | 11 | 22.33 | 498.62 | |
| Four Crop Combination | 25.00 | 35 | 60 | 3600 | |
| | 25.00 | 9 | 16 | 256 | 4068 |
| | 25.00 | 11 | 14 | 196 | |
| | 25.00 | 21 | 4 | 16 | |
| **Godda District** | | | | | |
| One Crop Combination | 100 | 48 | 52 | 2204 | 2204 |
| Two Crop Combination | 50.00 | 48 | 2 | 4 | 7303 |
| | 50.00 | 23 | 27 | 729 | |
| Three Crop Combination | 33.33 | 48 | -14.67 | 215.20 | |
| | 33.33 | 23 | 10.33 | 106.70 | 657 |
| | 33.33 | 15 | 18.33 | 335.98 | |
| Four Crop Combination | 25.00 | 48 | 23 | 529 | |
| | 25.00 | 23 | 2 | 4 | 733 |
| | 25.00 | 15 | 10 | 100 | |
| | 25.00 | 15 | 10 | 100 | |
| **Koderma District** | | | | | |
| One Crop Combination | 100 | 94 | 6 | 36 | 36 |

| | | | | | |
|---|---|---|---|---|---|
| Two Crop Combination | 50.00 | 94 | 44 | 1936 | 2000 |
| | 50.00 | 58 | 8 | 64 | |
| Three Crop Combination | 33.33 | 94 | -60.67 | 3680.84 | |
| | 33.33 | 58 | -24.67 | 608.60 | 4348.26 |
| | 33.33 | 41 | -7.67 | 58.82 | |
| Four Crop Combination | 25.00 | 94 | -69 | 4761 | |
| | 25.00 | 58 | -33 | 1089 | 6590 |
| | 25.00 | 41 | -16 | 256 | |
| | 25.00 | 47 | -22 | 484 | |
| **Giridih District** | | | | | |
| One Crop Combination | 100 | 105 | -5 | 25 | 25 |
| Two Crop Combination | 50.00 | 105 | -55 | 3025 | 4181 |
| | 50.00 | 16 | 34 | 1156 | |
| Three Crop Combination | 33.33 | 105 | 71.67 | 5136.58 | |
| | 33.33 | 16 | 17.33 | 300 | 5849 |
| | 33.33 | 13 | 20.33 | 413 | |
| Four Crop Combination | 25.00 | 105 | -80 | 6400 | |
| | 25.00 | 16 | 9 | 81 | 6626 |
| | 25.00 | 13 | 12 | 144 | |
| | 25.00 | 26 | -1 | 1 | |

| Dhanbad District | | | | | |
|---|---|---|---|---|---|
| One Crop Combination | 100 | 120 | 20 | 400 | 400 |
| Two Crop Combination | 50.00 | 120 | 70 | 4900 | 6196 |
| | 50.00 | 14 | 36 | 1296 | |
| Three Crop Combination | 33.33 | 120 | 86.67 | 7511.68 | |
| | 33.33 | 14 | 19.33 | 373.64 | 9204.32 |
| | 33.33 | 9 | 36.33 | 1319.86 | |
| Four Crop Combination | 25.00 | 120 | 35 | 9025 | |
| | 25.00 | 14 | 11 | 121 | 9763 |
| | 25.00 | 9 | 16 | 256 | |
| | 25.00 | 6 | 19 | 361 | |
| **Bokaro District** | | | | | |
| One Crop Combination | 100 | 200 | -100 | 10000 | 10000 |
| Two Crop Combination | 50.00 | 200 | -150 | 22500 | 23869 |
| | 50.00 | 87 | 37 | 1369 | |
| Three Crop Combination | 33.33 | 200 | -166.67 | 27778.88 | |
| | 33.33 | 87 | -53.67 | 2880.46 | 30864.68 |
| | 33.33 | 19 | 14.33 | 205.34 | |

| | | | | | |
|---|---|---|---|---|---|
| Four Crop Combination | 25.00 | 200 | -175 | 30625 | |
| | 25.00 | 87 | -62 | 3844 | 34866 |
| | 25.00 | 19 | 6 | 36 | |
| | 25.00 | 6 | 19 | 361 | |
| **Sahebganj District** | | | | | |
| One Crop Combination | 100 | 95 | 5 | 25 | 25 |
| Two Crop Combination | 50.00 | 95 | -45 | 2025 | 2169 |
| | 50.00 | 38 | 12 | 144 | |
| Three Crop Combination | 33.33 | 95 | -61.67 | 3803.18 | |
| | 33.33 | 38 | -4.67 | 21.80 | 3846.78 |
| | 33.33 | 29 | 4.33 | 21.80 | |
| Four Crop Combination | 25.00 | 95 | 70 | 4900 | |
| | 25.00 | 38 | -13 | 169 | 5089 |
| | 25.00 | 29 | -4 | 16 | |
| | 25.00 | 23 | 2 | 4 | |
| **Pakur District** | | | | | |
| One Crop Combination | 100 | 84 | 16 | 256 | 256 |
| Two Crop Combination | 50.00 | 84 | 34 | 4456 | 1997 |
| | 50.00 | 21 | 29 | 841 | |

| | | | | | |
|---|---|---|---|---|---|
| Three Crop Combination | 33.33 | 84 | -50 | 2500 | |
| | 33.33 | 21 | 12.33 | 152 | 3106.96 |
| | 33.33 | 12 | 21.33 | 454 | |
| Four Crop Combination | 25.00 | 84 | 59 | 3481 | |
| | 25.00 | 21 | 4 | 16 | 3730 |
| | 25.00 | 12 | 13 | 169 | |
| | 25.00 | 17 | 8 | 64 | |

regions are designed on the basis of a statistical assessment of land use patterns. The method was first elaborated by J.C. Weaver (1954, P. 175- 200).

The percentage of land occupied by each crop is ascertained for each sub area and the crops are ranked from greatest to lowest percentage. The actual percentages of crop land are compared to the theoretical percentage that would occur if particular crop combination were present (e.g., monoculture=100 percent in one crop; four crop combinations = 25 percent in each of four crops). The differences between the actual and theoretical distributions are compared using different possible theoretical distributions and for each comparison the result is quantified as:

$$\delta^2 = \sum \frac{\delta^2}{n}$$

'Where $\delta$ is the difference between the actual crop percentage and the appropriate theoretical percentage, and n is the number of crops in the combination tested. The minimum value of $\delta^2$ indicates the most representative crop combination. D. Thomas (1963) has modified the statistical equation to eliminate the division by n by incorporating all crops throughout the analysis of $\delta^2$, Weaver had only used the crops with ranks up to that appropriate to the theoretical combination being tested.' (Johnston, R.J, 1986, P84)

Using the table 6.4, table 6.5 has been calculated where we get difference between hypothetical and actual are. Then we get the square of the difference and the total value of difference.

**Table 6.6**
**Crop Combination Regions of Jharkhand**

| Crop Combination Region | Combination Crops | District Name | Percentage Area of Districts |
|---|---|---|---|
| One Crop Combination | Rice | Garhwa, Chatra, Hazaribagh, Koderma, | 80% |

| | | | |
|---|---|---|---|
| | | Deoghar, Dumka, Ramgarh, Pakur, Sahebganj, Jamtara, Lohardaga, Dhanbad, Gumla, Simdega, East Singhbhum, West Singhbhum, Giridih, Ranchi, Bokaro, Saraikela | |
| Two Crop Combination | Rice-Pulses | Palamu, Latehar | 10% |
| Three Crop Combination | Rice-Wheat-Maize | Godda | 5% |
| Four Crop Combination | Rice-Pulses-Maize-Wheat | Khunti | 5% |

In whole Jharkhand about 80 percent area of the state is covered by rice cultivation. In order of area covered, the most important crops are Rice, Maize, Wheat, Bajra, Marua, Pulses and Oil Seeds. These crops occupy about 99 percent of the total cropped area.

## One Crop Combination Region

80 percent of the region of Jharkhand which is covered by 20 districts are dominated by Rice cultivation. It means that in these districts Rice cultivation occupies the highest percentage of the total cropped area. It does not mean that there are no other crops being cultivated in these areas.

## Two Crop Combination Region

10 percent area of Jharkhand is dominated by two crops i.e., Pulses and Rice. Here Pulses are the first ranking crop and Rice is the second ranking crop.

## Three Crop Combination Region

Only 5 percent area, mostly the Godda district is the region dominated by three crops. Again, Rice is the first crop here but Maize and Wheat are also in the same status.

## Four Crop Combination Region

Khunti is a new district in Jharkhand. The area is small but four crops are grown here with the same status. Again,

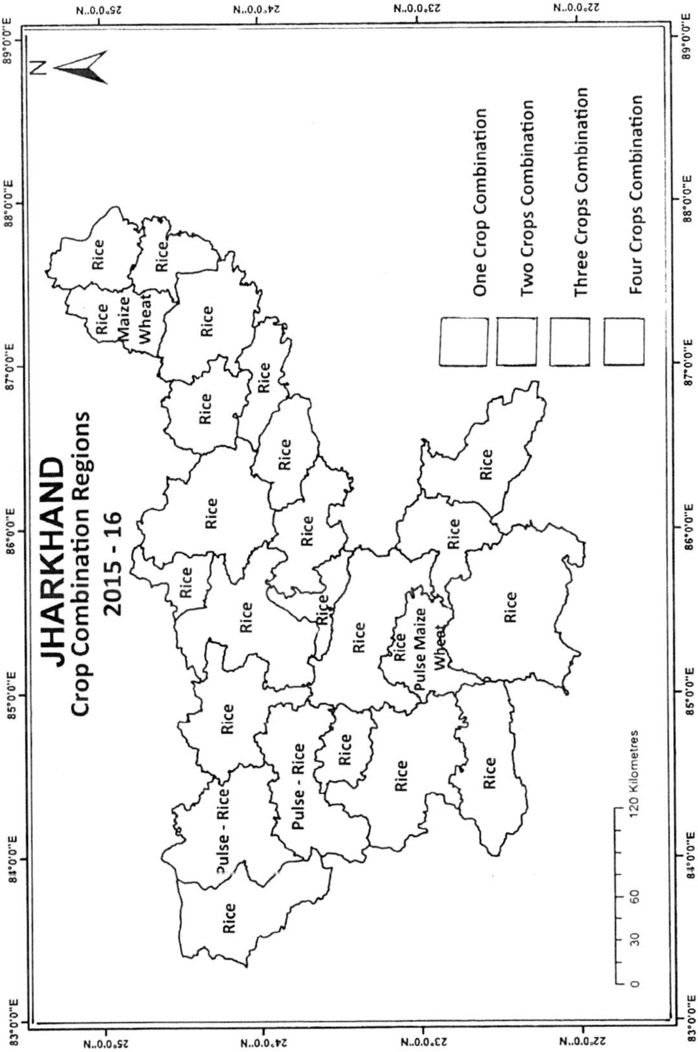


**Map 6.4: Crop Combination Regions (2015-16)**

Rice is in the first crop here. Second position is taken by Pulses, third by Maize and Wheat is the fourth one. The district occupies approximately 5 percent area of the state.

As discussed above, one can imagine if a region is growing four crops, it is rich and developed in agriculture. High yielding seeds, manuring system and irrigation facilities are sufficient here. Cropping diversity is high in this region. On the other hand, regions where rice is the first crop are still weak in diversity and scenario of almost whole Jharkhand is monoculture.

Majid Husain has depicted a map of crop combination regions in India for 1961-64 where Jharkhand has been mentioned as a monoculture region (Husain, 1996, P.234). Now it is a matter of surprise that even after 60 years, agricultural condition of Jharkhand is rather same as a state.

## Regions by Cropping Pattern (Relative Yield and Spread Method)

Cropping or agricultural pattern is the extent to which arable land under different agricultural activities can be put to use. These largely depend upon the socio-economic influents which determine the possibility of the enterprise the farmer chooses and the input intensity with which he farms. With the assured supply of water and availability of modern inputs - specially high - yielding varieties of seeds and commercial fertilizers - it becomes possible for the farmer to replace less profitable crops or enterprises with more profitable ones and also to enhance the intensity of the use of the available land by growing two or even three crops in the same field in a year (Singh, 1979, P.39-46) cropping pattern refers to the proportion of land under cultivation of different crops at different points of time. This indicates the time and arrangement of crops in a particular area. Any change in the cropping pattern would cause: change in the proportion of land under different crops as well as change in space sequence and time of crops. Cropping pattern is, however, a dynamic concept as it changes over space and time.

The cropping patterns of a region or areal unit may be

determined on the basis of areal strength of individual crops. The first, second and third ranking crops of an areal unit may be called as the dominant crops of that unit. These crops, if occupying more or less the same percentage of the total cropped area, shall be competing for area with each other and the farmer will decide which crop may fetch him more profit in a given year. Apart from the proportion of area under a particular crop, its relative yield also guides the suitability of that crop in a given geoclimatic and cultural setting. The relative yield index for the determination of suitability of crop may be calculated by applying the following formulas:

$$\text{Relative Yield Index} = \frac{\text{Mean yield of the crop in a component areal unit}}{\text{Mean yield of the total area}} \times 100$$

$$\text{Relative Spread Index} = \frac{\text{Area of the crop expressed as percentage of the total cultivated area in the areal unit}}{\text{Area of the crop expressed as percentage of the total cultivated land in the entire region}}$$

On the basis of these indices the suitability of the crop grown in a region may be ascertained. If the relative yield is below 90 percent, then it may be an inefficient crop and therefore should not be devoted more area to its cultivation. The area under each crop in a given region may be classified under four categories:

(i) High Yield, High Spread
(ii) High Yield, Low Spread
(iii) Low Yield, High Spread
(iv) Low Yield, Low Spread

## Computation of data

To compute the cropping pattern of Jharkhand we have collected the data of Maize, Rice, Wheat, Gram, Tuar, Rai/ Mustard. All these data have been calculated district wise and in respect of percentage of the area (Hectare) to the total area and percentage of production (Metric Tons) to the total production and thus we get average production(kg/hect), Average Production of area and relative yield index.

Similarly, to get the relative spread index we calculate the data of all districts as total area, total cropped area and ultimately, we find the relative spread index when we divide the total percentage of cropped area by the total percentage area of the state.

**Table 6.7**
**Relative yield index 2005-06 of Jharkhand**

| District | Total cropped area (Hect.) | Total Production (Metric Tons) | Total Average Production (Kg/ Hectare) | Average Production (Kg/ Hectare) of state | Relative Yield Index (In %) |
|---|---|---|---|---|---|
| | (Maize, Rice, Wheat, Gram, Tuar,Rai/Mustard) | | | | |
| Dumka | 82206 | 119195 | 1449 | 270 | 537 |
| Deoghar | 48152 | 16978 | 353 | 68 | 519 |
| Godda | 51493 | 82882 | 1610 | 393 | 409 |
| Pakur | 49804 | 65098 | 1307 | 360 | 363 |
| Sahebganj | 35117 | 42787 | 1218 | 237 | 514 |
| Jamtara | 13280 | 16572 | 1248 | 92 | 1356 |
| Hazaribagh | 46184 | 48794 | 1057 | 82 | 1289 |
| Kodarma | 11738 | 9870 | 841 | 66 | 127 |
| Chatra | 24374 | 11207 | 460 | 30 | 1533 |
| Giridih | 58663 | 32284 | 550 | 66 | 833 |
| Bokaro | 18433 | 9390 | 509 | 33 | 1542 |
| Dhanbad | 17718 | 19934 | 1125 | 95 | 1184 |
| Ranchi | 115602 | 67008 | 580 | 87 | 666 |
| Lohardaga | 28414 | 20274 | 714 | 136 | 525 |
| Gumla | 73781 | 96041 | 1302 | 180 | 723 |
| Palamu | 48024 | 24513 | 510 | 48 | 1062 |
| Garhwa | 33225 | 19774 | 595 | 49 | 1214 |
| Simdega | 40864 | 38464 | 941 | 102 | 922 |
| Latehar | 24493 | 19982 | 816 | 55 | 1483 |
| E. Singhbhum | 87533 | 68989 | 788 | 195 | 404 |
| W. Singhbhum | 119290 | 72751 | 610 | 101 | 603 |
| Saraikela-Kharsawan | 65384 | 44372 | 674 | 187 | 360 |

**Source:** Calculated by Author from statistical and evaluation Directorate Profile, 2006

**Table 6.8**
**Relative spread index of Jharkhand 2005-06**

| District | Total Area of District (000 Hectare) | Total Cropped Area (000 Hectare) | Percent to total cropped area of state | Percent of cropped area | Relative Spread Index (Col-5 / Col-4) |
|---|---|---|---|---|---|
| Dumka | 441.0 | 125.2 | 6.17 | 28.32 | 4.59 |
| Deoghar | 247.9 | 67.7 | 3.34 | 27.31 | 8.17 |
| Godda | 211.0 | 84.5 | 4.17 | 40.05 | 9.60 |
| Pakur | 180.6 | 63.1 | 3.11 | 34.91 | 11.22 |
| Sahebganj | 159.9 | 48.9 | 2.41 | 30.61 | 12.70 |
| Jamtara | 180.2 | 58.1 | 8.87 | 32.27 | 30.39 |
| Hazaribagh | 596.5 | 118.3 | 5.84 | 19.83 | 3.39 |
| Kodarma | 149.4 | 21.5 | 1.06 | 14.41 | 13.59 |
| Chatra | 370.6 | 50.9 | 2.51 | 13.75 | 5.47 |
| Giridih | 491.9 | 94.2 | 4.65 | 19.14 | 4.11 |
| Bokaro | 288.0 | 25.9 | 1.28 | 9.00 | 7.03 |
| Dhanbad | 208.9 | 40.0 | 1.97 | 19.14 | 4.11 |
| Ranchi | 769.8 | 272.6 | 13.46 | 35.41 | 2.63 |
| Lohardaga | 149.1 | 45.5 | 2.24 | 30.40 | 13.57 |
| Gumla | 532.1 | 169.3 | 8.35 | 31.82 | 3.81 |
| Palamu | 508.3 | 118.6 | 5.85 | 23.33 | 3.98 |
| Garhwa | 404.4 | 89.1 | 4.39 | 22.03 | 5.01 |
| Simdega | 375.6 | 86.9 | 4.29 | 23.15 | 5.39 |
| Latehar | 362.2 | 72.5 | 3.58 | 20.02 | 5.59 |
| E.Singhbhum | 353.3 | 146.8 | 7.25 | 41.57 | 5.73 |
| W. Singhbhum | 718.2 | 160.2 | 7.90 | 22.30 | 2.82 |
| Saraikela-Kharsawan | 237.2 | 65.3 | 3.22 | 27.52 | 8.54 |
| Jharkhand | 7971.40 | 2025.51 | | 25.41 | |

**Source:** Calculated by the author from Agricultural Data, Directorate of Agriculture, Govt. of Jharkhand.

**Table 6.9**
**Cropping pattern of Jharkhand (2005-06)**

| | Relative Yield Index (%) | | Relative Spread Index (%) | |
|---|---|---|---|---|
| High Value | 1542 | Bokaro | 13.59 | Kodarma |
| | 1533 | Chatra | 13.57 | Lohardaga |
| | 1483 | Latehar | 12.70 | Sahebganj |
| | 1356 | Jamtara | 11.22 | Pakur |
| | 1289 | Hazaribagh | 9.71 | Dhanbad |

| | | | | |
|---|---|---|---|---|
| | 1214 | Garhwa | 9.60 | Godda |
| | 1184 | Dhanbad | 8.54 | Saraikela |
| | 1062 | Palamu | 8.17 | Deoghar |
| | 922 | Simdega | 7.03 | Bokaro |
| | 833 | Giridih | 5.73 | E.Singhbhum |
| | 723 | Gumla | 5.59 | Latehar |
| Low Value | 666 | Ranchi | 5.47 | Chatra |
| | 603 | W. Singhbhum | 5.39 | Simdega |
| | 537 | Dumka | 5.01 | Garhwa |
| | 525 | Lohardaga | 4.59 | Dumka |
| | 519 | Deoghar | 4.11 | Giridih |
| | 514 | Sahebgunj | 3.98 | Palamu |
| | 409 | Godda | 3.81 | Gumla |
| | 404 | E. Singhbhum | 3.39 | Hazaribagh |
| | 363 | Pakur | 3.39 | Jamtara |
| | 360 | Saraikela | 2.82 | W.Singhbhum |
| | 127 | Kodarma | 2.63 | Ranchi |

On the availability of an alternative more efficient crop than the existing ones, new cropping pattern may emerge in a region. The cropping patterns may be intensified with the help of high yielding short duration varieties. Any cropping sequence to be adapted by the cultivators should be flexible. Hence, we may conclude that the area of the first pattern and the second pattern are very flexible for the cultivator to adopt the new cropping pattern.

| | |
|---|---|
| High Yield, High Spread | Bokaro, Latehar, Dhanbad |
| High Yield, Low Spread | Chatra, Jamtara, Hazaribagh, Garhwa,Palamu, Simdega, Giridih, Gumla |
| Low Yield, High Spread | Lohardaga, Deoghar, Sahebganj, Godda, E.Singhbum, Pakur, Saraikela, Koderma |
| Low Yield, Low Spread | Ranchi, E.Singhbum, Dumka |

## Previous Work

It would be relevant to mention in this chapter about the previous work done by different scholars of Geography regarding regionalization of Jharkhand area.

## 1947

Dr. P. Dayal had completed his Ph.D. work from University of London on the topic "Agriculture Geography of Bihar" In this work he has included a chapter on Agricultural regions. Dr. Dayal has demarcated four agricultural regions in Bihar - (i) The Wet Lowland 'dhanhar' tract, (ii) Bhith (including 'diara' tract, (iii) South Gangetic irrigated tract and (iv) Chotanagpur Plateau tract.

This fourth region is now Jharkhand. There is no sub-division for this region. It was an empirical technique to demarcate the region like Baker who had at first demarcated the agricultural belt of the USA in 1931. (The thesis of Dr. Dayal has been published by Rajesh publications, New Delhi in 2013).

## 1965

Dr. Enayat Ahmed has demarcated the agricultural regions of Bihar in his book "Bihar: Physical, Economic, Regional Geography". He has divided the portion of South-Bihar (Now Jharkhand) into six crop regions (1965, P.146). These regions are divided according to crop combination method. These regions are as under -

| No. In the division of Bihar | Name of the Region by Crops | Area by Districts |
|---|---|---|
| (2) | Rice-Maize-Gram | Santhal Pargana (old) |
| (8) | Rice-Maize-Gram | Palamu District (old) |
| (9) | Rice-Ragi-Fruits & Vegetables | Ranchi District (old) |
| (10) | Rice-Maize-Ragi | Hazaribagh District (old) |
| (11) | Rice-Maize-Fruits & Vegetables | Dhanbad District |
| (12) | Rice-Maize-Gram | Singhbhum District (old) |

## 2020

Present work is the next work after Dr. Enayat Ahmed.

Agricultural Regions of India has been demarcated by different scholars in different textbooks. It would be relevant

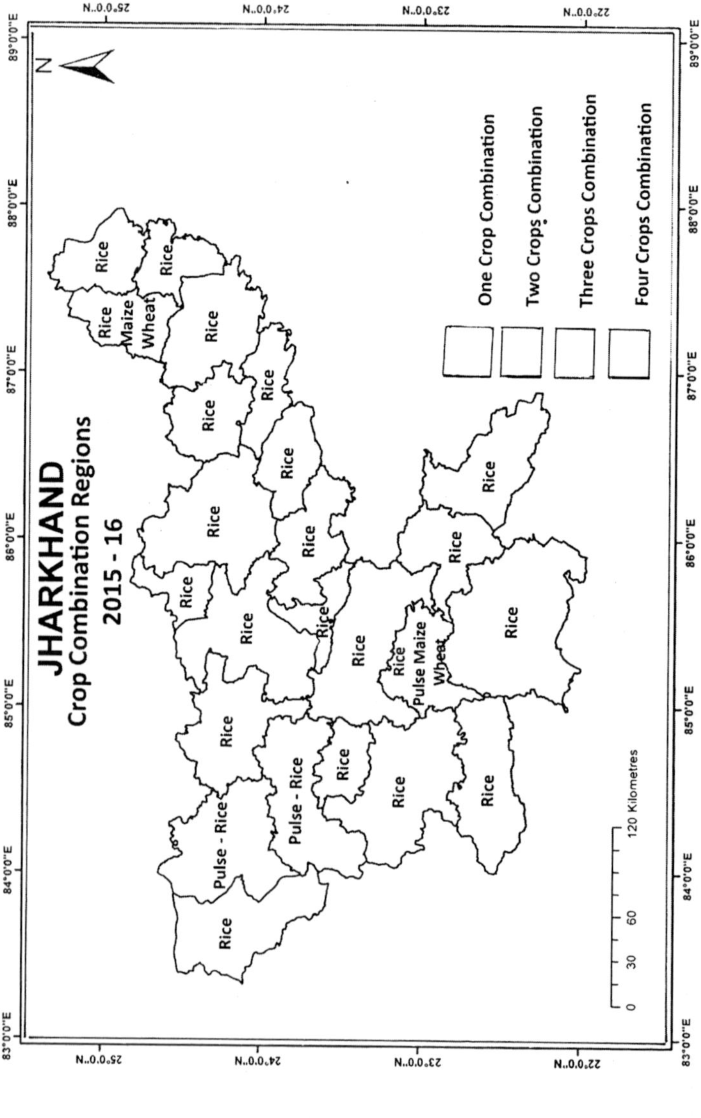


**Map 6.5: Region by Cropping Pattern**

to study the position of Jharkhand according to their division.

- M.S. Randhawa (1958) has demarcated India into five agricultural regions - 1. Temperate Himalayan Region, 2. Northern (Dry) Wheat Region, 3. Eastern Rice Region, 4. Western Wet (Malabar) region and 5.; Southern Coarser Grain (Medium Rainfall) Region. Here Jharkhand is under the third category i.e., Eastern Rice Region. (Randhawa, 1958, P. 73-75)
- Otokar Slampa (1968) has demarcated India into 13 agricultural regions. These are 1. Northern Mountain Region, 2. Northern Plain, 3. North-West Region, 4. North-East Region, 5. North-East Frontier region, 6. Central region, 7. Western Region, 8. Eastern Region, 9. Konkan Region, 10. Malabar and Kanara Region, 11. South-Eastern Region, 12. Lakshadweep Region and 13. Andaman-Nicobar Region. Among these Jharkhand is under the 8th category i.e., Eastern Region of Agriculture. (Slampa, 1968)
- P. Sengupta and Galina Sdasyuk (Eds) (1968) have classified the agricultural regions of India into four major divisions - 1. Himalayan Agricultural Zone, 2. Dry Agricultural Zone, 3. Sub-Humid Agricultural Zone and 4. The Wet Agricultural Zone. These are macro regions classified according to homogeneous climatic conditions. Afterwards they have classified eleven Meso regions according to physical divisions. These are further divided into sixty Micro regions according to crop combinations. Position of Jharkhand under Macro region is in 'Wet Agricultural Region'. Under Meso regions it is known as 'Peninsular Plateau Region' and under the Micro-Regions it is known as 'Chotanagpur and Orissa plateau agricultural region'. (Sengupta & Sdasyuk, 1968)
- B.L.C. Johnson (1979) has demarcated India into 15 main crop- combination regions. Among these Jharkhand has been demarcated as North-East plateau region where crop combination is Rice-Maize-Pulse. Northern part of Jharkhand has Rice-Wheat combination.
- R.L. Singh (1971) has divided India into First order,

Second order and third order crops. According to first order crops India has been divided into 8 Agricultural Regions. Second order crops have 27 regions and third order crops has 48 regions. According to this division Jharkhand has a crop combination of Rice-Maize/Millet-Oilseeds region.

- Jasbir Singh (1976) has demarcated India in 12 main crop combination regions and 60 sub-crop combination regions. Jharkhand is under the Rice region according to the 12 main crop combination of Jasbir Singh.
- National Commission for Agriculture in India (1976) has demarcated the 'Agro Climatic Regions' in India. This is based on Koppen's and Thornthwaite's classification. In this division, potential evapotranspiration, moisture index and thermal efficiency are the basis of demarcation. They have XV broader divisions of India in which Jharkhand falls under VII category which is known as the 'Eastern Plateau Region'. Again, they have divided India in 14 sub-divisions of physiographic units. Here Jharkhand has two sub-divisions - 3 and 4. 3rd is the Lower and Middle Ganga Plain and 4th is the Eastern and Central Plateau and Hills. Ultimately crop specialization has been demarcated according to the physical and climatic conditions. Jharkhand has VII category where crop specialization is Rice (Yojna, 1-15 October, 1990).

## References

Ahmad, Enayat (1965); Bihar: Physical, Economic, Regional Geography, Ranchi University, Ranchi.

Bhatia, S.S. (1965); Pattern of Crop Concentration and Diversifications in India, Economic Geography, 41.

Dayal, P (2013); Agriculture Geography of Bihar, Rajesh Publications, New Delhi.

Husain, Majid (1996); Systematic Agricultural Geography, Rawat Publications, Jaipur.

Johnson, B.L.C. (1979); India: Resource and Development. Anrold-Heinemann, London, First Indian Edition, 1980.

Johnson, R.J. (1986); The dictionary of Human Geography, National Publishing House, New Delhi.

Pokshishevskiy, V.V (1968); Foreword in Economic regionalization if India, Commissioner of India.

Randhawa, M.S. (1958); Agriculture and Animal Husbandry in India, ICAR, New Delhi.

Sengupta, P & Sdasyuk, Galina (1968); Principles and methods of Economic Regionalization, Census of India, 1961, Monograph Series, Vol-I, No.-8.

Singh, J & Dhillon S.S. (1984); Tata McGraw-Hill Publishing Co. Ltd, New Delhi.

Singh, R.L. (1971) (Eds); India: A Regional Geography, National Geographical Society of India, Varanasi.

Singh, V.R. (1979); A method for analyzing agricultural productivity, Transactions, Vol-1, Pune.

Slampa, O (1968); Agricultural Regions of India, Sbornik Ceskoslovenske Spolecsosti Zemepusne.

Thomas, D (1963); Agriculture in Wales during the Napoleonic War: a study in Geographical interpretation of historical sources, Cardiff, University of Wales Press.

Weaver, J.C. (1954); Crop Combination Region in the Middle West. Geographical Review, Vol-44, No.-2 (Apr. 1954) P 175-200, Published by Taylor & Francis Ltd, United States.

# 7

## Animal Husbandry

- Introduction
- Animal Husbandry
- Fisheries
- Dairying
- Livestock Products
- Livestock Development Programs

### Introduction

Livestock is commonly defined as domesticated animals raised in an agricultural setting to produce labor and commodities such as meat, eggs, milk, fur, leather and wool. Ever since the man started farming in organizational form, the need for draught animals for cultivation was increasingly felt and the domestication of animals started. Fodder, feed and crop by-products got increasingly used for animals kept for milk, wool, meat and draught purpose. The by-products, which otherwise don't find alternative use, would have been disposed of by burning causing more air pollution and global warming. The animal droppings proved to be helpful in enriching the soil of organic matter and nutrients. Crop and livestock have thus proved to be complementary to each other in many other ways as well such as:

- The resources having opportunity cost at certain point of time due to seasonality in crop operations get gainful employment for the livestock activities.
- The income from crops is generated after incurring expenditure for a couple of months. But from livestock, it is generally regular and thus family expenses get

regulated over the year without depending much on the sources of credit.

- The improvement in income of small farmers and landless poor families in the rural areas' poverty redressal program.
- The availability of milk, meat and eggs improves the human nutritional status in the rural areas.
- Livestock is an important capital asset and useful in certain non- market functions such as exchanges, milk, lassi and common grazing land etc. (Lekhi & Singh, 2018, P.480)

The process of livestock can be explained through the following figure

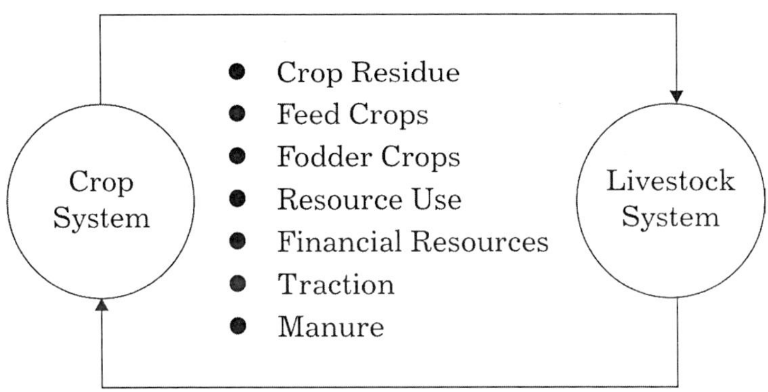


Livestock has been divided into two parts - (i). Animal Husbandry and (ii). Fisheries.

## Animal Husbandry

Animal husbandry has assumed a much broader role in the overall economy than so far envisaged as an integral part of expanding and diversified agriculture. This sector accounts for over 25 percent of gross value of agricultural output. India's vast livestock population offers tremendous potential for meeting domestic demands for milk, egg, meat, wool etc. This sector has also begun to be regarded as a source of new employments specially for marginal and small farmers and agricultural laborers by offering subsidiary occupation to supplement family income. It is not easy to assess the monetary value of cattle's contribution to the country's national income, because large proportion of produce is consumed by producers themselves instead of being marketed. Some ideas of actual as well as potential value of cattle to Jharkhand's agriculture can be secured from the following statement made by Animal Husbandry department, Govt. of India that "Jharkhand is now self- sufficient in milk production, producing 1894 thousand metric tons, Fish 118

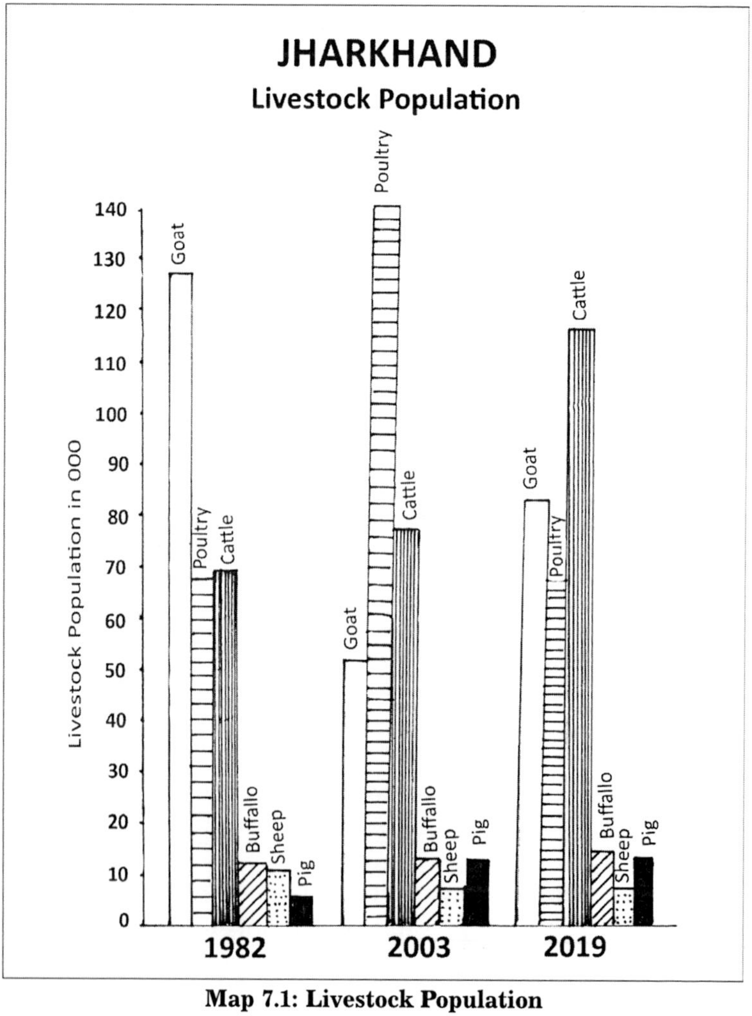


Map 7.1: Livestock Population

thousand metric tons and meat 55 thousand metric tons. Eggs production in Jharkhand is 510000 thousand in numbers."

**Table 7.1**
**Volume of Milk, Fish, Meat and Eggs in Jharkhand - 2017 (000 metric tons)**

| Volume of Milk, Fish, Meat and Eggs in Jharkhand - 2017 (000 metric tons) | |
|---|---|
| Milk | 1894 |
| Fish | 118 |
| Meat | 55 |
| Eggs | 510 (million) |

India possesses the largest number of cattle of any country in the world. Out of the world's cattle population of about 690 million animals, 215 million (or just one third) are located in India. Jharkhand possesses about 5.5% of the total animal population of India (2019). It was 6 percent during 2003 census. Trend of the livestock population growth in India and Jharkhand are in opposite directions. India has a growing status while Jharkhand has a decreasing animal population. Data in the below table will make the trend clear.

**Table 7.2**
**Census of Livestock in India (million)**

| Species | 1951 | 2003 | 2012 |
|---|---|---|---|
| Cattle | 155 | 185 | 191 |
| Buffaloes | 43 | 98 | 108 |
| Sheep | 39 | 61 | 65 |
| Goat | 47 | 124 | 135 |
| Pig | 4 | 14 | 12 |
| Poultry | 74 | 489 | 729 |
| Total Livestock | 293 | 485 | 512 |

**Table 7.3**
**Census of Livestock in Jharkhand (thousand)**

| Species | 1982 | 2003 | 2012 |
|---|---|---|---|
| Cattle | 6798 | 7659 | 11174 |
| Buffaloes | 1033 | 1343 | 1349 |

| | | | |
|---|---|---|---|
| Sheep | 975 | 680 | 784 |
| Goat | 12750 | 5031 | 8375 |
| Pig | 418 | 1108 | 1276 |
| Poultry | 6628 | 14429 | 6634 |
| Total Livestock | 28602 | 30260 | 29592 |

In the above-mentioned table, number of total livestock in Jharkhand shows a decline from 30.2 million in 2003 to 29.5 million in 2012.

Here we shall discuss all livestock population one by one because all category has its separate importance.

**Cattle**

Cattles play an important role in our agricultural economy. But their present position is far from satisfactory. Thus, the prevailing condition of the livestock is studied under the following heads.

**Cattle Population**

Cattle include Bulls, Ox and Cows all together. The productive value of cattle is not commensurate with its size. Nevertheless, the actual and potential value of cattle products is great. During the census of 1982, there were only 6798 thousand cattle in Jharkhand. It was enumerated 7659 thousand during the census of 2003 and during the census of 2019 it rose to 11174 thousand. It was a growth of 64 percent of cattle population in a duration of 40 years between 1982 to 2019. Though the population has decreased with respect to 2003. District wise distribution of cattle shows that the number is highest (974 thousand) in Giridih district. Second place is taken by Dumka with 748 thousand followed by Garhwa (702 thousand), Deoghar (693 thousand), Godda (685 thousand), Palamu (666 thousand) and Ranchi (636 thousand). These are the districts with the highest concentration of cattle. Chatra (511 thousand), Gumla (558 thousand) and Hazaribagh (548 thousand) are districts with above moderate distribution of cattle population.

A population between 500 to 300 thousand is considered as moderate population of cattle. Bokaro, Dhanbad, East Singhbhum, Jamtara, Latehar, Pakur, Saraikela, Simdega and

West Singhbhum districts are the area of moderate distribution of cattle.

Low population of cattle has been recorded in the districts of Koderma, Khunti, Lohardaga, Ramgarh and Sahebganj. Except Lohardaga and Sahebganj all are newly created districts. Seems like the Cattle improvement programs has not been implemented properly in these districts yet.

If we analyze the cattle scenario of the whole state, we notice that the northern districts of Jharkhand, mostly those bordering to Bihar have higher concentration of cattle. Godda, Garhwa are in this category of districts. The moment we start moving south, we find the districts have moderate or lower population of cattle. We also notice that the smaller districts of the state have low concentration of cattle as well. These districts need to focus more on cattle population.

**Problems with Cattle Population**

- In view of Royal Commission on Agriculture, in whatever respect Indian cattle may be lacking, they do not lack in number. Planning commission has observed the ratio of dry cows which are not in milk with those in milk differs very considerably. This is mainly due to the poor quality of cattle. So, the cows become un-economic when they are not into milk. This ratio is greater in Jharkhand. Here we find 200 dry cows for every 100 milk cows whereas in other parts of the country like Punjab, this ratio is about 150 and as low as 75. So, the main problem is - Uneconomic Cattle.
- **Cattle being Herbivorous:** It is well known that cattle are herbivorous animals. They convert plant and vegetable food into milk. It is also known that for efficient production of milk, the cattle need sufficient well blended food which must be satisfactory, both quantitatively as well as qualitatively. On the contrary a large number of Indian livestock as well as Livestock in Jharkhand are severely ill-fed and deficient in nutrition. As far as possible, they should be provided with green and succulent fodder. Planning commission has observed that the quantity of fodder

available is nearly three fourth of our requirement. This is mainly due to increase in the requirements of food for the human population, areas, where grazing was possible, have steadily diminished. Harbans Singh has rightly opined that in our country the natural grazing lands are classified as 'forests', 'culturable waste' and 'areas not available for cultivation' but are of limited or little value as a source of fodder supply due to stringency, un-useful herbage and absolute barrenness respectively. In Jharkhand permanent pasture and other grazing land was 461 thousand hectares but now it is only 110 thousand hectares.

- **Poor Nourishment:** Cattle should always be treated gently and with kindness. They should never be frightened. But in our country, cattle are housed under notoriously insanitary conditions and are shut up in low, dark and small rooms devoid of ventilation and drainage. They are often abominable in construction and emit foul suffocating odors and ammoniacal gases which irritate the throat, while floors are kutcha, rugged and uncomfortably damp. The result is that the animal being on the borderline of health, with no inherent vigor and stamina, fall on easy prey to contagious and non-contagious diseases and die of in thousands annually leaving their owners in despair.
- Poor Breeding: In Jharkhand cattle are short in height. They produce milk in very low quantity rather the quality is good enough. The inferior quality of cattle is also due to wrong breeding. At present due to promiscuous mating there is great degeneration in their quality. Good and healthy cattle are allowed to mix with the weak and degenerate types. In Jharkhand and other part of the country cattle roam about on the village commons or on the stable with mixed herds. Cows are thus covered by indifferent bulls; which leads to the progressive deterioration of their breed. In Jharkhand the cattle are 'non-descript' and do not belong to any defined breed.

Regarding cattle it is a significant fact that good cattle are generally found in dry areas and inferior quality cattle are found in heavy rainfall area. Calculations show that among the districts of Jharkhand cattle growth is always positive. (Table 7.1)

## Buffaloes

Buffalo is a large animal of the cattle family with curved long horns. Over the centuries species of buffaloes have been domesticated. Domestic buffaloes are equally important as cattle cow among the tribal population in Jharkhand. Buffaloes are capable of hauling heavy loads and surviving on a low-quality diet. In addition, they provide people with an essential supply of milk and hides and bones. There are eight indigenous buffalo breeds in India they are 1. Murrah (Haryana, Punjab), 2. Surti which is also known as Gujrati, Decani, Talabada are from Gujrat, 3. Jaffarabadi (Giir, Kutch, Jamnagar) from Gujrat, 4. Bhadawari (Agra, Etawah district of U.P.), 5. Nili Ravi (originated around river Rabi) and found in Sutlej valley in Firozpur district of Punjab and Sahiwal (Pakistan) of undivided India, 6. Mehsana, found in Mehsana district, Sabarkanda and Banaskanta district in Gujrat and adjoining Maharashtra state. 7. Nagpuri - also known as Elitchpuri or Barari from Nagpur, Akola, Amarawati district of Maharashtra; and 8. Toda - named after an ancient tribe toda of Nilgiri hills of South India.

### Population Growth

At present Jharkhand has registered about 1349 thousand population of buffaloes during 2019 census. It was 1343 thousand during 2003 and 1033 thousand during 1982. There is a growth of 31 percent in between 1982 and 2019 in Jharkhand. This growth is just half of the growth of the cattle or cows. Cattle population was 11174 thousand whereas the number of Buffaloes has been recorded 1349 thousand only during 2020.

District wise analysis of Buffaloes distribution shows highest numbers in Ranchi (125 thousand) followed by Giridih (112 thousand), Chatra (102 thousand), Godda (96), Palamu (94), Gumla (90), Garhwa (74) and Hazaribagh (70). The distribution is similar to the cattle but the density is just half of the cattle. Density of cattle is 140 cattle per sq. Km. but the density of Buffalo is only 17 Buffalo per sq. km. in the state. This shows that cattle and buffaloes both are rearing together but buffaloes are reared by the big cultivators who have large size land holdings.

Moderate distribution of buffaloes have been recorded in the districts of Dhanbad (33 thousand), Dumka (48 thousand), Khunti (33 thousand), Latehar (38), Sahebganj (47), Saraikela(33) and West Singhbhum (33).

Low population of Buffaloes have been recorded in the districts of Bokaro (29 thousand), Jamtara (20 thousand), Koderma (20), Lohardaga (17), Ramgarh (29) and Simdega (27).

When we compare the growth of cattle to buffaloes, we find that cattle growth is always positive but growth of buffaloes has been recorded negative in some districts like Bokaro, Dhanbad, Dumka, Hazaribagh, Jamtara, Latehar and Simdega. Exact reason is a matter of research by Animal Husbandry department but my assumption is the cultivators show more interest towards rearing cattle rather than buffaloes. Second reason is the fragmentation of land holdings and increasing their uneconomic size where need of buffalos' plough is not required. Another reason could be Buffaloes are not reared for meat purpose in Jharkhand.

## Sheep

There are about 14 breeds of sheep in India which can be divided into two distinct types namely, Wooly and Hairy. The wooly type produces wool fibers of fine or coarse quality, while the other just hair and are reared for mutton and manure. In India there are about 18 states producing wool and Jharkhand is placed just above Sikkim and Arunachal Pradesh among them. During 2006-07 about 146 tons of wool was produced in Jharkhand.

## Distribution of Sheep

High distribution of sheep population areas are West Singhbhum (113 thousand) followed by East Singhbhum (92 thousand), Saraikela (76 thousand) districts in Jharkhand.

Jharkhand has highest distribution of Sheep in southern area like Singhbhum and Saraikela districts.

Areas of moderate distribution are Pakur (47 thousand), Ranchi (40 thousand), Jamtara (38 thousand), Palamu and Khunti (35 thousand) districts. This is distributed over three different pockets, Santhal Pargana, Ranchi and Palamu.

Low concentration of sheep population is in the districts of Bokaro, Chatra, Deoghar, Dhanbad, Dumka, Garhwa, Giridih, Godda, Gumla, Hazaribagh, Koderma, Latehar, Lohardaga, Ramgarh, Sahebganj and Simdega districts.

Chatra, Koderma and Latehar are districts where sheep population were going to be extinct, which prompted the Jharkhand government to establish a major sheep breeding center in Chatra district in order to promote sheep rearing in this area.

When we study the growth of sheep population across the state, we find that only three districts have positive growth. These are Jamtara, Pakur and Saraikela districts. Rest all districts show a major decline of growth in sheep population in the state. (Table 7.3)

## Goat

There is a very popular proverb in the rural India that "Goats are the poor man's cow". Goats produce milk, meat, hides and wool. This is the main source of meat production in the country. Thus, goats become an integral part of the Indian agricultural economy. The important breeds of Indian goats fall under three broad groups 1. Himalayan Goats found in the hills of Kashmir, Punjab and UP. 2. Pashmina goats of Kashmir which produces very fine variety of hair which goes into the manufacture of quite expensive shawls, 3. The plain goats which provide some of the best mutton and milk. Some of the better-known breeds are Marwari, Konor, Jamnapari, Barbari, Mohsana, Kathiawari, Jalwari, Surti. Some hybrid varieties are Alpine, Nubian, Sanen, Togenberg, Angora etc. Varieties of goats rearing in Jharkhand are - Beetal Goat, Lahori Goat, Adult Sirohi Goat, Devgarhi Goat, Parbatsari Goat and Black Bengal Goat (Negi, 1971, P.99).

Around 80% of the global goat population is in developing countries. India ranks second in goat population with 135 million goats. At present 28 breeds of goats documented in India out of which Black Bengal is one of the most important meat producing breed found in Bengal, Bihar, Jharkhand, Orissa and North-Eastern part of India. (Mandal, 2018, P. 501-509).

## Population Distribution of Goats

District wise goat population distribution in Jharkhand recorded the highest concentration in Ranchi (785 thousand), followed by Giridih (730 thousand), Gumla (609 thousand) and West Singhbhum (528 thousand) districts respectively. These four districts have highest concentration of goats in the state.

Moderate concentration of goat population within a range of 500 to 300 thousand is recorded in Hazaribagh (477 thousand), Deoghar (473), Simdega (418), Garhwa (415), Dumka (407), East Singhbhum (359), Saraikela (365), Sahebganj (355), Palamu (352), Bokaro (347) and Chatra (301 thousand) respectively.

Low concentration has been recorded in Dhanbad (271), Godda (287), Jamtara (257), Khunti (267), Koderma (120), Latehar (274), Lohardaga (176) and Ramgarh district.

Overall total population of goats was 12750 thousand during 1982, it decreased to 5031 thousand during 2003 census and then again increased to 8375 thousand during 2019 census. This was a slight increase but it was still below the population size of 1982. This amounts to a -35% growth in between 1982 to 2019.

## Pig

As we know that pig has two types, one is domestic and other is wild pig. The domestic pig often called swine, hog or simply pig when there is no need to distinguish one from the other. Domestic pigs are farmed primarily for the consumption of their flesh, called 'Pork'. Pig is a mammal of the suidae family. In Britain the farm pig refers to all domestic swine, while in the united states it refers to younger swine not yet less than 82 Kg; others being called hogs. Following are the selected breeds of pigs - Duroc or Duroc-Jersey, Hampshire, Landrace, Spotted and Yorkshire (White). Among them only Duroc-Jersey is used for lard and all others are used for meat. Swine flu is a disease related to respiratory system of pigs that is caused by an influenza virus. This was first isolated from pig in 1930.

### Population Distribution of Pigs

Though piggery is small but present everywhere in Jharkhand. In the state pig rearing has increased steadily. During 1982 number of pigs in the state was 418 thousand which increased to 1108 thousand during 2003 and then 1276 thousand in 2019. This was an increase of 205 percent over this time duration. Among all other animal this was the highest growth number in the state. Pig population is conspicuous to the extent of having a share of 4.3 percent in the total livestock in Jharkhand.

District wise distribution of pig population shows two pockets of highest concentration in the state. First is Dumka (129 thousand) and Godda (105 thousand) in Santhal Pargana region and second is Gumla and Simdega (110 thousand) districts of Jharkhand.

Moderate concentration (50 to 100 thousand population) is recorded in three pockets of the state. One is Chatra and Latehar districts, second is Pakur and Sahebganj districts, third is Ranchi. Rest all districts have very low population of pigs.

### Poultry

The poultry industry in India has been so far village or cottage industry with no support from commercial hatcheries, feed industry or organized or supervised marketing of poultry products. New factors are now contributing to the development of the poultry industry in this country. The demand for poultry and eggs is increasing enormously due to general improvement in the economic levels of the people and greater appreciation of nutrition in general and food value of poultry products in particular, as eggs and poultry meat are one of the richest sources of animal protein.

There are only four pure Indian breeds of chicken available -Aseel, Chittagong (Malay), Kadaknath and Busra. Particularly in Jharkhand, 'Jharsim' is a local variety developed by Birsa Agricultural University, Ranchi. The name Jharsim is derived from 'Jhar' for Jharkhand and 'Sim' meaning hen in the local tribal dialect. Apart from the above there are some more varieties - Kamrupa (Assam), Pratapdhan (Rajsthan), Kari Shyama (Kadaknath cross, locally known as 'Kalamasi', Madhya Pradesh), Hitcari (Naked neck cross), Upkari (frizzle cross), Grampriya etc.

### Distribution of Poultry Population

22 percent share of the total livestock population of Jharkhand belongs to poultry. This is the highest share among the livestock population of the state. However, the growth is minimal (0.09%). Total number of poultry production in the state during 1982 was 6628 thousand which increased to 14429 thousand during 2003 census. During 2019 census the number came down to 6634 thousand in the state.

District wise population reveals the highest concentration in Ranchi district (975 thousand) followed by West Singhbhum (775 thousand). Only these two districts comprise of 26 percent of the total population of poultry in the state.

Moderate distribution of poultry can be seen in the East Singhbhum (447 thousand), Hazaribagh (416 thousand), Dhanbad (370 thousand), Gumla (342 thousand) and Saraikela (319 thousand) districts.

Below 300 thousand or low concentration of poultry is seen in districts like Bokaro, Chatra, Deoghar, Dumka, Garhwa, Godda, Jamtara, Koderma, Latehar, Lohardaga, Palamu and Sahebganj districts.

When we study the growth of poultry, we find a decreasing tendency in about eleven districts of Jharkhand.

However, few districts have recorded a growth of more than 200 percent in the same period of time. For example, West Singhbhum has recorded a growth of 226 percent, 219 percent in Saraikela, 197 percent in Dhanbad, 89 percent in East Singhbhum and 67 percent in Hazaribagh districts. Still the overall growth in the state stands at 0.009 percent only.

**Table 7.4**
**District wise Cattle Population in Jharkhand (Figs in 000)**

| District | 1982 | 2003 | 2019 | Growth in % |
|---|---|---|---|---|
| Bokaro | 312 | 327 | 401 | 29 |
| Chatra | 254 | 331 | 511 | 101 |
| Deoghar | 288 | 396 | 693 | 141 |
| Dhanbad | 221 | 309 | 350 | 58 |
| Dumka | 439 | 488 | 748 | 70 |
| East Singhbhum | 395 | 276 | 413 | 5 |
| Garhwa | 334 | 357 | 702 | 110 |
| Giridih | 533 | 588 | 974 | 83 |
| Godda | 269 | 324 | 685 | 155 |
| Gumla | 408 | 438 | 558 | 37 |
| Hazaribagh | 410 | 484 | 548 | 34 |
| Jamtara | 208 | 247 | 432 | 103 |
| Khunti | - | - | 280 | - |
| Koderma | 88 | 146 | 184 | 109 |
| Latehar | 248 | 253 | 348 | 40 |
| Lohardaga | 116 | 123 | 146 | 26 |
| Pakur | 210 | 225 | 387 | 84 |
| Palamu | 409 | 500 | 666 | 63 |
| Ramgarh | - | - | 120 | - |
| Ranchi | 574 | 671 | 636 | 11 |
| Sahebganj | 234 | 196 | 263 | 12 |
| Saraikela | 168 | 223 | 305 | 82 |
| Simdega | 281 | 324 | 403 | 43 |
| West Singhbhum | 399 | 394 | 471 | 18 |
| Jharkhand | 6798 | 7659 | 11174 | 64 |

**Source:** Animal Husbandry Department, Govt. of India

**Table 7.5**
**District wise Buffaloes Population in Jharkhand (Figs in 000)**

| District | 1982 | 2003 | 2019 | Growth in % |
|---|---|---|---|---|
| Bokaro | 52 | 61 | 29 | -44 |
| Chatra | 56 | 95 | 102 | 82 |
| Deoghar | 37 | 54 | 54 | 46 |
| Dhanbad | 37 | 59 | 33 | -11 |
| Dumka | 57 | 49 | 48 | -16 |
| East Singhbhum | 41 | 35 | 59 | 44 |
| Garhwa | 54 | 59 | 74 | 37 |
| Giridih | 89 | 119 | 112 | 26 |
| Godda | 35 | 64 | 96 | 174 |
| Gumla | 61 | 62 | 90 | 48 |
| Hazaribagh | 90 | 143 | 70 | -22 |
| Jamtara | 36 | 26 | 20 | -44 |
| Khunti | - | - | 33 | - |
| Koderma | 19 | 37 | 20 | 5 |
| Latehar | 40 | 36 | 38 | -5 |
| Lohardaga | 17 | 12 | 17 | 00 |
| Pakur | 27 | 31 | 70 | 159 |
| Palamu | 66 | 98 | 94 | 42 |
| Ramgarh | - | - | 29 | - |
| Ranchi | 86 | 155 | 125 | 45 |
| Sahebganj | 30 | 47 | 47 | 46 |
| Saraikela | 18 | 29 | 31 | 72 |
| Simdega | 42 | 21 | 27 | -36 |
| West Singhbhum | 43 | 51 | 31 | 28 |
| Jharkhand | 1033 | 1343 | 1349 | 31 |

**Source:** Animal Husbandry Department, Govt. of India

**Table 7.6**
**District wise Sheep Population in Jharkhand (Figs in 000)**

| District | 1982 | 2003 | 2019 | Growth in % |
|---|---|---|---|---|
| Bokaro | 46 | 34 | 16 | -65 |
| Chatra | 06 | 01 | 0.5 | -92 |
| Deoghar | 48 | 37 | 26 | -46 |

| | | | | |
|---|---|---|---|---|
| Dhanbad | 33 | 33 | 18 | -45 |
| Dumka | 73 | 50 | 29 | -60 |
| East Singhbhum | 127 | 55 | 92 | -27 |
| Garhwa | 32 | 36 | 25 | -21 |
| Giridih | 79 | 22 | 12 | -85 |
| Godda | 45 | 04 | 05 | -89 |
| Gumla | 35 | 10 | 08 | -77 |
| Hazaribagh | 11 | 10 | 03 | -72 |
| Jamtara | 35 | 51 | 38 | 9 |
| Khunti | - | - | 35 | - |
| Koderma | 02 | 07 | 0.4 | -80 |
| Latehar | 24 | 10 | 0.9 | -99 |
| Lohardaga | 10 | 02 | 02 | -80 |
| Pakur | 34 | 17 | 47 | 38 |
| Palamu | 39 | 50 | 35 | -10 |
| Ramgarh | - | - | 4 | - |
| Ranchi | 50 | 87 | 40 | -20 |
| Sahebganj | 39 | 03 | 04 | -89 |
| Saraikela | 54 | 62 | 76 | 41 |
| Simdega | 24 | 11 | 14 | -42 |
| West Singhbhum | 129 | 98 | 113 | -12 |
| Jharkhand | 975 | 680 | 784 | -20 |

**Source:** Animal Husbandry Department, Govt. of India

**Table 7.7**

**District wise Goat Population in Jharkhand (Figs in 000)**

| District | 1982 | 2003 | 2019 | Growth in % |
|---|---|---|---|---|
| Bokaro | 193 | 207 | 346 | 79 |
| Chatra | 170 | 221 | 301 | 77 |
| Deoghar | 186 | 200 | 473 | 154 |
| Dhanbad | 139 | 214 | 271 | 95 |
| Dumka | 285 | 213 | 407 | 43 |
| East Singhbhum | 323 | 230 | 389 | 20 |
| Garhwa | 147 | 139 | 415 | 182 |
| Giridih | 333 | 372 | 730 | 119 |
| Godda | 174 | 185 | 289 | 66 |
| Gumla | 2748 | 284 | 609 | -78 |

| | | | | |
|---|---|---|---|---|
| Hazaribagh | 275 | 356 | 477 | 73 |
| Jamtara | 134 | 155 | 257 | 92 |
| Khunti | - | - | 267 | - |
| Koderma | 59 | 91 | 120 | 103 |
| Latehar | 109 | 170 | 274 | 151 |
| Lohardaga | 783 | 85 | 176 | -77 |
| Pakur | 136 | 136 | 310 | 128 |
| Palamu | 180 | 232 | 352 | 96 |
| Ramgarh | - | - | 210 | - |
| Ranchi | 3867 | 643 | 785 | -79 |
| Sahebganj | 151 | 172 | 355 | 135 |
| Saraikela | 137 | 191 | 365 | 166 |
| Simdega | 1894 | 244 | 418 | -77 |
| West Singhbhum | 327 | 290 | 528 | 61 |
| Jharkhand | 12750 | 5031 | 8375 | -34 |

**Source:** Animal Husbandry Department, Govt. of India

**Table 7.8**
**District wise Pig Population in Jharkhand (Figs in 000)**

| District | 1982 | 2003 | 2019 | Growth in % |
|---|---|---|---|---|
| Bokaro | 26 | 43 | 17 | -35 |
| Chatra | 21 | 47 | 74 | 252 |
| Deoghar | 05 | 29 | 30 | 500 |
| Dhanbad | 19 | 61 | 36 | 89 |
| Dumka | 07 | 109 | 129 | 1743 |
| East Singhbhum | 14 | 25 | 22 | 57 |
| Garhwa | 17 | 27 | 31 | 82 |
| Giridih | 46 | 51 | 53 | 15 |
| Godda | 04 | 47 | 105 | 2525 |
| Gumla | 45 | 74 | 108 | 140 |
| Hazaribagh | 35 | 71 | 45 | 28 |
| Jamtara | 03 | 50 | 37 | 1133 |
| Khunti | - | - | 41 | - |
| Koderma | 07 | 06 | 03 | -57 |
| Latehar | 12 | 44 | 53 | 341 |
| Lohardaga | 13 | 10 | 08 | -38 |
| Pakur | 04 | 69 | 98 | 2350 |
| Palamu | 21 | 56 | 42 | 100 |
| Ramgarh | - | - | 18 | - |

| | | | | |
|---|---|---|---|---|
| Ranchi | 64 | 123 | 68 | 06 |
| Sahebganj | 04 | 58 | 68 | 1600 |
| Saraikela | 06 | 16 | 42 | 600 |
| Simdega | 31 | 60 | 110 | 254 |
| West Singhbhum | 14 | 32 | 38 | 171 |
| Jharkhand | 418 | 1108 | 1276 | 205 |

**Source:** Animal Husbandry Department, Govt. of India

**Table 7.9**

**District wise Poultry Population in Jharkhand (Figs in 000)**

| District | 1982 | 2003 | 2019 | Growth in % |
|---|---|---|---|---|
| Bokaro | 223 | 600 | 279 | 25 |
| Chatra | 154 | 280 | 152 | 1.3 |
| Deoghar | 294 | 323 | 134 | -54 |
| Dhanbad | 158 | 702 | 370 | 197 |
| Dumka | 448 | 598 | 284 | -37 |
| East Singhbhum | 236 | 959 | 447 | 89 |
| Garhwa | 174 | 304 | 154 | -11 |
| Giridih | 381 | 430 | 221 | -42 |
| Godda | 274 | 473 | 215 | -21 |
| Gumla | 780 | 705 | 343 | -56 |
| Hazaribagh | 249 | 919 | 416 | 67 |
| Jamtara | 211 | 440 | 183 | -13 |
| Khunti | - | - | - | - |
| Koderma | 53 | 96 | 53 | 00 |
| Latehar | 130 | 400 | 165 | 26 |
| Lohardaga | 222 | 254 | 112 | -49 |
| Pakur | 215 | 656 | 319 | 48 |
| Palamu | 214 | 396 | 189 | -12 |
| Ramgarh | - | - | - | - |
| Ranchi | 1097 | 2105 | 975 | -11 |
| Sahebganj | 239 | 559 | 255 | 7 |
| Saraikela | 100 | 831 | 319 | 219 |
| Simdega | 538 | 563 | 273 | -49 |
| West Singhbhum | 238 | 1836 | 775 | 226 |
| Jharkhand | 6628 | 14429 | 6634 | 0.09 |

**Source:** Animal Husbandry Department, Govt. of India

**Table 7.10**
**District wise Total Livestock Population in Jharkhand**
**(Figs in 000)**

| District | 1982 | % to total | 2003 | % to. total | 2019 | % to total |
|---|---|---|---|---|---|---|
| Bokaro | 852 | 2.9 | 1272 | 4.2 | 1088 | 3.6 |
| Chatra | 661 | 2.3 | 975 | 3.2 | 1145 | 3.8 |
| Deoghar | 858 | 2.9 | 1039 | 3.4 | 1410 | 4.7 |
| Dhanbad | 607 | 2.1 | 1378 | 4.5 | 1078 | 3.6 |
| Dumka | 1309 | 4.5 | 1507 | 4.9 | 1645 | 5.5 |
| East Singhbhum | 1136 | 3.9 | 1580 | 5.2 | 1422 | 4.8 |
| Garhwa | 758 | 2.6 | 922 | 3.0 | 1401 | 4.7 |
| Giridih | 1461 | 5.1 | 1582 | 5.2 | 2102 | 7.1 |
| Godda | 801 | 2.8 | 1097 | 3.6 | 1395 | 4.7 |
| Gumla | 4077 | 14.2 | 1571 | 5.1 | 1716 | 5.7 |
| Hazaribagh | 1070 | 3.7 | 1983 | 6.5 | 1559 | 5.2 |
| Jamtara | 627 | 2.1 | 999 | 3.3 | 967 | 3.2 |
| Khunti | - | - | - | - | 956 | 3.2 |
| Koderma | 228 | 0.7 | 383 | 1.2 | 380 | 1.2 |
| Latehar | 563 | 1.9 | 913 | 3.0 | 878 | 2.9 |
| Lohardaga | 1161 | 4.0 | 495 | 1.6 | 461 | 1.5 |
| Pakur | 626 | 2.1 | 1134 | 3.7 | 1231 | 4.1 |
| Palamu | 929 | 3.2 | 1332 | 4.4 | 1378 | 4.6 |
| Ramgarh | - | - | - | - | 381 | 1.2 |
| Ranchi | 5738 | 20.0 | 3784 | 12.5 | 2629 | 8.8 |
| Sahebganj | 697 | 2.4 | 1035 | 3.4 | 992 | 3.3 |
| Saraikela | 483 | 1.6 | 1352 | 4.4 | 1138 | 3.8 |
| Simdega | 2810 | 9.8 | 1223 | 4.0 | 1245 | 4.2 |
| West Singhbhum | 1150 | 4.0 | 2701 | 8.9 | 1956 | 6.6 |
| **Jharkhand** | **28602** | | **30260** | | **29592** | |
| **INDIA** | | | **485000** | | **535780** | |
| **% to India** | | | **6%** | | **5.5%** | |

**Source:** Animal Husbandry Department, Govt. of India

**Table 7.11**

**Number and Percentage of Livestock to the total of the Districts for 2019 in Jharkhand (Figs in 000)**

| District | Total Live-stock | No. of Cattle | No of Buffalo | No of Sheep | No of Goat | No of Pig | No of Poultry | Cattle % | Buffalo % | Sheep % | Goat % | Pig % | Poultry % |
|---|---|---|---|---|---|---|---|---|---|---|---|---|---|
| Bokaro | 1088 | 401 | 29 | 16 | 346 | 17 | 279 | 37 | 3 | 2 | 32 | 3 | 26 |
| Chatra | 1145 | 511 | 102 | 0.5 | 301 | 74 | 152 | 45 | 9 | .04 | 26 | 6 | 13 |
| Deoghar | 1410 | 693 | 54 | 26 | 473 | 30 | 134 | 49 | 4 | 2 | 34 | 2 | 10 |
| Dhanbad | 1078 | 350 | 33 | 18 | 271 | 36 | 370 | 32 | 3 | 2 | 25 | 3 | 34 |
| Dumka | 1645 | 748 | 48 | 29 | 407 | 129 | 284 | 45 | 3 | 2 | 25 | 8 | 15 |
| East Singhbhum | 1422 | 413 | 59 | 92 | 389 | 22 | 447 | 29 | 4 | 6 | 27 | 2 | 31 |
| Garhwa | 1401 | 702 | 74 | 25 | 415 | 31 | 154 | 50 | 5 | 2 | 30 | 2 | 11 |
| Giridih | 2102 | 974 | 112 | 12 | 730 | 53 | 221 | 46 | 5 | 0.6 | 35 | 3 | 10 |
| Godda | 1395 | 685 | 96 | 5 | 289 | 105 | 215 | 49 | 7 | 0.3 | 21 | 8 | 15 |
| Gumla | 1716 | 558 | 90 | 8 | 609 | 108 | 343 | 33 | 5 | 0.5 | 35 | 6 | 20 |
| Hazaribagh | 1559 | 548 | 70 | 3 | 477 | 45 | 416 | 35 | 4 | 0.2 | 30 | 3 | 27 |
| Jamtara | 967 | 432 | 20 | 38 | 257 | 37 | 183 | 45 | 2 | 4 | 26 | 4 | 19 |
| Khunti | 1051 | 280 | 33 | 35 | 267 | 41 | 395 | 26 | 3 | 3 | 25 | 4 | 38 |
| Koderm a | 380 | 184 | 20 | 0.4 | 120 | 3 | 53 | 48 | 5 | 0.1 | 31 | .8 | 14 |
| Latehar | 878 | 348 | 38 | .09 | 274 | 53 | 165 | 40 | 4 | .01 | 31 | 6 | 19 |
| Lohardaga | 461 | 146 | 17 | 02 | 176 | 8 | 112 | 31 | 3 | 0.4 | 38 | 2 | 24 |
| Pakur | 1231 | 387 | 70 | 47 | 310 | 98 | 319 | 31 | 6 | 4 | 25 | 8 | 82 |
| Palamu | 1378 | 663 | 94 | 35 | 352 | 42 | 189 | 48 | 7 | 3 | 26 | 3 | 14 |

| | | | | | | | | | | | | | |
|---|---|---|---|---|---|---|---|---|---|---|---|---|---|
| Ramgarh | 720 | 120 | 29 | 4 | 210 | 18 | 339 | 17 | 4 | 0.6 | 29 | 3 | 47 |
| Ranchi | 2629 | 636 | 125 | 40 | 785 | 68 | 975 | 24 | 5 | 2 | 30 | 3 | 37 |
| Sahebganj | 992 | 263 | 47 | 4 | 355 | 68 | 255 | 25 | 5 | 0.4 | 36 | 7 | 26 |
| Saraikela | 1138 | 305 | 31 | 76 | 365 | 42 | 319 | 27 | 3 | 7 | 32 | 4 | 28 |
| Simdega | 1245 | 403 | 27 | 14 | 418 | 110 | 273 | 32 | 2 | 1 | 33 | 9 | 22 |
| West Singhbhum | 1956 | 471 | 31 | 113 | 528 | 38 | 775 | 24 | 2 | 6 | 27 | 2 | 40 |
| Jharkhand | 29592 | 11174 | 1349 | 784 | 8375 | 1276 | 6634 | 38 | 5 | 3 | 28 | 4 | 22 |

**Source:** Animal Husbandry Department, Govt. of India

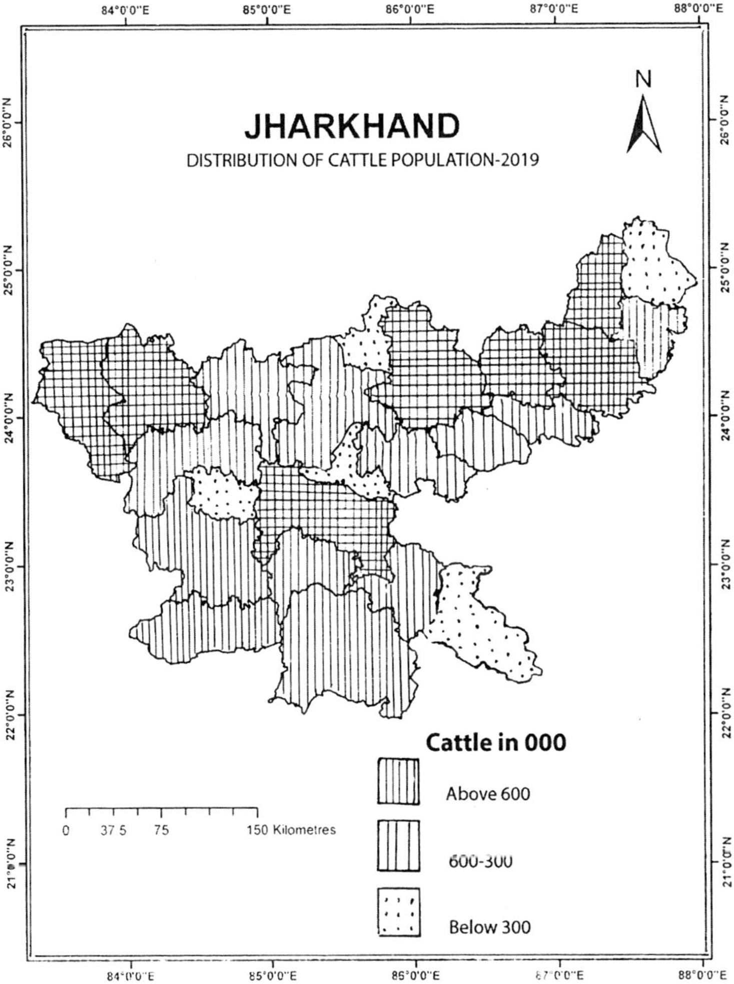


Map 7.2: Distribution of Cattle Population (2019)

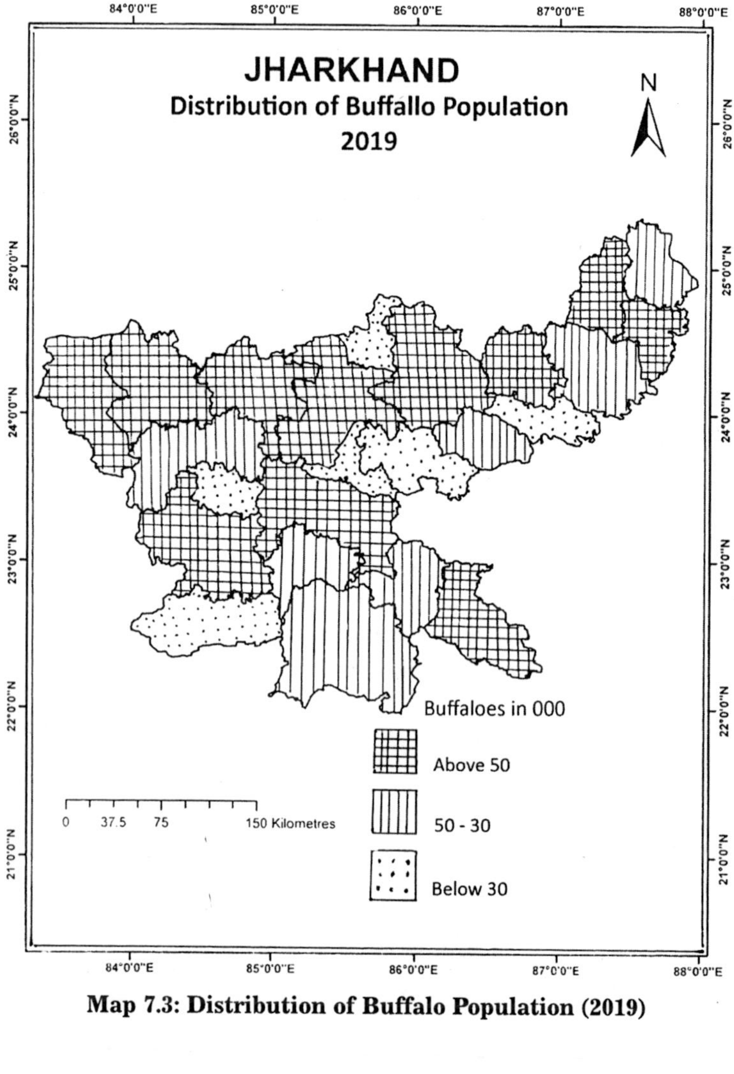


**Map 7.3: Distribution of Buffalo Population (2019)**

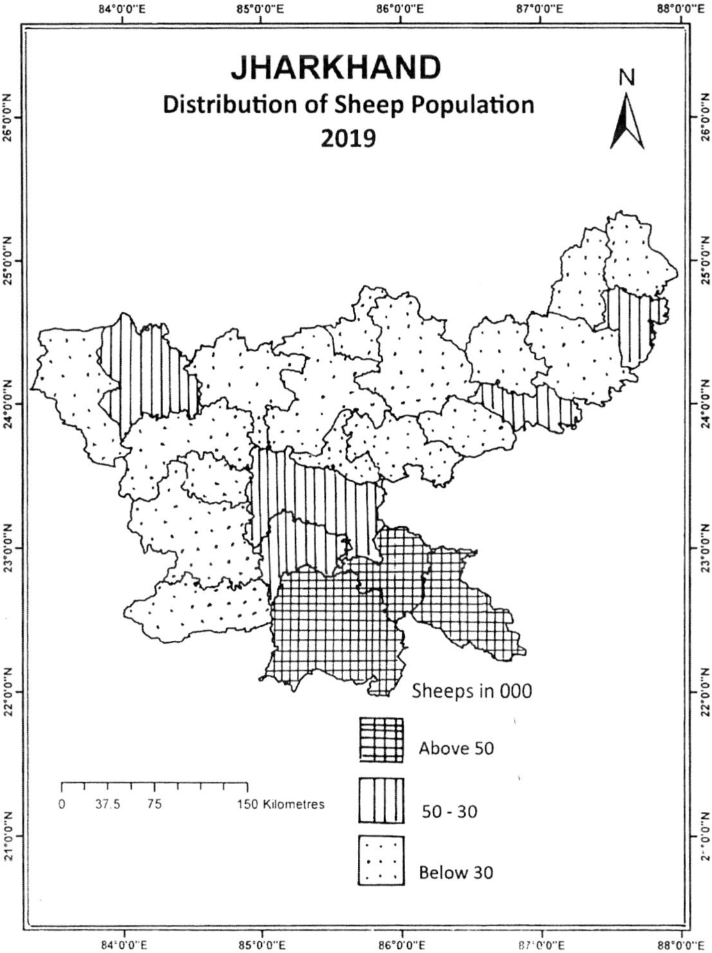


**Map 7.4: Distribution of Sheep Population (2019)**

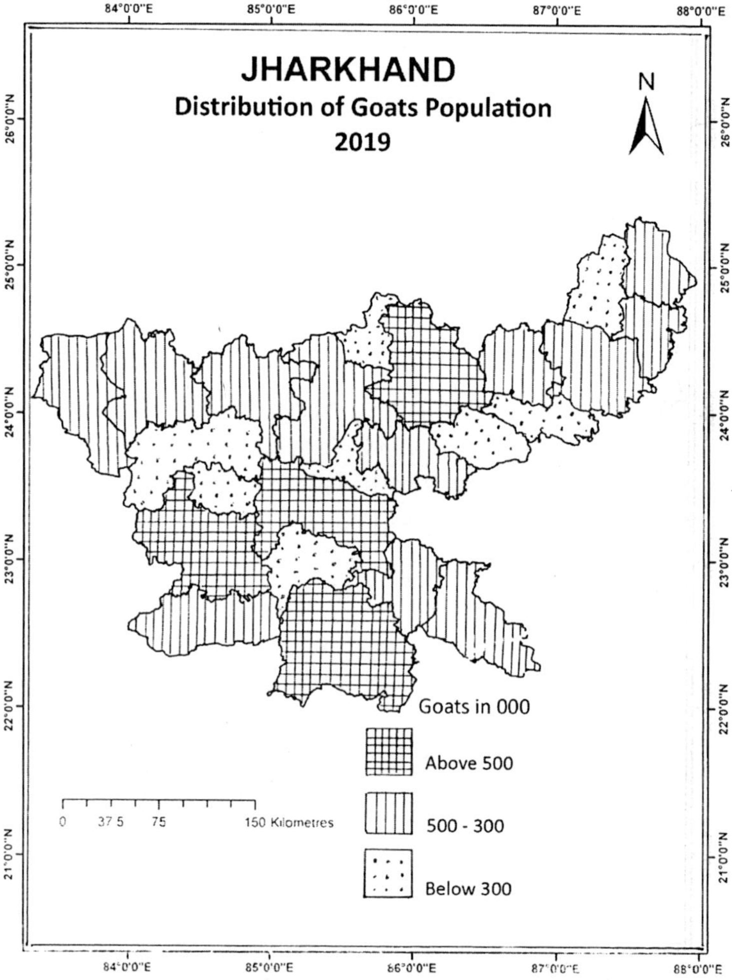


**Map 7.5: Distribution of Goat Population (2019)**

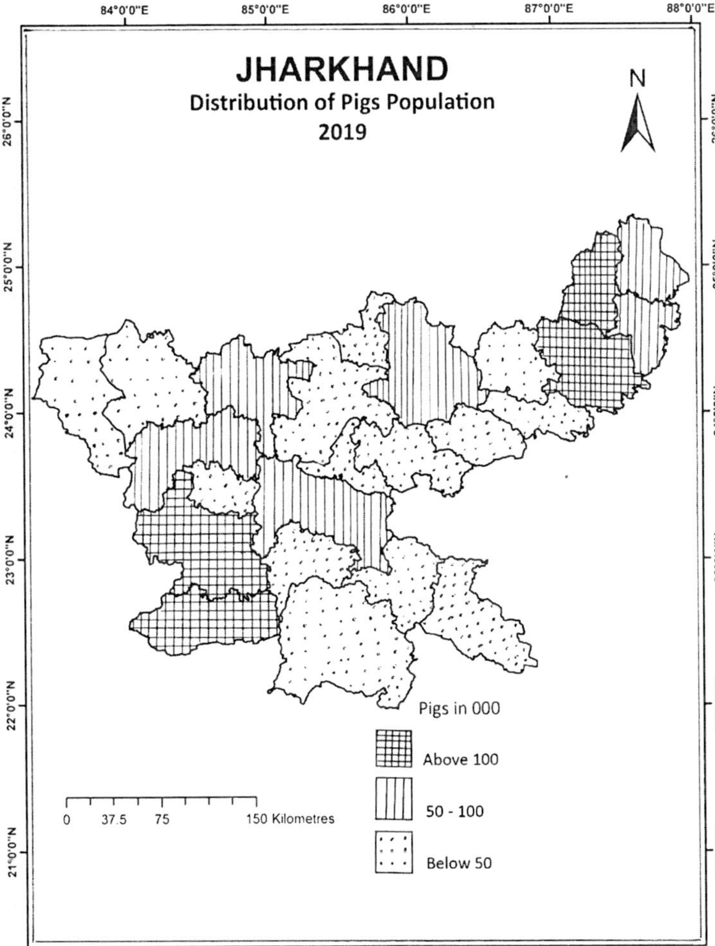


**Map 7.6: Distribution of Pig Population (2019)**

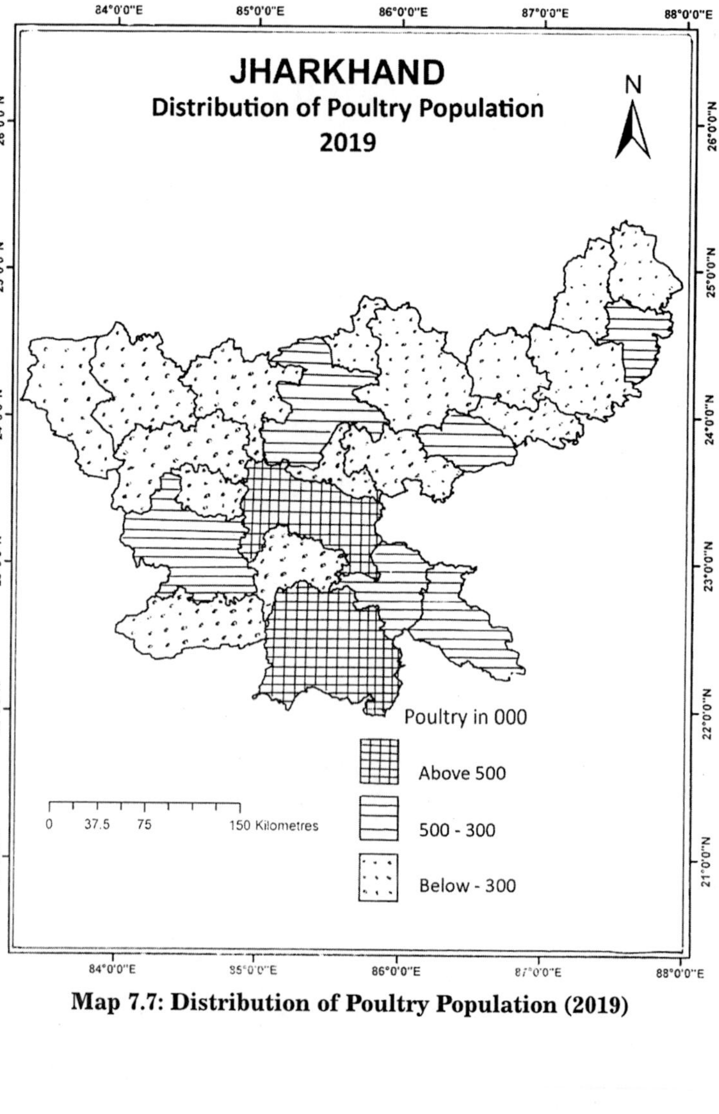


**Map 7.7: Distribution of Poultry Population (2019)**

**Table 7.12**
**Combination of Livestock in Jharkhand**

| Number of Livestock in combination | Percent of Hypothetical Livestock Combination | Percent of Actual Livestock in Descending order | Difference between hypothetical and actual % of livestock (d) | Square of Difference ($d^2$) | Total of Square and Name and Number of Combination |
|---|---|---|---|---|---|
| **Bokaro District** | | | | | |
| One Livestock Combination | 100 | 37 | 63 | 3969 | 3969 |
| Two Livestock Combination | 50.00 | 37 | 13 | 169 | 493 |
| | 50.00 | 32 | 18 | 324 | |
| Three Livestock Combination | 33.33 | 37 | 3.67 | 13.46 | 68.94 (Cattle-Goat-Poultry) |
| | 33.33 | 32 | 1.33 | 1.76 | |
| | 33.33 | 26 | 7.33 | 53.72 | |
| Four Livestock Combination | 25.00 | 37 | 12 | 114 | 648 |
| | 25.00 | 32 | 7 | 49 | |
| | 25.00 | 26 | 1 | 1 | |
| | 25.00 | 3 | 22 | 484 | |
| **Chatra District** | | | | | |
| One Livestock Combination | 100 | 45 | 55 | 3025 | 3025 |

| | | | | | |
|---|---|---|---|---|---|
| Two Livestock Combination | 50.00 | 45 | 5 | 25 | 601 |
| | 50.00 | 26 | 24 | 576 | (Cattle- Goat) |
| Three Livestock Combination | 33.33 | 45 | 11.67 | 136.18 | |
| | 33.33 | 26 | 7.33 | 53.72 | 603.2 |
| | 33.33 | 13 | 20.33 | 413.30 | |
| Four Livestock Combinatio n | 25.00 | 45 | 20 | 400 | |
| | 25.00 | 26 | 1 | 1 | 801 |
| | 25.00 | 13 | 12 | 144 | |
| | 25.00 | 9 | 16 | 256 | |
| **Deoghar District** | | | | | |
| One Livestock Combination | 100 | 49 | 51 | 2601 | 2601 |
| Two Livestock Combination | 50.00 | 49 | 1 | 1 | 257 |
| | 50.00 | 34 | 16 | 256 | (Cattle- Goat) |
| Three Livestock Combination | 33.33 | 49 | 15.67 | 245.54 | |
| | 33.33 | 34 | 0.67 | 0.448 | 490.26 |
| | 33.33 | 10 | 23.33 | 244.28 | |
| Four Livestock Combination | 25.00 | 49 | 24 | 576 | |
| | 25.00 | 34 | 9 | 81 | 1323 |
| | 25.00 | 10 | 15 | 225 | |
| | 25.00 | 4 | 21 | 441 | |

| | | | | | |
|---|---|---|---|---|---|
| **Dhanbad District** | | | | | |
| One Livestock Combination | 100 | 34 | 66 | 4356 | 4356 |
| Two Livestock Combination | 50.00 | 34 | 16 | 256 | 580 |
| | 50.00 | 32 | 18 | 324 | |
| Three Livestock Combination | 33.33 | 34 | 1.33 | 1.76 | 72.908 |
| | 33.33 | 32 | 1.33 | 1.768 | (Poultry-Cattle- |
| | 33.33 | 25 | 8.33 | 69.38 | Goat) |
| Four Livestock Combination | 25.00 | 34 | 9 | 81 | |
| | 25.00 | 32 | 7 | 49 | 578 |
| | 25.00 | 25 | 0 | 0 | |
| | 25.00 | 3 | 22 | 448 | |
| **Dumka District** | | | | | |
| One Livestock Combination | 100 | 45 | 55 | 3025 | 3025 |
| Two Livestock Combination | 50.00 | 45 | 5 | 25 | 277 |
| | 50.00 | 25 | 25 | 252 | (Cattle- Goat) |
| Three Livestock Combination | 33.33 | 45 | 11.67 | 136.18 | |
| | 33.33 | 25 | 8.33 | 69.38 | 541.54 |
| | 33.33 | 15 | 18.33 | 335.98 | |

| | | | | | |
|---|---|---|---|---|---|
| Four Livestock Combination | 25.00 | 45 | 20 | 400 | 789 |
| | 25.00 | 25 | 0 | 00 | |
| | 25.00 | 15 | 10 | 100 | |
| | 25.00 | 8 | 17 | 289 | |
| **East Singhbhum District** | | | | | |
| One Livestock Combination | 100 | 31 | 69 | 4761 | 4761 |
| Two Livestock Combination | 50.00 | 31 | 19 | 361 | 802 |
| | 50.00 | 29 | 21 | 441 | |
| Three Livestock Combination | 33.33 | 31 | 2.33 | 5.42 | 77.88 |
| | 33.33 | 29 | 4.33 | 18.74 | (Poultry-Cattle- |
| | 33.33 | 27 | 7.33 | 53.72 | Goat) |
| Four Livestock Combination | 25.00 | 31 | 6 | 36 | |
| | 25.00 | 29 | 4 | 16 | 421 |
| | 25.00 | 27 | 2 | 8 | |
| | 25.00 | 6 | 19 | 361 | |
| **Garhwa District** | | | | | |
| One Livestock Combination | 100 | 50 | 50 | 2500 | 2500 |
| Two Livestock Combination | 50.00 | 50 | 00 | 00 | 400 |
| | 50.00 | 30 | 20 | 400 | (Cattle- Goat) |

| | | | | | |
|---|---|---|---|---|---|
| Three Livestock Combination | 33.33 | 50 | 17.33 | 300.32 | |
| | 33.33 | 30 | 3.33 | 11.08 | 810 |
| | 33.33 | 11 | 22.33 | 498.62 | |
| Four Livestock Combination | 25.00 | 50 | 25 | 625 | |
| | 25.00 | 30 | 5 | 25 | 1246 |
| | 25.00 | 11 | 14 | 196 | |
| | 25.00 | 5 | 20 | 406 | |
| | | **Giridih District** | | | |
| One Livestock Combination | 100 | 46 | 54 | 2916 | 2916 |
| Two Livestock Combination | 50.00 | 46 | 54 | 2916 | 3141 |
| | 50.00 | 35 | 15 | 225 | |
| Three Livestock Combination | 33.33 | 46 | 13.33 | 177.68 | 727.38 |
| | 33.33 | 35 | 2.33 | 5.42 | (Cattle-Goat- |
| | 33.33 | 10 | 23.33 | 544.28 | Poultry) |
| Four Livestock Combination | 25.00 | 46 | 21 | 441 | |
| | 25.00 | 35 | 10 | 100 | 1166 |
| | 25.00 | 10 | 15 | 225 | |
| | 25.00 | 5 | 20 | 400 | |
| | | **Godda District** | | | |
| One Livestock Combination | 100 | 49 | 51 | 2601 | 2601 |

| | | | | | |
|---|---|---|---|---|---|
| Two Livestock Combination | 50.00 | 49 | 1 | 1 | 842 |
| | 50.00 | 21 | 29 | 841 | |
| Three Livestock Combination | 33.33 | 49 | 16.33 | 266.66 | |
| | 33.33 | 21 | 12.33 | 152.02 | 755 |
| | 33.33 | 15 | 18.33 | 335.98 | |
| Four Livestock Combination | 25.00 | 19 | 6 | 36 | 441 |
| | 25.00 | 21 | 4 | 16 | (Cattle-Poultry- |
| | 25.00 | 15 | 10 | 100 | Goat-Pig) |
| | 25.00 | 8 | 17 | 289 | |
| **Gumla District** | | | | | |
| One Livestock Combination | 100 | 35 | 65 | 4225 | 4225 |
| Two Livestock Combination | 50.00 | 35 | 5 | 225 | 514 |
| | 50.00 | 33 | 17 | 289 | |
| Three Livestock Combination | 33.33 | 35 | 2.33 | 5.42 | 183 |
| | 33.33 | 33 | 0.33 | 0.10 | (Goat-Cattle- |
| | 33.33 | 20 | 13.33 | 177.68 | Poultry) |
| Four Livestock Combination | 25.00 | 35 | 10 | 100 | |
| | 25.00 | 33 | 8 | 64 | 550 |
| | 25.00 | 20 | 5 | 25 | |
| | 25.00 | 6 | 19 | 361 | |

| | | | | | |
|---|---|---|---|---|---|
| | | **Hazaribagh District** | | | |
| One Livestock Combination | 100 | 35 | 65 | 4225 | 422 |
| Two Livestock Combination | 50.00 | 35 | 5 | 225 | 625 |
| | 50.00 | 30 | 20 | 400 | |
| Three Livestock Combination | 33.33 | 35 | 2.33 | 5.42 | 57 |
| | 33.33 | 30 | 3.33 | 11:08 | (Cattle-Goat- |
| | 33.33 | 27 | 6.33 | 40.06 | Poultry) |
| Four Livestock Combination | 25.00 | 35 | 10 | 100 | |
| | 25.00 | 30 | 5 | 25 | 570 |
| | 25.00 | 27 | 2 | 4 | |
| | 25.00 | 4 | 21 | 441 | |
| | | **Jamtara District** | | | |
| One Livestock Combination | 100 | 45 | 55 | 3025 | 3025 |
| Two Livestock Combination | 50.00 | 45 | 5 | 25 | 601 |
| | 50.00 | 26 | 24 | 576 | |
| Three Livestock Combination | 33.33 | 45 | 12.33 | 152 | 410 |
| | 33.33 | 26 | 7.33 | 53 | (Cattle-Goat- |
| | 33.33 | 19 | 14.33 | 205 | Poultry) |

| | | | | | |
|---|---|---|---|---|---|
| Four Livestock Combination | 25.00 | 45 | 20 | 400 | |
| | 25.00 | 26 | 1 | 1 | 878 |
| | 25.00 | 19 | 6 | 36 | |
| | 25.00 | 4 | 21 | 441 | |
| | | **Khunti District** | | | |
| One Livestock Combination | 100 | 38 | 62 | 3844 | 3844 |
| Two Livestock Combination | 50.00 | 38 | 12 | 144 | 720 |
| | 50.00 | 26 | 24 | 576 | |
| Three Livestock Combination | 33.33 | 38 | 5.33 | 28 | 151 |
| | 33.33 | 36 | 7.33 | 54 | (Poultry-Cattle- |
| | 33.33 | 25 | 8.33 | 69 | Goat) |
| Four Livestock Combination | 25.00 | 38 | 13 | 169 | |
| | 25.00 | 26 | 1 | 1 | 611 |
| | 25.00 | 25 | 00 | 0 | |
| | 25.00 | 4 | 21 | 441 | |
| | | **Koderma District** | | | |
| One Livestock Combination | 100 | 48 | 52 | 2704 | 2704 |
| Two Livestock Combination | 50.00 | 48 | 2 | 4 | 845 |
| | 50.00 | 31 | 29 | 841 | |

| | | | | | |
|---|---|---|---|---|---|
| Three Livestock Combination | 33.33 | 48 | 15.33 | 235 | 614 |
| | 33.33 | 31 | 2.33 | 5 | (Cattle-Goat- |
| | 33.33 | 14 | 19.33 | 374 | Poultry) |
| Four Livestock Combination | 25.00 | 48 | 23 | 529 | |
| | 25.00 | 31 | 6 | 36 | 1086 |
| | 25.00 | 14 | 11 | 121 | |
| | 25.00 | 5 | 20 | 400 | |
| **Latehar District** | | | | | |
| One Livestock Combination | 100 | 40 | 60 | 3600 | 3600 |
| Two Livestock Combination | 50.00 | 40 | 10 | 100 | 461 |
| | 50.00 | 31 | 19 | 361 | |
| Three Livestock Combination | 33.33 | 40 | 13.33 | 178 | 388 |
| | 33.33 | 31 | 2.33 | 5 | (Cattle-Goat- |
| | 33.33 | 19 | 14.33 | 205 | Poultry) |
| Four Livestock Combination | 25.00 | 40 | 15 | 225 | |
| | 25.00 | 31 | 6 | 36 | 658 |
| | 25.00 | 19 | 6 | 36 | |
| | 25.00 | 6 | 19 | 361 | |
| **Loahrdaga District** | | | | | |
| One Livestock Combination | 100 | 38 | 62 | 3844 | 3844 |

| | | | | | |
|---|---|---|---|---|---|
| Two Livestock Combination | 50.00 | 38 | 12 | 144 | 505 |
| | 50.00 | 31 | 19 | 361 | |
| Three Livestock Combination | 33.33 | 38 | 5.33 | 28 | 120 |
| | 33.33 | 31 | 2.33 | 5 | (Goat-Cattle- |
| | 33.33 | 24 | 9.33 | 87 | Poultry) |
| Four Livestock Combination | 25.00 | 38 | 13 | 169 | |
| | 25.00 | 31 | 6 | 36 | 690 |
| | 25.00 | 24 | 1 | 1 | |
| | 25.00 | 3 | 22 | 484 | |
| **Pakur District** | | | | | |
| One Livestock Combination | 100 | 82 | 22 | 484 | 484 (Poultry) |
| Two Livestock Combination | 50.00 | 82 | 32 | 1024 | 1465 |
| | 50.00 | 31 | 21 | 441 | |
| Three Livestock Combination | 33.33 | 82 | 49.33 | 2433 | 2507 |
| | 33.33 | 31 | 2.33 | 5 | |
| | 33.33 | 25 | 8.33 | 69 | |
| Four Livestock Combination | 25.00 | 82 | 57 | 3249 | |
| | 25.00 | 31 | 6 | 36 | 674 |
| | 25.00 | 25 | 00 | 00 | |
| | 25.00 | 8 | 17 | 289 | |

| | | | | | |
|---|---|---|---|---|---|
| **Palamu District** | | | | | |
| One Livestock Combination | 100 | 48 | 52 | 2704 | 2704 |
| Two Livestock Combination | 50.00 | 48 | 2 | 4 | 580 |
| | 50.00 | 26 | 24 | 576 | (Cattle- Goat) |
| Three Livestock Combination | 33.33 | 48 | 15.33 | 235 | |
| | 33.33 | 26 | 7.33 | 54 | 663 |
| | 33.33 | 14 | 19.33 | 374 | |
| Four Livestock Combination | 25.00 | 48 | 23 | 529 | |
| | 25.00 | 26 | 1 | 1 | 715 |
| | 25.00 | 14 | 11 | 121 | |
| | 25.00 | 7 | 8 | 64 | |
| **Ramgarh district** | | | | | |
| One Livestock Combination | 100 | 47 | 53 | 2809 | 2809 |
| Two Livestock Combination | 50.00 | 47 | 3 | 9 | 450 |
| | 50.00 | 29 | 21 | 441 | (Poultry-Goat) |
| Three Livestock Combination | 33.33 | 47 | 14.33 | 205 | |
| | 33.33 | 29 | 4.33 | 19 | 490 |
| | 33.33 | 17 | 16.33 | 266 | |

| | | | | | |
|---|---|---|---|---|---|
| Four Livestock Combination | 25.00 | 47 | 22 | 484 | |
| | 25.00 | 29 | 4 | 16 | 1005 |
| | 25.00 | 17 | 8 | 64 | |
| | 25.00 | 4 | 21 | 441 | |
| **Ranchi District** | | | | | |
| One Livestock Combination | 100 | 37 | 77 | 5929 | 5929 |
| Two Livestock Combination | 50.00 | 37 | 13 | 169 | 569 |
| | 50.00 | 30 | 20 | 400 | |
| Three Livestock Combination | 33.33 | 37 | 4.33 | 19 | 117 |
| | 33.33 | 30 | 3.33 | 11 | (Poultry-Goat- |
| | 33.33 | 24 | 9.33 | 87 | Cattle) |
| Four Livestock Combination | 25.00 | 37 | 12 | 144 | |
| | 25.00 | 30 | 5 | 25 | 570 |
| | 25.00 | 24 | 1 | 1 | |
| | 25.00 | 5 | 20 | 400 | |
| **Sahebganj District** | | | | | |
| One Livestock Combination | 100 | 36 | 76 | 5776 | 5776 |
| Two Livestock Combination | 50.00 | 36 | 14 | 196 | 772 |
| | 50.00 | 26 | 24 | 576 | |

| | | | | | |
|---|---|---|---|---|---|
| Three Livestock Combination | 33.33 | 36 | 3.33 | 11 | 134 |
| | 33.33 | 26 | 7.33 | 54 | (Goat-Poultry- |
| | 33.33 | 25 | 8.33 | 69 | Cattle) |
| Four Livestock Combination | 25.00 | 36 | 11 | 121 | |
| | 25.00 | 26 | 1 | 1 | 446 |
| | 25.00 | 25 | 0 | 0 | |
| | 25.00 | 7 | 18 | 324 | |
| **Saraikela District** | | | | | |
| One Livestock Combination | 100 | 32 | 68 | 4624 | 4624 |
| Two Livestock Combinatio n | 50.00 | 32 | 18 | 324 | 808 |
| | 50.00 | 28 | 22 | 484 | |
| Three Livestock Combination | 33.33 | 32 | 1.33 | 2 | 70 |
| | 33.33 | 28 | 5.33 | 28 | (Goat-Poultry- |
| | 33.33 | 27 | 6.33 | 40 | Cattle) |
| Four Livestock Combination | 25.00 | 32 | 7 | 49 | |
| | 25.00 | 28 | 3 | 9 | 386 |
| | 25.00 | 27 | 2 | 4 | |
| | 25.00 | 7 | 18 | 324 | |
| **Simdega District** | | | | | |
| One Livestock Combination | 100 | 33 | 77 | 5929 | 5929 |

| | | | | | |
|---|---|---|---|---|---|
| Two Livestock Combination | 50.00 | 33 | 17 | 289 | 613 |
| | 50.00 | 32 | 18 | 324 | |
| Three Livestock Combination | 33.33 | 33 | 0.33 | 0.1 | 109 |
| | 33.33 | 32 | 1.33 | 2 | (Goat-Cattle- |
| | 33.33 | 22 | 10.33 | 107 | Poultry) |
| Four Livestock Combination | 25.00 | 33 | 8 | 64 | |
| | 25.00 | 32 | 7 | 49 | 394 |
| | 25.00 | 22 | 5 | 25 | |
| | 25.00 | 9 | 16 | 256 | |
| **West Singhbhum District** | | | | | |
| One Livestock Combination | 100 | 40 | 60 | 3600 | 3600 |
| Two Livestock Combination | 50.00 | 40 | 10 | 100 | 629 |
| | 50.00 | 27 | 23 | 529 | |
| Three Livestock Combination | 33.33 | 40 | 13.33 | 178 | 305 |
| | 33.33 | 27 | 6.33 | 40 | (Poultry-Goat- |
| | 33.33 | 24 | 9.33 | 87 | Cattle) |
| Four Livestock Combination | 25.00 | 40 | 15 | 225 | |
| | 25.00 | 27 | 2 | 4 | 591 |
| | 25.00 | 24 | 1 | 1 | |
| | 25.00 | 6 | 19 | 361 | |

**Livestock Regions**

| | | |
|---|---|---|
| One Livestock Region | Poultry | Pakur |
| Two Livestock Region | Cattle-Goat | Chatra, Deoghar, Dumka, Garhwa, Palamu |
| | Poultry-Goat | Ramgarh |
| Three Livestock Region | Cattle-Goat-Poultry | Bokaro, Giridih, Hazaribagh, Jamtara, Koderma, Latehar |
| | Poultry-Cattle-Goat | Dhanbad, East Singhbhum, Khunti |
| | Goat-Cattle-Poultry | Gumla, Lohardaga, Simdega |
| Four Livestock Region | Poultry-Goat-Cattle | Ranchi, West Singhbhum |
| | Goat-Poultry-Cattle | Sahebganj, Saraikela |
| | Cattle-Goat-Poultry-Pig | Godda |

**Livestock Combination in Jharkhand**

As we have studied about crop-combination in an earlier chapter (Six), here we shall study about livestock combination. This is a very important aspect in the study of Animal Husbandry. Since the components of crops and livestock are intimately related, combination analysis is inevitable. The relative position and strengths of different livestock units can be better comprehended with combination analyses. From established livestock combinations one can infer the geography of climate and technological variables in particular and socio-economic factors in general. (Singh & Dhillon) The farmers keep several types of livestock for their various needs and it is a common practice to raise many types rather than a single class of livestock is not unusual. To get a comprehensive idea about the distribution pattern of livestock in Jharkhand

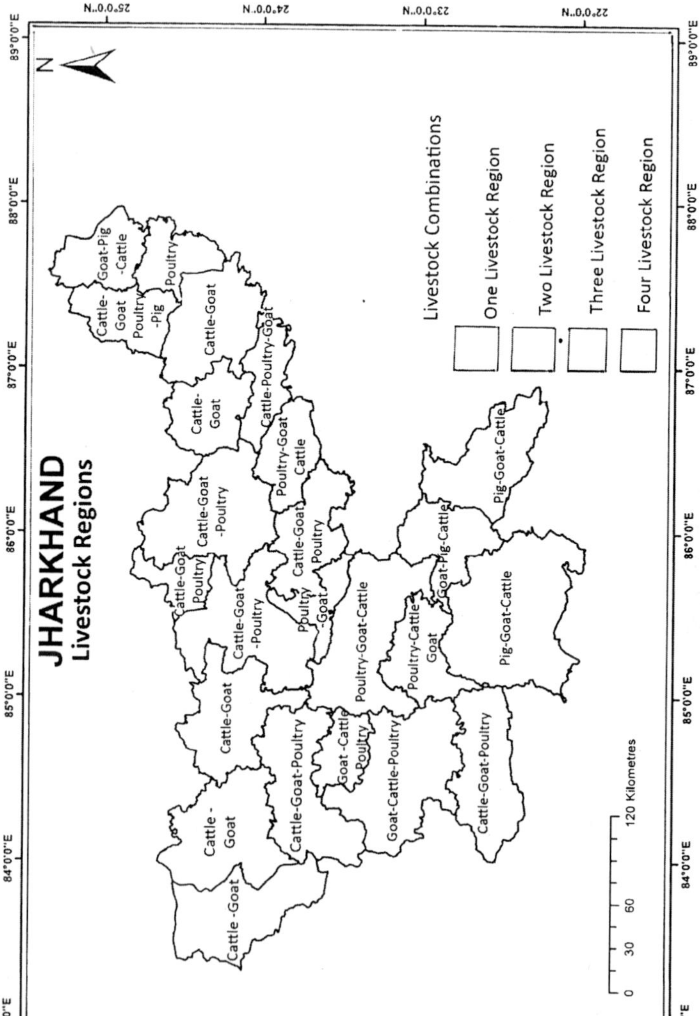


**Map 7.8: Livestock Regions**

we have prepared a combination of livestock using the method of crop combination. First, we prepared table 7.11 calculating the percentage of livestock to the total of district livestock. Here we have considered only six types of livestock for calculations, these are - Cattle, Buffalo, Sheep, Goat, Pig and Poultry. Table 7.12 represents the livestock combination for all the districts in the state We observed four types of combinations in the calculations - One Livestock combination, Two Livestock combination, Three Livestock combination and Four Livestock combination. These four combinations have revealed nine types of sub groups of livestock which can be called livestock regions. One livestock region is Pakur district where only Poultry is prevalent. Though other types of livestock are available in this region but poultry is the most prominent livestock combination for this district. Two Livestock combination has two types - one is "Cattle-Goat" combination which is spread over Chatra, Deoghar, Dumka, Garhwa and Palamu districts while other combination "Poultry-Goat" which is found only in Ramgarh district. Three livestock combination has five types of combinations - "Cattle-Goat-Poultry" is spread over Bokaro, Giridih, Hazaribagh, Jamtara, Koderma and Latehar districts. Another group is "Poultry-Cattle-Goat" which is predominant in Dhanbad, East Singhbhum and Khunti. Third sub-group is "Goat-Cattle-Poultry" which is found in Gumla, Lohardaga and Simdega districts. Fourth sub-group which is "Poultry-Goat-Cattle" is prevalent in Ranchi and West Singhbhum districts. Last sub-group is "Goat-Poultry-Cattle" which is found in Sahebganj and Saraikela districts. In this category, first sub-group "Cattle- Goat-Poultry" is similar to state average.

Four livestock combination - "Cattle-Goat-Poultry-Pig" is a major group of Jharkhand which consists of four livestock. It shows a vast presence of livestock in the district. This combination is spread over only in Godda district. Interestingly enough, the livestock combination in major parts of Jharkhand are uniform i.e., large spreads are covered either by 'Cattle- Goat" or "Poultry-Goat" and Poultry-Cattle" except a few districts. State shows three livestock combination in the major parts i.e., Cattle Goat and Poultry.

When we look into the northern part of the Jharkhand state, the combination of "Cattle-Goat" is prominent, whereas in the southern part of the state has "Goat-Cattle" or "Poultry-Goat" or "Poultry-Cattle" combination in prominence. In a different perspective we get Cattle as first livestock in only Palamu, Jamtara, Koderma, Latehar, Giridih, Godda, Hazaribagh, Dumka, Garhwa, Deoghar, Bokaro and Chatra districts. These are twelve districts in number which cover around 50 percent area of the state. Rest 50 percent register the first livestock as either Goat or Poultry. These districts are Ranchi, Sahebganj, Saraikela, Simdega, Lohardaga, Pakur, Ramgarh, Khunti, Gumla, East Singhbhum, West Singhbhum and Dhanbad. If we deep dive into the reasons of the differences in first animal we find it is mostly culture related. Northern part of Jharkhand is influenced heavily by Bihar which is predominantly agriculture based and vegetarian while the southern part bordering with West-Bengal and Orissa are prominently non-vegetarian.

## Fisheries

In India chief sources of fishing are coastal margins of the sea, river, estuaries and backwaters for marine and estuarine fish and rivers, canals, tanks, inundated tracts etc. for fresh water fish. Thus, there are three sources of fishing - **Inland fishing, Sea Fishing** and **Estuarine and Backwater Fishing**.

Inland fisheries constitute fresh fish from rivers, ponds, canals, tanks, irrigation canals, inundated tracts etc. These are mostly found in Assam, Orissa, Madya Pradesh, Bihar, Jharkhand and all other states in the country. Assam leads in this category of fisheries to the rest of the country. The state of West Bengal, Bihar, Jharkhand and Assam taken together account for 71.6 percent of the total fresh water fish in India. The production of Inland fish has gone up from 7.5 lakh tons in 1950-51 to 125 lakh tons in 2017-18. In Jharkhand fish production was only 14 thousand tons during 2001 and now it has grown to 2.33 lakh tons in 2020. It is about 1.2 percent to the production of India. (Jaiprakash & Singh, R.P., 2016)

Fisheries in Jharkhand are developing with a good pace and is one of the sectors which is attracting rural youth in a big way. Fisheries resources in state are limited but considerable efforts are being made for their restoration, renovation and area expansion. Jharkhand fishery department has several schemes for the development of fisheries in the state. Some of the main schemes are - Training programs for fish culturists, Seed production by trained seed growers. New ponds are being made by farmers with the help of Jharkhand government. 3 lakh per acre subsidy, fish marketing by co-operative society, farmers can apply for fish sale center to 'Jhascofish', cage farming facility is available in different reservoirs like Maithan, Chandil, Tenughat, Konar, Masanjor, Tilaiya for the FCC members, fish feeds are available for farmers in Ranchi, Chandil, Dhanbad, Bokaro, Ved Vyas awas yojana is for providing houses for poor fishermen, group accident insurance scheme for active fishermen (Department of Animal Husbandry & Cooperative, Jharkhand Government).

**Table 7.13**
**Aquaculture resources in Jharkhand**

| Water Area Description | Number | Area (Hectares) |
|---|---|---|
| Private Tanks | 129863 | 63053.11 |
| Govt. Tanks | 17959 | 15956.17 |
| Reservoir | 435 | 129990 |
| Check Dam and Ahar | 9360 | 10056 |
| Coal Pits and mines | 1741 | 9880 |
| Rivers | - | 1800 Kms |
| MNREGA Wells | 204000 | |
| Dhoba | 135279 | |
| RFF | 962 | 1870 Units |
| Fisheries Cooperation Committee | 488 | |
| No of Members | 32635 | |

| | |
|---|---|
| No of Members (Reservoirs) | 15688 |
| No of Fishermen | 205920 |
| Voluntary Matsya Mitra | 7500 |
| No of Installed Cages | 5937 |
| Fish Production in Cages | 3-4 tons per cage |
| Pucca Turtle Habitat | 22354 |
| Accidental Insurance Medication Payment (till 2018-19) | 42 |
| **Fish Production** | **Metric Tons (2020)** |
| Fish Production (Year 2019-20) | 223000 |
| Demand (Year 2020-21) | 240000 |
| Present Production Capacity | 2570 Kg/ ha/year |
| National Capacity | 3000 Kg/ ha/year |
| Fish Seed Production in 2019-2020 | 1000 crore |
| Total Fish Seed Production till now | 1056.92 Crore |
| **Hatchery** | **Number** |
| Departmental | 17 |
| Private | 107 |
| Portable Hatchery | 128 |
| Fish Field (in different district) | 88 |

**Source:** Departmental website of Jharkhand Fisheries (jharkhandfisheries.org)

**Table-7.14**
**Fish Production in Jharkhand (in lakh tons)**

| Year | Production (in lakh tons) |
|---|---|
| 2011-12 | 0.92 |
| 2012-13 | 0.97 |
| 2013-14 | 1.05 |

| | |
|---|---|
| 2014-15 | 1.06 |
| 2015-16 | 1.16 |
| 2016-17 | 1.45 |
| 2017-18 | 1.90 |

**Source:** Handbook on fisheries statistics 2018

**Table 7.15**
**Fish released for development of fisheries sector during 2010-11 to 2017-18 in Jharkhand**

| Year | Rs (In Lakh) |
|---|---|
| 2011-12 | 256.33 |
| 2012-13 | 251.50 |
| 2013-14 | 313.11 |
| 2014-15 | 554.81 |
| 2015-16 | 297.72 |
| 2016-17 | 589.12 |
| 2017-18 | 2300.20 |
| Total | 4637.79 |

**Source:** Handbook on fisheries statistics 2018

**Table 7.16**
**Inland Fish Landings by Species in Jharkhand-2017**

| Species | Quantity (in tons) |
|---|---|
| 1. Major Carps (Catla, Rohu, Marigal) | 161800 |
| 2. Minor Carps | 1520 |
| 3. Exotic Carps (Common, Silver, Grass) | 17200 |
| 4. Murrels (Ophiocephalus spp) | 00 |
| 5. Catfish (Wallago attu, Pangasius) | 9500 |
| 6. Other Freshwater Fish | 00 |
| Total | 190020 |

**Source:** Handbook on fisheries statistics 2018

## Resources

Resources are the basic thing for development of any industry in any region. Jharkhand Government has improved the infrastructure and resources for the development of fisheries in the state - natural and cultural. As fish rearing resource Jharkhand has - Well developed private tanks, Government Tanks, Reservoirs, Check dams, Ahar, Coal Pits and Mines, Rivers, Dhoba in abundant numbers. Table 7.13 shows the numbers of fish rearing resources. There are 1.29 lakh private tanks, 17.95 thousand government tanks. Throughout the state the length of the rivers run up to 1800 Kms, 435 Reservoirs, 9360 Check Dams and Ahar, 1.35 lakh

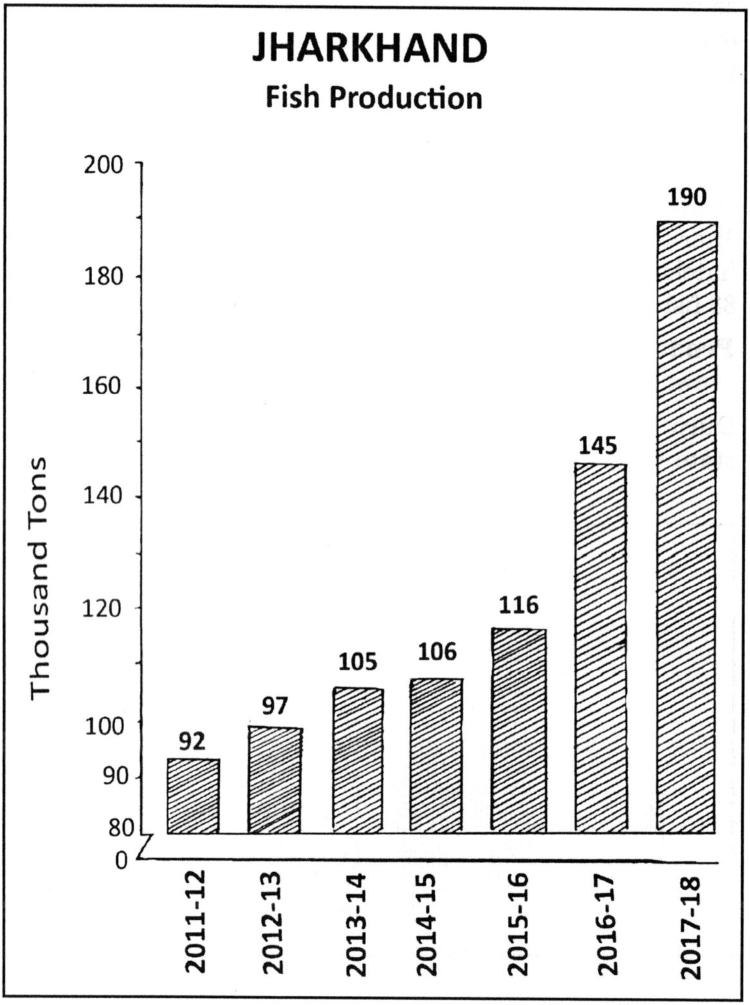


Map 7.9: Fish Production

Dhoba, 2.04 lakh MNREGA wells. There are 488 Fisheries Cooperation Committees with 32.63 thousand members are associated with them. 15.68 thousand members are associated with the reservoirs. Number of fishermen in the state is around 2.05 lakh. There are 7500 voluntary Matsya Mitra members as well in the state. There are 11 districts in the state where different fish cultivation activities are being carried out with the help of government agencies. These districts are Ranchi, Dhanbad, Koderma, Saraikela, Deoghar, Ramgarh, Gumla, Jamshedpur (East Singhbhum), Pakur and Dumka. The activities mainly include seed production, cage culture, fish feed, wholesale market, ornamental fisheries etc. Hatchery is a facility where eggs are hatched under artificial conditions, especially those of fish or poultry. Jharkhand has 17 government hatcheries and 107 private hatcheries. Portable hatcheries are 128 in number. There are 88 fish field across the state in Jharkhand.

## Production

Table 7.14 shows the fish production details in Jharkhand. During 2011- 12 the fish production was 0.92 lakh tons in the state which doubled in the duration of six years and was

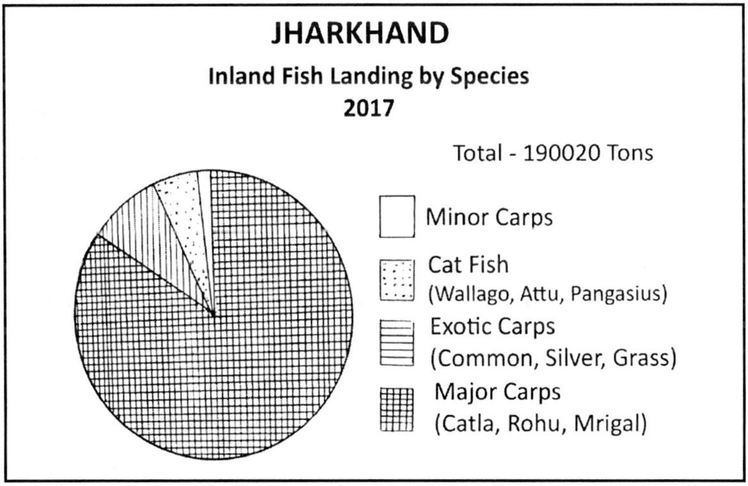


**Map 7.10: Inland Fish Landings by Species**

recorded 1.90 lakh tons during 2017-18.

Table 7.16 details various species of fish in Jharkhand. There are mainly six types of fish available in Jharkhand. These are Major Carp (Catla, Rohu, Marigal), Minor Carp, Exotic Carps, Murrels, Catfish and other fresh water fish.

## Dairying

History of dairy development in India can be divided into two phases—pre and post-Operation Flood. History of pre-operation flood starts from 1913 when British rule established military dairy farms to ensure the supply of milk and butter to colonial army at Allahabad. This was the first dairy farm followed by Bangalore, Karnal and Ooty. However, these did not have much impact on civilian consumers. Growth of the urban area and growth of population in these urban areas, consumers had to depend on milk vendors who kept cattle in the urban area and sold their milk door to door. As a result, several cattle sheds came into existence in different cities, which led to environmental problems. Vendors wanted to increase their income, they started lactation and consequently cattle developed sterility problems which led to reduce the number of calving. Once the cattle became unproductive, they were sold to slaughterhouses.

Onset of second world war gave momentum to private dairy in big cities of the country with some modernized processing facilities like Polson, Keventers and Express dairy. Now the modernization of dairy became priority for the government after first five-year plan in 1951. Second five-year plan was based on it and government implemented programs like Integrated Cattle Development Project (ICDP) and Key Village Scheme (KVS) and several other milk-production stimulating actions. As a result, during 1951 to 1970 the annual growth rate in milk production was about 1 percent.

The cooperative movement to organized dairy development was started in 1960 and National Dairy Development Board (NDDB) was established by an act of parliament in 1965. Govt of India approved the OFP (Operation Flood Program) in 1969 and in 1970 government established a public sector company The Indian Dairy Cooperation (IDC). OFP was setup with the objective to enhance milk production and to increase rural income. OFP was implemented in three phases - The first phase (1970-1980) was inclined to se up dairy cooperative, best milkshed and linking them with four main cities of Bombay, Calcutta, Delhi and Madras and thus 18 of India's premier milksheds were linked with consumers in these four metropolitan cities.

Operation Flood Program (OFP) phase 2 (1981-85) integrated the Indian dairy association assisted dairy development projects. Now milksheds increased from 18 to 136 and 290 Urban markets expanded the outlet of milk. 'Raksha' Vaccine was developed by Hyderabad Research Center to protect from diseases of cattle.

OFP Phase 3 (1985-1996) enabled dairy co-operatives to expand, strengthen the infrastructure and to increase the volume of milk. It was concentrated for veterinary first-aid health care service, feed and artificial insemination service for co-operative members.

India currently has 190516 village dairy co-operatives federated into 177 milk unions and 15 federations that procured on an average 507.69 lakh kg of milk per day. These village dairy co-operatives have nearly 16.9 million farmers as members (www.nddb.org, 2018-19).

It is a well-known fact that Animal Husbandry and milk production is an integral part of the diversified agricultural system. The animal husbandry sector plays a crucial role in the rural economy by providing gainful employment particularly to small, marginal farmers and agricultural laborers in the state of Jharkhand. Jharkhand is such a state where unemployment and poverty is higher than other states. Animal Husbandry helps to eradicate these chronic problems. It provides gainful employment and source of income and livelihood to landless labors and additional source of income to families through employment of women. In agrarian economy of Jharkhand, livestock plays an important role in contribution to income generation, self-employment, tribal development and women empowerment.

Agriculture and livestock rearing are a back-breaking occupation not rewarded by big returns because of unproductive land, climate factor and poor breed of livestock (yielding 0.5 to 1 liters of milk per day) in Jharkhand. At the same time, milk production activity is an important enterprise to supplement income and reduce underemployment of small and marginal farmers and agricultural laborers. It also helps in improving the nutritional standard of the people. It has been realized that dairy development could be used as a tool for bringing about socio-economic change among the rural farming community by providing milk production as a subsidiary occupation (Department of Agriculture, Brochure).

There is a tremendous potential for milk production activities in the state, as the agro-climatic conditions of this state are ideally suited for cross- breed milch cattle with moderate production capacity specially of 50% exotic inheritance from Holstein Friesian. There is adequate market potential for sale of milk and milk products in this state due to a number of industrially developed area.

In the state, rural milk trade practices are not established and milk marketing network is not much developed and milk is produced mostly for household consumption and local marketing. There is scope for co- operative development and milk route development through institutional arrangements with milk processing plant.

In order to meet the requirement of milk and milk products in the state around 6 to 8 lakh liter of milk is being procured from neighboring states daily by organized as well as non-organized sector players. At the same time to make the state self-sufficient in milk production, the breed improvement, feed and fodder development program and other productivity enhancement programs are intensively implemented. As a result, the milk production in rural area is increasing significantly. (Choudhary and Marandi, 2017, P 12-13).

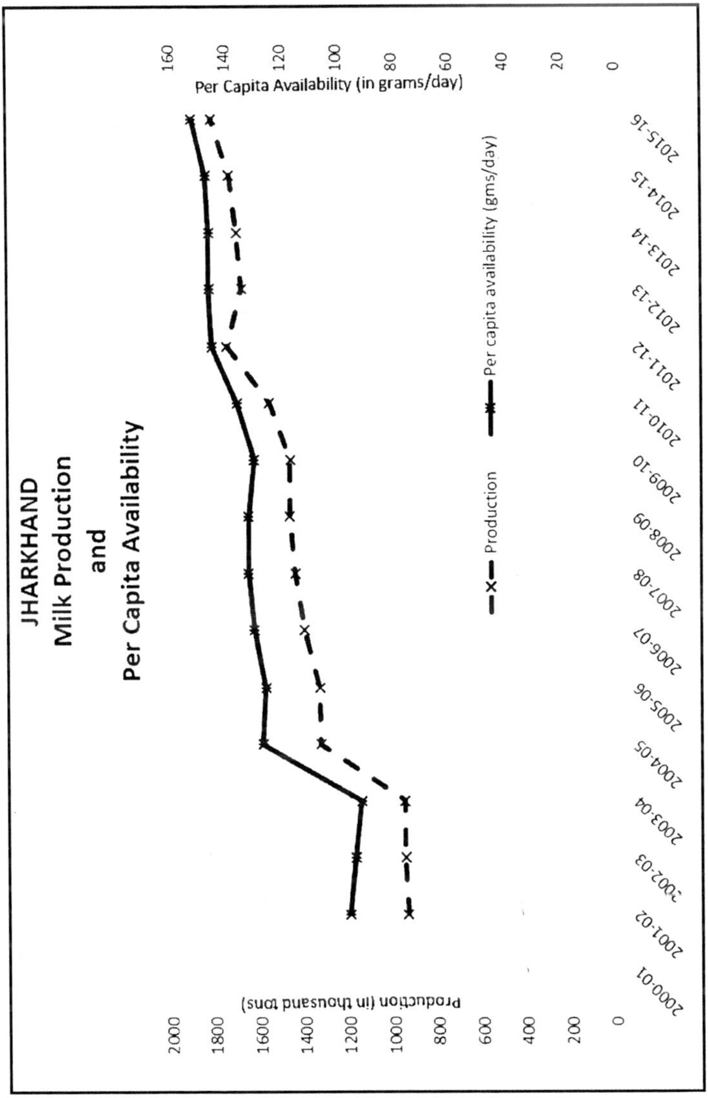


**Map 7.11: Milk Production and Availability**

## Milk Production and per capita availability of Milk in Jharkhand (2001 to 2015 -16)

The year wise milk production and per capita availability of milk are presented in table 7.17. An analysis of this table reveals that total milk production was increased to 1812.38 thousand tones in 2015-16 from 910 thousand tones in 2000-01 accounting for 99.16 per cent increase during 2000-01 to 2015-16. So, we can say that Jharkhand has been achieved just double in milk production during 2000-01 to 2015-16. In this state, per capita availability of milk was also increased to 152 gms per day in 2015-16 from 96 gms/day in 2001-02 accounting for 58.33 per cent increased during 2001-02 to 2015-16.

**Table 7.17**

**Year wise Milk Production and per capita availability of milk in Jharkhand (2000-01 to 2015-16) (In 000 tones)**

| Years | Production | Per capita availability (gms/day) |
|---|---|---|
| 2000-01 | 910 | - |
| 2001-02 | 940 | 96 |
| 2002-03 | 952 | 94 |
| 2003-04 | 954 | 92 |
| 2004-05 | 1330 | 127 |
| 2005-06 | 1335 | 126 |
| 2006-07 | 1401 | 130 |
| 2007-08 | 1442 | 132 |
| 2008-09 | 1466.35 | 132 |
| 2009-10 | 1463.00 | 130 |
| 2010-11 | 1555.64 | 136 |
| 2011-12 | 1745.00 | 145 |
| 2012-13 | 1679.00 | 146 |
| 2013-14 | 1699.83 | 146 |
| 2014-15 | 1733.72 | 147 |
| 2015-16 | 1812.38 | 152 |

## District wise Milk Production in Jharkhand

The district wise total milk production in the state of Jharkhand is presented in table 7.18. Analysis of this table

**Table 7.18**

**District wise Milk Production in Jharkhand, (2009-10 to 2015-16) (in 000 MT)**

| District | 2009-10 (%) | 2010-11 (%) | 2011-12 (%) | 2012-13 (%) | 2013-14 (%) | 2014-15 (%) | 2015-16 (%) |
|---|---|---|---|---|---|---|---|
| Bokaro | 65.80 | 70.75 | 70.99 | 75.61 | 76.50 | 84.54 | 101.57 |
| | (4.50) | (4.54) | (4.50) | (4.50) | (4.50) | (4.88) | (5.60) |
| Chatra | 79.68 | 84.33 | 85.66 | 89.90 | 90.90 | 87.91 | 86.88 |
| | (5.45) | (5.42) | (5.42) | (5.35) | (5.35) | (5.07) | (4.79) |
| Deoghar | 83.69 | 89.18 | 91.22 | 97.55 | 98.81 | 99.59 | 110.14 |
| | (5.72) | (5.73) | (5.77) | (5.81) | (5.81) | (5.74) | (6.08) |
| Dhanbad | 84.36 | 91.09 | 90.36 | 95.51 | 96.52 | 109.95 | 130.16 |
| | (5.77) | (5.86) | (5.72) | (5.69) | (5.68) | (6.34) | (7.18) |
| Dumka | 78.55 | 83.26 | 86.39 | 93.11 | 94.43 | 90.82 | 86.53 |
| | (5.37) | (5.35) | (5.46) | (5.54) | (5.56) | (5.24) | (4.77) |
| Garhwa | 77.41 | 81.90 | 83.12 | 87.07 | 88.01 | 84.80 | 79.39 |
| | (5.30) | (5.26) | (5.26) | (5.18) | (5.18) | (4.89) | (4.38) |
| Giridih | 112.53 | 119.79 | 122.49 | 131.02 | 132.71 | 134.34 | 128.17 |
| | (7.69) | (7.70) | (7.75) | (7.80) | (7.81) | (7.75) | (7.07) |
| Godda | 76.38 | 80.69 | 82.20 | 86.32 | 87.29 | 83.22 | 74.76 |
| | (5.22) | (5.19) | (5.20) | (5.14) | (5.14) | (4.80) | (4.13) |
| Gumla | 64.78 | 67.98 | 70.07 | 74.43 | 75.37 | 70.93 | 66.88 |
| | (4.43) | (4.37) | (4.43) | (4.43) | (4.43) | (4.09) | (3.69) |

| | | | | | | | |
|---|---|---|---|---|---|---|---|
| Hazaribagh | 79.80 | 84.90 | 86.31 | 92.02 | 93.16 | 96.45 | 101.20 |
| | (5.45) | (5.46) | (5.46) | (5.48) | (5.48) | (5.56) | 5.58) |
| Jamtara | 39.00 | 41.48 | 43.05 | 46.62 | 47.29 | 46.42 | 47.23 |
| | (2.87) | (2.67) | (2.72) | (2.78) | (2.78) | (2.68) | (2.61) |
| Khunti | 26.65 | 28.20 | 28.96 | 30.95 | 31.35 | 30.95 | 31.10 |
| | (1.82) | (1.81) | (1.83) | (1.84) | (1.84) | (1.79) | (1.72) |
| Koderma | 30.61 | 32.70 | 33.10 | 35.15 | 35.56 | 36.97 | 37.40 |
| | (2.09) | (2.10) | (2.09) | (2.09) | (2.09) | (2.13) | (2.06) |
| Latehar | 42.34 | 44.58 | 45.29 | 47.40 | 47.91 | 45.87 | 39.04 |
| | (2.89) | (2.87) | (2.86) | (2.82) | (2.82) | (2.65) | (2.15) |
| Lohardaga | 19.47 | 20.78 | 21.01 | 22.42 | 22.69 | 24.34 | 25.58 |
| | (1.33) | (1.34) | (1.33) | (1.34) | (1.34) | (1.40) | (1.41) |
| Pakur | 48.34 | 50.86 | 52.39 | 55.49 | 56.18 | 52.20 | 49.00 |
| | (3.30) | (3.27) | (3.31) | (3.30) | (3.31) | (3.01) | (2.70) |
| Palamu | 104.9 | 110.70 | 112.62 | 117.61 | 118.87 | 111.07 | 94.31 |
| | (7.17) | (7.12) | (7.12) | (7.00) | (6.99) | (6.41) | (5.20) |
| W. Singhbhum | 33.02 | 34.64 | 36.03 | 39.18 | 39.76 | 39.72 | 42.14 |
| | (2.26) | (2.23) | (2.28) | (2.33) | (2.34) | (2.29) | (2.33) |
| E. Singhbhum | 60.85 | 66.04 | 65.73 | 70.80 | 71.66 | 86.21 | 116.73 |
| | (4.16) | (4.25) | (4.16) | (4.22) | (4.22) | (4.97) | (6.44) |
| Ramgarh | 30.45 | 32.65 | 32.41 | 34.27 | 34.63 | 39.62 | 45.89 |
| | (2.08) | (2.10) | (2.05) | (2.04) | (2.03) | (2.28) | (2.53) |

| | | | | | | | |
|---|---|---|---|---|---|---|---|
| Ranchi | 102.01 | 109.57 | 108.46 | 114.59 | 115.75 | 136.60 | 179.43 |
| | (6.97) | (7.04) | (6.86) | (6.82) | (6.81) | (7.88) | (9.90) |
| Sahebganj | 59.59 | 63.41 | 64.14 | 67.53 | 68.30 | 66.36 | 62.63 |
| | (4.07) | (4.08) | (4.06) | (4.02) | (4.02) | (3.83) | (3.45) |
| Saraikela | 30.81 | 32.53 | 33.76 | 36.72 | 37.27 | 37.62 | 40.45 |
| | (2.11) | (2.09) | (2.14) | (2.20) | (2.19) | (2.17) | (2.23) |
| Simdega | 31.85 | 33.52 | 35.11 | 38.21 | 38.80 | 37.23 | 35.64 |
| | (2.18) | (2.15) | (2.22) | (2.28) | (2.28) | (2.15) | (2.00) |
| **Total** | **1463** | **1555.64** | **1580.98** | **1679.58** | **1699.83** | **1733.73** | **1812.37** |
| | **(100%)** | **(100%)** | **(100%)** | **(100%)** | **(100%)** | **(100%)** | **(100%)** |

**Source:** Jharkhand Economic Survey, 2015-16 & 2016-17

reveals that total milk production was increased to 1812.37 thousand MT in 2015-16 from 1463.00 thousand MT in 2009-10 accounting for 23.88 per cent increased during 2009-10 to 2015-16. So, we can say that total milk production in Jharkhand has been continuously increased during above mentioned period. Moreover, district wise analysis reveals that Ranchi has highest milk production, which increased to 179.44 thousand MT in 2015-16 from 102.02 thousand MT in 2009-10 followed by Dhanbad with 54.29 per cent increased, Giridih with 13.90 per cent increased, Singhbhum (E) with 91.83 per cent increased and Deoghar with 31.62 per cent increased during 2009-10 to 2015-16. Thus, in Lohardaga district of Jharkhand, the total production of milk was lowest (25.59 thousand MT) in 2015-16 among all the 24 districts of Jharkhand state but its milk production was increased to 25.58 TMT in 2015-16 from 19.47 TMT in 2009-10 accounting for 31.38 per cent increased during 2009-10 to 2015-16.

### District wise Cattle and Buffalo Population and per capita availability of Milk

The district wise cattle and buffalo population and per capita availability of milk are presented in table 7.19. An analysis of this table reveals that total livestock population and total breedable cattle and buffalo population in the Jharkhand was estimated to be 9916025 and 2957194 respectively. Its means that about 29.82 per cent of total livestock population was breedable. The district of Giridih has highest breedable population of cattle and buffalo (2,46814) accounting for 8.34 per cent of total state's breedable cattle and buffalo population followed by Palamu with 7.50 per cent, Deoghar with 6.22 per cent and Dumka with 6.01 per cent. Whereas, Giridih has also highest livestock population of state livestock population accounted for 8.25 per cent followed by Palamu with 6.34 per cent, Dumka with 6.07 per cent and Gumla with 5.61 per cent. However, Gumla has lowest number of livestock and breedable population accounted for 1.08 per cent and 1.38 per cent respectively of state total livestock and breedable population. Moreover, Chatra district of Jharkhand has highest availability of milk

Table 7.19

District wise Cattle and Buffalo Population and per capita availability of Milk

| District | Livestock Population | | | Breedable Cattle and Buffalo (%) | Annual Milk Production (MT)(%) | Per Capita Availability in gms |
|---|---|---|---|---|---|---|
| | Cow (%) | Buffalo (%) | Total (%) | | | |
| Bokaro | 311748 | 37455 | 3949203 | 112248 | 101570.88 | 135 |
| | 3.57 | 3.16 | 3.52 | 3.80 | 5.80 | |
| Chatra | 404132 | 83289 | 487421 | 156672 | 86889.57 | 228 |
| | 4.63 | 7.02 | 4.92 | 5.30 | 4.79 | |
| Deoghar | 507325 | 43260 | 550585 | 184103 | 110149.62 | 202 |
| | 5.81 | 3.65 | 5.55 | 6.22 | 6.08 | |
| Dhanbad | 304360 | 32292 | 336652 | 118429 | 130169.75 | 133 |
| | 3.48 | 2.72 | 3.40 | 4.00 | 7.18 | |
| Dumka | 562389 | 39108 | 601497 | 177880 | 86539.17 | 179 |
| | 6.44 | 3.30 | 6.07 | 6.01 | 4.77 | |
| Garhwa | 392257 | 60397 | 452654 | 149858 | 79395.81 | 164 |
| | 4.49 | 5.10 | 4.56 | 5.07 | 4.38 | |
| Giridih | 718845 | 98930 | 817775 | 246814 | 128175.99 | 143 |
| | 8.23 | 8.34 | 8.25 | 8.35 | 7.07 | |
| Godda | 389048 | 60186 | 449234 | 163812 | 74762.77 | 156 |
| | 4.46 | 5.07 | 4.53 | 5.54 | 4.13 | |

| | | | | | | |
|---|---|---|---|---|---|---|
| Gumla | 486668 | 69968 | 556636 | 133958 | 66889.55 | 179 |
| | 5.57 | 5.90 | 5.61 | 4.53 | 3.69 | |
| Hazaribag h | 446400 | 106784 | 553184 | 159596 | 101203.27 | 160 |
| | 5.11 | 9.00 | 5.58 | 5.40 | 5.58 | |
| Jamtara | 297107 | 20325 | 317432 | 95071 | 47230.53 | 163 |
| | 3.40 | 1.71 | 3.20 | 3.21 | 2.61 | |
| Khunti | 232346 | 24158 | 256504 | 70424 | 31107.55 | 160 |
| | 2.66 | 2.04 | 2.59 | 2.38 | 1.72 | |
| Koderma | 163259 | 30584 | 19843 | 58617 | 37403.42 | 143 |
| | 1.87 | 2.58 | 1.95 | 2.00 | 2.06 | |
| Latehar | 277380 | 37798 | 315178 | 83498 | 39042.22 | 147 |
| | 3.18 | 3.19 | 3.18 | 2.82 | 2.15 | |
| Lohardaga | 141135 | 15160 | 156295 | 38039 | 25588.74 | 152 |
| | 1.62 | 1.28 | 1.58 | 1.29 | 1.41 | |
| Pakur | 378246 | 57051 | 435297 | 125819 | 49001.89 | 149 |
| | 4.33 | 4.81 | 4.39 | 4.25 | 2.70 | |
| Palamu | 540055 | 88732 | 628787 | 221806 | 94316.44 | 133 |
| | 6.20 | 7.48 | 6.34 | 7.50 | 5.20 | |
| W. Singhbhum | 415993 | 32643 | 448636 | 103316 | 42147.67 | 77 |
| | 4.77 | 2.75 | 4.52 | 3.49 | 2.33 | |
| E. Singhbhum | 340457 | 19484 | 359941 | 104044 | 116734.44 | 139 |
| | 3.90 | 1.64 | 3.63 | 3.52 | 6.44 | |

| | | | | | | |
|---|---|---|---|---|---|---|
| Ramgarh | 124421 | 36546 | 160967 | 40839 | 45889.98 | 132 |
| | 1.43 | 3.08 | 1.62 | 1.38 | 2.53 | |
| Ranchi | 446146 | 100800 | 546946 | 143705 | 179439.77 | 169 |
| | 5.11 | 8.50 | 5.52 | 4.86 | 9.90 | |
| Sahebganj | 261982 | 53655 | 315637 | 123226 | 62632.04 | 149 |
| | 3.00 | 4.52 | 3.18 | 4.17 | 3.45 | |
| Saraikela | 239298 | 19633 | 258931 | 59093 | 40451.26 | 104 |
| | 2.74 | 1.66 | 2.61 | 2.00 | 2.23 | |
| Simdega | 349086 | 17704 | 366790 | 86327 | 35644.03 | 163 |
| | 4.00 | 1.50 | 3.70 | 2.91 | 2.00 | |
| Total | 8730083 | 1185942 | 9916025 | 2957194 | 1812375.68 | |
| | 100% | 100% | 100% | 100% | 100% | |

**Source:** Jharkhand Economic Survey, 2015-16 & 2016-17.

per capita (228 gms/day) followed by Dumka and Gumla both (179 gms/day), Ranchi with 169 gms/day and Garhwa with 164 gms/day.

## Annual outlays and Expenditures under Dairy Development (11th Five Year Plan)

Annual outlays and expenditures under dairy development in Jharkhand is presented in table 7.20. An analysis of this table reveals that total amount of outlay and expenditure by state plan was Rs. 25375 lakh and Rs. 22516 lakhs during 11th Five Year Plan. Around 88.73 per cent of total outlay was expended during same period. Under CS/CSS schemes, around Rs. 500.90 lakh was released and its expenditure was Rs. 483.82 lakh accounted for 96.59 per cent of total grant released during same period of time.

**Table 7.20**

**Annual Outlays and Expenditures during 11th Five Year Plan under Dairy Development in Jharkhand (Rs. In Lakh)**

| Year | State Plan | | CS/CSS | |
|---|---|---|---|---|
| | Outlay | Expenditure | Grant Released | Expenditure |
| 2007-08 | 3965.00 | 3453.33 | 107.64 | 107.64 |
| 2008-09 | 4600.00 | 4415.21 | 93.50 | 89.25 |
| 2009-10 | 4710.00 | 4609.45 | 19.76 | 6.93 |
| 2010-11 | 5500.00 | 4674.41 | 280.00 | 143.4375 |
| 2011-12 | 6600.00 | 5364.04 | - | 136.5625 |
| Total | 25375.00 | 22516.44 | 500.90 | 483.82 |

**Source:** Dept. of Animal Husbandry & Fisheries, Annual Plan, 2012-13 of Jharkhand

## District wise infrastructure Facilities for Milk Production, Procurement, Processing and Marketing in Jharkhand

The district wise infrastructure facilities for milk; production, procurement, processing and marketing is presented in table. An analysis of this table reveals that total number of Gokul Gram Vikas Kendra in the state of

Jharkhand was estimated to 268. Out of which, 52 was in Ranchi district, as the highest number of total states Gokul Gram Vikas Kendra followed by Deoghar with 26, Hazaribagh with 23 and Giridih with 16 Gokul Vikas Kendra. The total number of Bulk milk cooler in the state of Jharkhand was estimated to 20. Out of which, Ranchi has highest 11 number of bulk milk cooler while Palamu, Hazaribagh and Deoghar each has two bulk milk coolers. The total number of AMCU/ PDMCU in the state of Jharkhand was recorded to be 136. Ranchi has highest number of AMCU/PDMCU (48) followed by Palamu with 16, Dumka and Deoghar each has 14 AMCU/ PDMCU and Lohardaga has 12 AMCU/PDMCU. However, some districts have no any such types of facilities. So, it should be effort by cooperative/state government to establish such types of facilities in each district of Jharkhand to empower the milk producers.

The total number of dairy cattle development center in the state of Jharkhand was 1010. Out of which, Palamu has highest number of cattle development center followed by Hazaribagh, Giridih, Godda and Deoghar each has 65. The total number of milk processing plant/milk chilling plant in the state was 9. It means 9 districts each has one chilling plant while remaining district has no any such types of facilities.

## Dairy Co-operatives in the State

The dairy cooperative in the state of Jharkhand is presented in table 2.6. An analysis of this table reveals that total number of DCS in the state was 60 and producer member was 1000 in 2015-16. The capacity of milk procurement in the state of Jharkhand was increased to 61.00 thousand kg/day in 2015-16 from 14.00 thousand kg/day in 2014-15 while liquid milk marketing capacity in the state was decreased to 304 thousand liters per day in 2015-16 from 308 thousand liters per day in 2014-15 accounting for 1.29 per cent decreased during this period.

## Veterinary Facilities in the State

The total number of class-I veterinary hospital was 424,

mobile veterinary hospital (04), provincial veterinary hospital (23), cattle breeding farms (03), bull mother farm (01) and AI centers (433) managed by department of animal husbandry & dairy, Govt. of Jharkhand.

## Requirement and availability of Feed and Fodder in Jharkhand and India (2011-12)

The requirement and availability of feed and fodder in India and Jharkhand are presented in table 2.9 and reveals that the requirement and availability of dry fodder in Jharkhand were 20.09 MT and 35.54 MT respectively which indicates 15.45 MT of dry fodder was surplus in the state of Jharkhand while that was 163 MT deficits in India. The requirement and availability of green fodder in the state were 40.18 MT and 21.32 MT respectively which indicates 18.86 MT green fodders was deficit in the state while deficit of green fodder in India was 79 MT. Moreover, requirement and availability of concentrate in the state were recorded to be 4.02 MT and 1.80 MT respectively which indicates 2.22 MT deficits while deficit of concentrate in India was 30 MT.

**Table 7.21**
**Requirement and Availability of Feed and Fodder in Jharkhand and india (2011-12) (in MT)**

| Particular | Jharkhand | | India | |
|---|---|---|---|---|
| | Requirement | Availability | Requirement | Availability |
| Dry Fodder | 20.09 (4.82) | 35.54 (14.04) | 416.00 (100) | 253.00 (100) |
| Green Fodder | 40.18 (18.09) | 21.32 (14.90) | 222.00 (11) | 143.00 (100) |
| Concentrate Feed | 4.02 (7.58) | 1.80 (7.82) | 53.00 (100) | 23.00 (100) |

**Source:** State Animal Husbandry Department.
(Figures in bracket indicate percentage share of Jharkhand in India)

## Animal wise Production of Milk in the State (2009-10 to 2013-14)

The animal wise production of milk in the state of Jharkhand is presented in table 7.22 and reveals that total milk production in the state was increased to 1699.83

thousand MT in 2013-14 from 1462.61 thousand MT in 2009-10 accounted for 16.22 per cent increased during 2009-10 to 2013-14. The total milk production of cattle was increased to 1017 thousand MT in 2011-12 from 765.86 thousand MT in 2009-10 accounted for 32.79 per cent increased whereas that of cattle was decreased to 978.44 thousand MT in 2013-14 from 1017 thousand MT in 2011-12. So, in the year of 2011-12, maximum milk production was recorded. The total production of buffalo milk was varied between 620.81 thousand MT to 636.04 thousand MT during 2009-10 to 2013-14. The total production of goat milk was also increased to 85.35 thousand MT in 2013-14 from 75.94 thousand MT in 2009-10. Milk production of goat in the Jharkhand was continuously increased except in 2010-11 year. However, the percentage share of cattle milk, buffalo milk and goat milk to total state milk production in 2013-14 were recorded to 57.56 per cent, 37.42 per cent and 5.02 per cent respectively.

**Table 7.22**
**Animal wise Production of Milk in the State (2009-10 to 2013-14)**

| Year | Cow Milk | | | Buffalo Milk | Goat Milk | Total |
|---|---|---|---|---|---|---|
| | Exotic/CB | Indigenous Cow | Total | | | |
| 2009-10 | 121.04 | 644.82 | 765.86 | 620.81 | 75.94 | 1462.51 |
| 2010-11 | 147.00 | 695.20 | 842.24 | 647.13 | 65.42 | 1554.80 |
| 2011-12 | 123.00 | 894.00 | 1017.00 | 648.00 | 80.00 | 1745.00 |
| 2012-13 | 136.00 | 826.21 | 962.21 | 633.05 | 83.74 | 1679.00 |
| 2013-14 | 137.34 | 841.10 | 978.44 | 636.04 | 85.35 | 1699.83 |

**Source:** State/UT Animal Husbandry Department

## Animal wise Breedable Population of Jharkhand and India

The animal wise breedable population of Jharkhand and India are presented in table 7.23 and reveals that as per census of livestock, 2012; total number of livestock population in Jharkhand was 99.16 lakh. Out of which, 29.58 lakh was breedable accounted for 29.83 per cent to the total livestock population. Out of total livestock population (99.16 lakh), cattle and buffalo population were 87.30 lakh and 11.86 lakh accounted for 88.04 per cent and 11.96 per cent respectively. Therefore, the percentage share of total livestock and breedable of Jharkhand in all-India were estimated to 3.31 per cent and 2.49 per cent respectively.

**Table 7.23**
**Animal wise Breedable Population of Jharkhand and India (No. in Lakhs)**

| Breed | Jharkhand | | India | |
|---|---|---|---|---|
| | Total | Breedable | Total | Breedable |
| Indigenous Cow | 84.74 | 24.34 | 1511.72 | 481.00 |
| Crossbreed Cow | 2.56 | 1.35 | 397.32 | 194.20 |
| Total Cow | 87.30 | 25.69 | 1909.04 | 675.20 |
| Buffalo | 11.86 | 3.89 | 1087.00 | 510.54 |
| Total | 99.16 | 29.58 | 2996.04 | 1185.74 |

**Source:** Dept of Animal Husbandry, Jharkhand, Livestock Census- 2012

## Percentage share of Livestock in Jharkhand and India

The percentage share of livestock of Jharkhand in India is presented in table 7.24 and reveals that there was 3.52 per cent share of livestock of Jharkhand in India, poultry (1.85%), bovines' male (6.06%) bovine female (2.22%) and total bovines (3.3%).

**Table 7.24**
**Percentage share of Livestock in Jharkhand and India (2012)**

| Particulars | Jharkhand (In Nos) | Country (In Nos) | % Share of the Country |
|---|---|---|---|
| Total Livestock | 18052746 | 512057301 | 3.52 |
| Total Poultry | 13559528 | 729209320 | 1.85 |
| Bovines (Male) | 5107916 | 84187200 | 6.06 |
| Bovines (Female) | 4808109 | 215793953 | 2.22 |
| Total Bovines | 9916025 | 299981153 | 3.3 |

**Source:** 19th Livestock Census 2012

## Livestock Population, Milk Production and per capita of milk in sample districts.

An analysis of its may been seen in table 7.25 and reveals that per capita availability of milk among selected districts was highest (202 gms) in Deoghar followed by Ranchi with 169 gms and Hazaribagh (160 gms). Livestock population was highest in Hazaribagh district (553144) among sample districts followed by Deoghar (550585) and Ranchi (546946) whereas the population of breedable animal was highest in Deoghar district followed by Hazaribagh and Ranchi district. However, Lohardaga district among selected districts was lowest in all regards as in livestock population, breedable population of animal and per capita availability of milk.

**Table 7.25**

**Livestock Population, Milk Production and per capita of milk in sample districts**

| District | Livestock Population (In Nos) | | | Breedable Cattle and Buffalo | Annual Milk Production (MT) | Per Capita Availability (in gms) |
|---|---|---|---|---|---|---|
| | Cow | Buffalo | Total | | | |
| Deoghar | 507325 | 43260 | 550585 | 184103 | 110149.62 | 202 |
| Hazaribagh | 446400 | 106784 | 553144 | 159596 | 101203.27 | 160 |
| Ranchi | 446146 | 100800 | 546946 | 143705 | 179439.77 | 169 |
| Lohardaga | 141135 | 15160 | 156295 | 38039 | 25588.74 | 152 |

**Source:** As per Census. 2012-13, Govt. of Jharkhand

## Animal's Health Infrastructure

Animal's health infrastructure facilities in the sample districts of Jharkhand are presented in table 7.26 and analysis reveals that Ranchi district of Jharkhand state was laced with all types of health infrastructure for animals as compared to other districts in Jharkhand. However, Ranchi has 424 veterinary hospitals, 04 mobile veterinary hospital, 03 cattle breeding farms and 432 artificial insemination centers. BAIF has been performing well in Ranchi to provide all types of facilities for dairy development in the state of Jharkhand.

## Table 7.26
## Animal's health infrastructure facilities in the sample districts of Jharkhand

| Item | Ranchi Dist. (In Nos.) | Hazaribagh (In Nos.) | Lohardaga (In Nos.) | Deoghar (In Nos.) |
|---|---|---|---|---|
| Class - I, Veterinary | 424 | 01 | NA | NA |
| Mobile Veterinary Hospital | 04 | 27 (C-1 Vet.disp.) | NA | NA |
| Provincial Veterinary | 23 | - | NA | NA |
| Cattle Breeding Farms | 03 | - | NA | NA |
| Bull mother Farm | 01 | - | NA | NA |
| Gokul Gram Vikas Centre | 04 | - | NA | NA |
| No. of State-run Poultry | 02 | - | NA | NA |
| No. of State-run Pig Farm | 06 | - | NA | NA |
| No. of State-run Goat Farm | 01 | - | NA | NA |
| No. of AI Centers Managed by the Department | 405 | 07 | NA | NA |
| No. of Dairy Cattle development center | 410 | - | NA | NA |
| No. of Milk chilling centers/ dairies | 16 | - | NA | NA |
| No. of Biological Production Unit | 01 | - | NA | NA |

**Source:** Dept. of Animal Husbandry, Jharkhand

Infrastructure facilities of milk marketing in Jharkhand is presented in table 7.26 and reveals that total number of villages coverage was increased to 990 in 2012-13 from 130 in 2008-09. Also, household coverage of milch animal holding was increased to 41500 in 2012-13 from 4500 in 2008-09. Total milk procurement was also increased to 90 TLPD in 2012-13 from 08 TLPD in 2008-09. Total liquid milk sold was also increased to 80 TLPD in 2012-13 from 08 TLPD in 2008-09 due to high demand of milk and milk product with increasing population of the state.

**Table 7.26**
**Infrastructure facilities of milk marketing in Jharkhand**

| Description | Unit | 2008-09 | 2009-10 | 2010-11 | 2011-12 | 2012-13 |
|---|---|---|---|---|---|---|
| Coverage of Village | No | 130 | 360 | 580 | 850 | 920 |
| Milch animal Holding household | No | 4600 | 11800 | 20850 | 32450 | 41500 |
| Total milk procurement | TLPD | 08 | 21 | 40 | 65 | 90 |
| Liquid milk sale | TLPD | 08 | 21 | 36 | 58 | 80 |
| AMCU positioning at village level | No | 10 | 30 | 52 | 80 | 88 |
| Cattle induction | No | 2000 | 2000 | 2000 | 2000 | 2000 |

**Source:** Deptt. of Animal Husbandry, Jharkhand

An overview on animal husbandry and dairy in Jharkhand is presented in table 7.26 and reveals that the total milk production was 1.33 MT in Jharkhand with 30.00 per cent of marketable surplus. Total number of DCS was 1490 with 48.75 thousand of milk producer members while average milk collection per day was 20.65. The processing and chilling capacity of milk by COMPFED was estimated to 185 thousand liters while milk marketing through dairies and chilling centers was estimated 230.45 thousand liters per day.

**Table 7.27**
**An overview on animal husbandry and dairy in Jharkhand**

| Items | Achievements |
|---|---|
| Breedable population of Milch cattle | 1.90 M |
| Breedable buffalo population (in lakh) | 0.39 |
| Productivity (kg/day/animal) | 1.59 |
| Per capita availability of milk (in grams) | 140 |
| Milk production (MT) | 1.33 |
| Marketable surplus 30 per cent of total milk production (MT) | 0.34 |
| Total Nos. Of milk producers' cooperative societies | 1490 |
| Total producers' members (in '000) | 48.75 |
| Average milk collection per day through societies ('000 liters) | 20.65 |
| Processing and chilling capacity (In '000 liters) managed by District Co-operative | 45 |
| Processing & chilling capacity (In '000 liters) Managed by COMPFED | 185 |
| Per day milk marketing through dairies & chilling centers (in '000 liters) | 230.45 |

**Source:** Dept. of Animal Husbandry, Jharkhand (Census 2001)

Achievement of dairy development during 10th Five Year Plan in Jharkhand is presented in table 2.17 and reveals that the achievement in regards to milk production, induction of milch cattle and strengthening of DCS in the state of Jharkhand was estimated to 100 per cent during 10th Five Year Plan. However, the achievement and target of milk production were increased to 14.52 & 14.52 lakh MT respectively in 200607 from 9.51 lakh MT and 13.36 lakh MT in 2002-03.

Infrastructure of dairy cooperatives in the state of Jharkhand is presented in table 2.18 and reveals that the total number of PDCS was 1490, member of dairy (48750), district cooperative milk union (12) and milk processing and chilling center was 14 with capacity of 45000 TLPD. (Choudhary & Marandi, 2017, P. 13-18)

## Livestock Products

### Milk

As we have studied in the earlier pages that cattle, buffalo and goat are the main sources of milk in India. During 2009-10 Jharkhand has produced about 1462 thousand tons on milk. This increased to 1554.80 tons of milk during 2010-11 and 1745 thousand tons in 2011-12. During 2012-13 the production of milk was recorded 1679 thousand metric tons which grew to 1699 thousand metric tons during 2013-14. However, we find that the per capita milk availability in Jharkhand is just half of our country. Census of India has mentioned the per capita availability of milk based on the state estimates and projected population as on 1st March. The estimates are as under

**Table 7.28**
**Per Capita Milk Availability in India and Jharkhand (grams/day)**

| | 2009-10 | 2010-11 | 2011-12 | 2012-13 | 2013-14 |
|---|---|---|---|---|---|
| India | 273 | 281 | 290 | 299 | 307 |
| Jharkhand | 130 | 136 | 145 | 146 | 146 |

**Source:** Calculated based on State Estimates of Production and Projected Population based on 2001, Published by Department of Animal Husbandry, Govt. of India.

In India, Milk availability during 2009-10 was 273 grams per day. It grew to 281 grams per day during 2010-11 and 290 grams during 2011-12. Census of 2012-13 and 2013-14 recorded the milk availability as 299 grams per day and 307 grams per day respectively.

In Jharkhand per capita availability of milk was recorded during the census of 2009-10, 2010-11, 2012-13 and 2013-14 as 130 grams per day, 136 grams per day, 145 grams per day, 146 grams per day and 146 grams per day respectively.

National Institute of Nutrition, Hyderabad recommended through a manual of dietary guidelines for Indians (2011) that RDA (Recommended Dietary Amount) for milk is 300 gram per day per capita. Jharkhand is still lacking 154 grams/ day. During 2013-14 data shows that now India is a self-sufficient producer as per the RDA.

**Table 7.29**
**Livestock Produced Items in Jharkhand (2013-14)**

| Items | 2005-06 | 2009-10 | 2010-11 | 2011-12 | 2012-13 | 2013-14 |
|---|---|---|---|---|---|---|
| Wool (tons) (Ewest+Lamb +Ram) | 146 | 136.17 | 149.33 | 201.68 | 158.89 | 156.13 |
| Eggs (Crore) | 71.1 | 39.25 | 41.53 | 43.13 | 42.39 | 44.45 |
| Meat (000 tons) | 43 | 46.76 | 43.77 | 45.94 | 44.52 | 45.45 |
| Poultry | - | 3.36 | 4.29 | 5.00 | 128.84 | 2.48 |
| Pig | - | 15.55 | 13.26 | 13.00 | 15.84 | 14.43 |
| Goat | - | 24.95 | 23.86 | 25.11 | 19.69 | 16.41 |
| Sheep | - | 2.90 | 2.36 | 2.38 | 1.34 | 1.11 |
| Buffalo | - | - | - | - | - | 5.00 |
| Cattle | - | - | - | - | - | - |
| Milk (000 tons) Total | - | 1462.61 | 1554.80 | 1745.00 | 1679.00 | 1699.83 |
| Cow Milk | - | 765.86 | 842.24 | 1017.00 | 962.21 | 978.44 |
| Buffalo Milk | - | 620.81 | 647.13 | 648.00 | 633.05 | 636.04 |
| Goat Milk | - | 75.94 | 65.42 | 80.00 | 83.74 | 85.35 |

## Meat

National Nutrition Monitoring Bureau (NNMB) surveys indicate that daily intake of all foods except cereals and millets (396 g) in Indian households is lower than the recommended dietary allowances (RDA). The specific amount of meat is not mentioned in the RDA chart while the general recommendation is 100 grams per day per capita and twice in a week for red meat. Number of Non-Vegetarian population is unknown and therefore it is not possible to calculate whether production in state or India is sufficient or not. Here we only place the data for meat production in Jharkhand as 45.5 thousand metric tons during 2013-14. Data in table reveals that highest production (16.41 thousand tons of meat in Jharkhand is procured by goat followed by pig (14.43

thousand tons), Poultry (2.48 thousand tons) and sheep (1,11 thousand tons). Data of Buffalo meat (5 thousand tons) is available only for 2013-14 only and for cattle it is not available.

## Eggs

Poultry development is one of the most important activities in the livestock sector. The total egg production in India was at 1832 million in the year 1950-51 and since then the production of eggs continues to rise over the period of time. There has been steady increase in production of eggs up to the year 1999-2000. From the year 2000 onwards, the production of eggs improved substantially and it reached 74752 million in the year 2013-14.

In Jharkhand total egg production was at 71.1 crore in the year 2005-06. Production of egg decreased to 39.25 crore in 2009-10. Again, increased to 41.53 crore during 2010-11 and 44.45 crore during 2013-14 census.

In India egg availability is 61 per capita per annum whereas it is 14 per capita per annum in the state of Jharkhand. It is definitely behind the national production by a heavy margin.

## Wool

Wool producing animals are Ewe, Lamb and Ram. Number of sheep during 2012 census was about 65069.19 thousand in India and only 582.93 thousand in Jharkhand. It was enumerated total 784 thousand during 2019 census which shows a significant growth of 35 percent in Jharkhand. Sheep population was decreasing in Jharkhand by 2003 to 2012 but 2019 census has recorded an increase in sheep population. This shows a positive growth of this animal. In Jharkhand a good number of sheep rearing practice prevails in Saraikela, Pakur, Jamtara and East & West Singhbhum and Garhwa districts. Though Jharkhand is not in a good position in wool production, but it has produced about 146 tons of wool during 2005-06 and in 2013-14 it has recorded about 156.13 tons of wool production. This shows a gradual increase in wool production in the state.

## Livestock Development Program

Government of India has implemented several livestock development programs through the animal husbandry department. Following are the highlights:

- **Intensive Cattle Development Program:** The program envisages the setting up of number of intensive cattle development blocks. Already, 22 intensive cattle development projects have been set up. In addition, 19 central semen collection stations, 105 regional artificial insemination centers and 1152 stocking centers have been opened for dairy expansion activities, castration of scrub bulls and disease control have been started in all project areas. It is also proposed to organize the distribution of balanced feed to livestock owners in areas covered by this program.
- **National Livestock Mission:** There are following Sub-Missions under this mission -
  1. Fodder and Feed Development
  2. Skill Development
  3. Insurance of Milch Cattle for their development.
- Key Village Scheme: Through this scheme livestock improvement is being pursued by state government. This scheme provides for castration of scrub bulls, artificial insemination, pasture development, cultivation of fodder are the main schemes.
- **Gosadan Scheme:** The scheme was worked out by cattle preservation and development committee in 1948. It envisages the segregation of all old, infirm, unproductive and uneconomic cattle from area where active cattle development program to be taken up by Go-Sadan located in wasteland, in forests and other places where cattle grazing facilities exist but not utilized. This is mainly to avoid their further multiplication and the resultant damage to crops.
- **Goshala Development Scheme:** It was considered that the cattle development movement might receive wide public support specially from private charitable institutions such as Goshala. Each Goshala has been

provided by government with certain animals of improved breed for the production of superior bulls, which will be further distributed for breeding purpose and also for increasing the production of milk.

Apart from all this, certain other measures have been taken for development of livestock by state government:

- Two Milch Cattle distribution plan on 90 percent subsidy.
- Four milch cattle distribution to tana bhagat family on hundred percent subsidy.
- 5 to 50 milch cattle distribution on 33.33 percent to unemployed graduate, progressive dairy farmers under Kamdhenu Dairy Scheme.
- Feed and fodder development scheme.
- Distribution of balanced feed in subsidy rate.
- Support to progressive farmers of dairy for old shed repairing, milking machine, paneer and khoya making unit, biofertilizer unit, cow mat, fodder farming etc.
- Bio-Gas Unit establishment.
- High breed calf rearing program
- Breed reformation and productivity development program
- Farmers awareness program and dairy farming training programs.
- Milk Pooling point.
- Gokul gram Vikash center etc.

## References

Choudhary, Rambalak & Marandi, Rosline Kusum (2017); Assessment of the status of dairying in the Eastern states and potential to improve socio-economic status of milk producer and convergence of all central and state schemes at district level in Jharkhand- Agro-economic Research Center for Bihar & Jharkhand, T.M Bhagalpur University Bhagalpur (Report No. 45/2017).

Department of Agriculture Brochure (Govt. of Jharkhand) Dairying in Jharkhand, Jharkhand State cooperative milk producer.

Dietary Manual (2011); Dietary Guidelines for Indians National Institute of Nutrition. Indian Council of Medical Research, Hyderabad.

Jaiprakash & Singh, R.P. (2016); Growth performance of fish production in Jharkhand. Agro Economist-An International Journal, Renu Publications.

Lekhi, R.K. & Singh, Joginder (2018); Agricultural Economics, An Indian Perspective, Kalyani Publishers, New Delhi.

Mandal, Banani & Others (2018); Chromosomal Study of Black Bengal Goats of Jharkhand, International Journal of Current Microbiology, Applied Science, Vol-7 (Special Issue), P. 501-509.

Negi, Balbir Singh (1971); Economic Geography of India, Kitab Mahal, Allahabad.

Singh, J & Dhillon, S.S. (1984); Agricultural Geography, Tata Mc Graw Hill Publishing Ltd, New Delhi.

Singh, Jasbir (1976); An Agricultural Geography of Haryana, Vishal publications, University Campus, Kurukshetra University, Haryana.

# 8

# Problems and Prospects of Agriculture

## Problems and Prospects of Agriculture

In a less developed country like India, agriculture has been considered as the primary occupation. Agricultural practices followed by most of the peasants are not managed on commercial lines. Rather, it is still maintained at the level of subsistence farming so as to meet the requirement of food for the entire family throughout the year. Although about 69 percent of the working population are engaged in agriculture in the country and the percentage is around 63 for Jharkhand, still this sector is not yet organized in a rational manner. Most of the working units are still maintained at the household or the family level instead of commercially viable organized farming units. This trend has resulted in a poor show in agricultural practice in the country and in Jharkhand too. In a sense agricultural sector in India as well as in Jharkhand has been rejected to countless problems but we shall discuss here the ones related to the topic of this work.

**Chapter One** deals with several physical and cultural aspects of agriculture. Jharkhand is geologically constructed from granitic and gneissic formations structured with schists and at some places archean lava and sand stones over the surface. Hence, we get two main groups of soil here - Pedocals (rich in lime) and Pedalfer (rich in aluminum). The former is found in areas with less than relatively less moisture while the latter Pedalfers are a characteristic of high moisture regions. (Ahmad, 1965, P.54). These soils are low to very low in available Phosphorous and Sulphur, medium in available Nitrogen and Potassium status and deficient in available Boron in general. Soil area of about 1.6 million hectares (19% TGA) is acidic in nature. The soil of the state is also affected with the problem of soil-erosion. There is slight erosion in 52% area and moderated erosion in about 36% of the area.

Agriculture is associated with climate. The state of Jharkhand is under the Agroclimatic Zone VII i.e., Eastern Plateau and Hilly Region. State receives annual rainfall of 1200-1600 mm. Climate is dry semi humid to humid semi-arid type. Hence, key characteristics of this region are

1. Upland soil has low water retaining capacity
2. Late arrival and early cessation of monsoon. Uneven rainfall.
3. Drying of tanks and wells by February.
4. Poor soil fertility.

Irrigation is also important for monsoon dependent agricultural area. Here irrigated land against net area sown is only 13 percent. About 80 percent area is drought prone and 7 percent flood prone. Most of the rain water is lost as run off losses. But in Jharkhand a great prospect is its undulating surface and receiving a reasonable amount of rainfall which provides an opportunity to arrest run off losses by creating water retention structures. This not only checks erosion of surface soil leading to siltation of dams/water reservoirs but also increases irrigation potential and increases sub-surface moisture, which is essential for good agriculture in Jharkhand.

Population density measures the pressure of population on per-unit area of land. In the state overall density of population was only 76 person/sq. km in 1951. It became 122 in 1951 census and has grown to 274 during 1991 census. During 2001 and 2011 census, density of Jharkhand was recorded 338 persons/sq. km and 414 persons/sq. km respectively.

**Table 8.1**
**Population Density of India and Jharkhand (1901-2011)**

| Census Year | 1901 | 1911 | 1921 | 1931 | 1941 | 1951 | 1961 | 1971 | 1981 | 1991 | 2001 | 2011 |
|---|---|---|---|---|---|---|---|---|---|---|---|---|
| India | 77 | 82 | 81 | 90 | 103 | 117 | 142 | 177 | 216 | 273 | 325 | 368 |
| Jharkhand | 76 | 85 | 85 | 99 | 111 | 122 | 146 | 178 | 221 | 274 | 338 | 414 |

When we compare to the density of India it remains almost same during 1901 to 1991. After this census population density of Jharkhand began to grow rapidly and now it is higher than the density of the country.

Agricultural density is the measurement of the pressure of agricultural population over per unit of agricultural land. It was 3.67 persons/hectare during 2001 and it has grown to 5.30 persons/hectare during 2011 census. Growth of pressure over agricultural land is increasing over the agricultural land and that is the part of a problem.

Working force composition is showing an increase of marginal workers which is a presentation of almost unemployed population in the state. Number of populations is decreasing in cultivation, agricultural laborers and household industries. It means the workers are concentrating in 'Other Workers' group which is almost service sector group. This is a major cause of distress and anxiety in the state.

Land Use of Jharkhand clearly shows the increasing fallow land and decreasing net area sown year by year. Similarly, the major portion of 'Other Uncultivable Land' (Cultivable waste land + Permanent pasture & other grazing land + Land under miscellaneous trees) is also decreasing year by year in Jharkhand. This is also a major problem in agriculture of Jharkhand because this was the comfort zone for animal husbandry in Jharkhand. Forest area has decreased from 3186 thousand hectare in 1955-56 to 2239 hectares in 2015-16.

**Chapter two** is related to land tenure and size of holding. The land tenure system which is practiced in Jharkhand is full of defects. Insecurity of tenancy was a big problem for the tenants. CNTA and SPTA was enforced for the protection of land of tribal and SC people but still money lenders are active as ever and the alienation of land has not been totally stopped. (Singh, 1987). With the gradual breakdown of the tribal land system due to the invasion of the non-tribals in the region, these tribal tenures are gradually becoming extinct, and now a majority of the holdings are ordinary raiyati land. Under the raiyat a system of subletting is also prevalent in the region. Problems of the region has increased more after the separation of the state from Bihar. Cost of land has increased tremendously but tribals are not allowed to sell it freely at the market price. They are bound to sale to tribals only and hence they are not getting the fair market price. Return from agriculture is not enough. Fallow land is increasing due to less interest in agriculture or non-availability of labor to the cultivators. Now a days a surprising practice is prevailing in the entire region that without execution of sale of tribal land, non-tribals avail their position by the consent of tribal land owners violating the CNT act. This is very much similar to Benami Transfer of land.

Fragmentation of holdings is a major problem in agriculture sector. In Jharkhand, the average size of a holding has declined from 3165 thousand hectares in 2010-11 to 3091 in 2015-16. It is a decline of about 14.8 thousand hectares per year. Similarly, number of holdings has recorded an increase of 18.8 holdings per year. Number of marginal holding (less than 1 hectare) was 1848 thousand during 2010-11 and it raised to 1962 thousand holdings in 2015-16 but number of medium and large holdings are same in number. Marginal holdings were 68.4 percent during 2010-11 whereas the percentage of 1915-16 was very high (69.9 percent). Thus, we clearly see that about 70 percent holdings of the area are very small/ marginal (less than 1 hectare). Thus, the size of agricultural holding is quite uneconomic, small and fragmented. There is continuous sub- division and fragmentation of agricultural land due to increasing pressure of population and breakdown of the joint family system and also due to forced selling of the land for meeting debt repayment option. Thus, the size of holding has been declining year by year leading to the increase in number of marginal and small holdings.

This increasing trend of sub-division and fragmentation has resulted in inefficient use of the land. Moreover, it has led to the problem of increasing the volume of capital required for farming with low productivity.

Inequality in land distribution is a major problem in

agriculture in Jharkhand and in the country. Though we have not studied the land distribution in details it is a well-known fact that it is a major constraint in the development of the agricultural system. A considerable degree of concentration of land holdings is in the hands of rich landlords, farmers and money lenders not only in Jharkhand but in the whole country. Vast majority of farmers are operating a very small and uneconomic size of holding. This results in higher costs per unit.

**Chapter Three** has discussed about farm technology. Farmers in Jharkhand have been using traditional methods of agriculture. These methods are inefficient. Technology of agriculture is also inefficient and orthodox. In recent years some rich farmers have started to adopt improved implements like steel ploughs, seed drills, barrows, hoes, water pump, tractors etc. Most of the farmers are relying on centuries old wooden ploughs and other implements. Though some are using improved implements but their number is very limited. Hence, Farm Power Availability (FPA) as observed by the Agriculture Department puts Jharkhand in the IIIrd category of FPA. Third category means the state has FPA between 1.726 and 1.00 KW/Hectare. It is less than the national average of India which was 1.94 KW/Hectare during 2012-13. It is clear indication that Jharkhand is still using very less instruments of agriculture which are operated by power.

Prospect in this regard has been mentioned by the agriculture department in their final report that the availability of adequate farm power for mobile and stationary farm operations should be increased from the present level of about 1.21 KW/hectare to about 1.54 and 2.65 KW/hectare to about 1.54 and 2.65 KW/hectare in Jharkhand by 2020 and 2030 respectively.

District wise study of FPA shows an alarming problem in the state as only three districts (Bokaro, Koderma and Giridih) have high availability of farm power. Rest all districts come under less Farm Power Availability in the state.

Migration and trafficking of working population from

Jharkhand is a serious and chronic problem. This directly affects the agriculture of the state. Decade after decade the agricultural population is decreasing. It was 67.1 percent during 1991 and decreased to 58.2 percent in 2001. Again during 2011 it decreased to 48.9 percent. Reasons behind this sharp decrease is either migration or trafficking in Jharkhand. The economic survey of India has revealed that Jharkhand lost close to 5 million of its population between 2001 and 2011 due to migration. More than 5 percent of working population migrates annually to other states in search of better employment opportunities, education or because of loss of traditional livelihood. The net outflow of the working age population is highest among the states in the country. Seasonal migration from Jharkhand is also well known to all in the state.

Jharkhand Economic Survey 2016-17 P.302 mentions that Jharkhand has emerged as a major source area for Intra-country trafficking in India. Women and children, especially adolescent girls are the major victims of trafficking. The sectors of exploitation include - Domestic labor, bride trafficking, Brick Kiln workers, sexual exploitation, carpet industry and trafficking en-route orphanages. New initiatives are being taken by the Jharkhand government to combat human trafficking. The victims of trafficking generally belong to either cultivator or agricultural laborer's family; naturally the agriculture of the state gets adversely affected.

Seed requirement as well as fertilizer requirement during Kharif season in Jharkhand increased to a highest point but supply of both are very poor. There is a wide gap between the requirement and distribution. Below table makes it clear though it has been discussed earlier in detail.

Almost every crop has a huge amount of deficit. Paddy (C) has the highest seed deficit of 129420.77 Quintals. It is followed by Paddy (H) with 71940.50 Q and then maize with 61908.8 Quintals of seed deficit. Urad, Moong and Jowar showed a deficit of 7309.72, 1470 and 1611 Q respectively. Overall, there is a huge deficit of Kharif seed distribution in the state. Similarly, the trend of fertilizer requirement and distribution can be seen in table 8.3 where shortage is continuously prevailing. This is an immediate and undeviating problem for the farmers and agriculture of Jharkhand. The plausible explanation for the gap could be the short fall in the supply of the fertilizers, however there are some exceptional years where the supply has exceeded the requirement and consumption.

**Table 8.2**
**Kharif Seed Distribution (2016-17)**

| Name of Crop | Distributed Quantity (in Q) | Requirement (in Q) |
|---|---|---|
| Paddy (C) | 88559.83 | 217800 |
| Paddy) H) | 3050.50 | 75000 (NSC) |
| Maize | 891.20 | 62800 |
| Urad | 478.28 | 7788 |
| Moong | 180.00 | 1650 |
| Jowar | 310.00 | 192 |
| Total | 93469.81 | 366959 |

**Source:** Directorate of Agriculture, Govt of Jharkhand

**Table 8.3**
**Trend of Fertilizer requirement and consumption in Jharkhand (in tons)**

| Year | Kharif | | | Rabi | | |
|---|---|---|---|---|---|---|
| | Requirement | Supply | Consumption | Requirement | Supply | Consumption |
| 2009-10 | 250000 | 175013 | 169672 | 140000 | 138555 | 148503 |
| 2010-11 | 261000 | 211782 | 156025 | 160000 | 85786 | 85798 |
| 2011-12 | 298000 | 197168 | 196744 | 229000 | 684894 | 88557 |
| 2012-13 | 350000 | 186056 | 165131 | 249000 | 120080 | 112758 |
| 2013-14 | 280000 | 153554 | 135092 | 265000 | 68176 | 74997 |
| 2014-15 | 151760 | 2302272 | 146397 | 206000 | 91352 | 89701 |
| 2015-16 | 215000 | 240354 | 223247 | 266000 | 47653 | 93088 |
| 2016-17 (Up to 31-10-2016) | 330000 | 227707 | 200795 | 165500 | 47653 | 11019 |

**Source:** Directorate of Agriculture, Govt of Jharkhand

**Chapter Four** has discussed in detail on cropping intensity and irrigation intensity. Cropping intensity is the percentage of gross cropped area to net sown area and irrigation intensity is the percent of net irrigated area to net sown area. Both are the percentage of net sown area. Thus, one can imagine the importance of net sown area.

> We researchers are facing a terrifying problem of reliable data. An example–Jharkhand Economic Survey 2016-17 has discussed the Agriculture of Jharkhand in Chapter Six. Page number 128 is related to Land use where Table 6.3 cites the detaisl of land Utilization. Table shows Net Sown Area as 25.75 lakh hectares Current Fallow & Other Fallow Land is 15.02 lakh hactare. They represent 28.08 and 119.58 percent of land respectively. This data has been taken from the Annual Plan (2016-17) of the Govt. of Jharkhand. When we refer the Agricultural Statistics at a Glance 2018, Govt. of India for the same data, there is a major gap between them. Here the Net Area Sown is 14.51 lakh hectare and Fallow Land (current and other) is 24.62 lakh hectare. Their percentage is 18.2 and 30.8 respectively. Calculations of percentage mentioned in Annual plan or Jharkhand Economic Survey do not match at all.

Here the main problem being faced by Jharkhand is raising of fallow land and decreasing of net area sown year by year continuously. In Jharkhand cultivable land/culturable land/Arable land is 4320 thousand hectares in which only 1281 thousand hectares are net sown area and 2513 thousand hectares are fallow land. Ratio of fallow land is far more than the double of the net sown area. During 1955-56 to 1989-90 net sown area was about 25 to 30 percent and now it is only 18.2 percent. Fallow land in the respective year was 15.5 to 23.8 percent and now it has increased to 30.8 percent. It is a big problem for Jharkhand. Why is Fallow land in rising and Net sown area in a shrinking mode!

Experts and planners are indicating towards the intensity of cropping. It is rising gradually year after year. It was 117.9 in 2012-13, it became 130.6 in 2015-16 and now it is 142.4 in 2018-19. They want to point towards the view that intensity of cropping is rising and hence don't bother about the fallow land and net sown area. They also point towards the increasing pattern of "Area sown more than once" which was 251 thousand hectares only in 2012-13 and now it is 544 in 2018-19. Certainly, it is a matter of research to find out its causes.

Somehow the reply of the above question is hidden in the trend of irrigation intensity. There is detailed temporal data of irrigation intensity (1955-56 to 2016-17) in chapter three. During 1955-56 it was 10 percent and then year after year it went on decreasing till 1995-96. After emergence of Jharkhand, it began to rise and now it is 16.20 percent in 2016-17. As the irrigation intensity rises the cropping intensity will rise too. When we compare to the data of the country, the cropping intensity of India was 136 during 2018-19 and irrigation intensity of the same year was 68.8. Thus, the level of the state is very poor than the national potential.

**Chapter Five** deals with cropping pattern and diversification. Cropping pattern which shows the proportion of the area under different crops at a definite point of time, is an important indicator of development and diversification of the sector. Food crops and non-food crops or cash crops are produced by the agricultural sector of the country or state. As the prices of the cash crops are becoming more and more attractive, more and more land has been diverted from the production of the food crops into cash crops or commercial crops. This has been creating the problem of food crisis in the region.

**Table 8.4**
**Per head/per day food crops (cereals and pulses) in Jharkhand**

| Year | Cereal Production (Metric Tons) | Pulses Production (Metric Tons) | Per head/ per day (in gram) Cereal | Per head/ per day (in gram) Pulses | Requirement of cereals (in grams) | Requirement of Pulses (in grams) |
|---|---|---|---|---|---|---|
| 1989-90 | 1773000 | 52000 | 219 | 7 | 460 | 40 |
| 1990-91 | 1751000 | 52000 | 220 | 6.5 | 460 | 40 |
| 2001-02 | 3057000 | 165000 | 311 | 16.90 | 460 | 40 |

| | | | | | | |
|---|---|---|---|---|---|---|
| 2002-03 | 2464000 | 160000 | 250 | 16.26 | 460 | 40 |
| 2003-04 | 2582000 | 208000 | 262 | 21.14 | 460 | 40 |
| 2004-05 | 2301000 | 190000 | 233 | 19.31 | 460 | 40 |
| 2015-16 | 3128000 | 547000 | 260 | 45 | 460 | 40 |

**Source:** Jharkhand Ka Bhugol, Ram Kumar Tiwari, Bihar Hindi Granth Academy, Patna

Production of rice and pulses in Jharkhand was not sufficient as per the requirement of the population till date. Our calculation on regional basis show that it is 260 grams per head per day while required amount is 460 grams per head per day. Pulse production was also not sufficient till 2004-05. But production of 2015-16 shows 45 grams per head per day which is the sufficient amount. But still the state has not recorded any surplus production.

It is plausible to see the data of Govt of India regarding daily availability of rice per capita across India for 2015, 2016, 2017, 2018, 2019, production is 186, 184, 183, 189, 189 grams per day per capita.

Rice research station, Hazaribagh has rightly observed that Jharkhand state falls under the category of low and unstable in maximum districts of Jharkhand. Causes are - small holdings, holdings are scattered at different places/ location, farmers are not able to generate their own resource like open wells, small private or community ponds and check dams to collect ground water and surface water for irrigation to save the crops from drought. Farmers also reported that drought, erratic rainfall, weed infestation and soil erosion in upland and cases of worms and submergence in lowland are the other major problems.

During the course of this work, we have noticed some other constraints or problems of agriculture in the state. These are - Instability and fluctuation of weather, orthodox farming techniques, inadequate use of inputs, poor conditions of agricultural laborers, absence of scientific crop rotation, instability in agricultural prices, unorganized agricultural marketing, high degree of agricultural debt and uneconomic use of crop residue etc.

It would be worthy to mention some indirect constraints or problems of agriculture which effect the whole country and therefore Jharkhand. These are - Rural Development Program, Agricultural Education, Food Policy, Land Policy etc.

**Chapter Six** has a discussion regarding the regionalization of the area. Three methods have been adopted for regionalization - Cop concentration, crop combination and cropping pattern. Crop concentration means the variations in the density of any crop in an area or region at a given time. We have discussed and demarcated the rice region, wheat region and maize region of Jharkhand on the basis of district level calculations. Here we would like to highlight all those areas which have very low concentration index so that the planners can ameliorate the agricultural conditions of those regions of concerned crops.

Crop concentration of rice has been worked out for the period of 2004-05 and 2017-18. During 2004-05 low concentration of rice area was Palamu, Latehar, Garhwa, Chatra and Koderma. During 2009-10 there were only four districts where the rice concentration was less. They are Palamu, Latehar, Garhwa and Lohardaga. During 2017-18 there were no districts under this low concentration index category. Certainly, production of rice has increased with a good pace. However, constrain is low productivity and it is still not sufficient to meet the daily availability per capita for the state.

Maize is the second largest crop in the state. Its concentration index is very low in Gumla, Dhanbad, Khunti, Hazaribagh, Ranchi, Saraikela, Simdega and West Singhbhum districts. Here crop concentration index for maize is below the average for the state.

Wheat is the third largest crop in Jharkhand. Only three districts (Garhwa, Godda and Koderma) have recorded highest concentration of wheat whereas ten districts have recorded concentration index below average of the state. These districts are Bokaro, Dhanbad, East Singhbhum,

Gumla, Khunti, Lohardaga, Ramgarh, Ranchi, Saraikela and West Singhbhum.

Crop combination analysis reveals that only one district (Khunti) has for crop combination (RPMW) whereas only one district (Godda) has three crop combination (RMW) and two districts (Palamu and Latehar) have two crop combination (PR). Rest all the districts have single crop combination (Rice). In this group about 10 districts are included which covers about 80 percent of the area of entire Jharkhand.

Cropping pattern has been related by the yield and spread index. Jharkhand has been demarcated into four regions - High Yield, High Spread, High Yield, Low Spread, Low Yield, Low Spread region. Here Ranchi, East Singhbhum and Dumka have been demarcated under Low Yield, Low Spread region. This requires amelioration in Agriculture.

**Chapter Seven** is related to animal husbandry. Animal husbandry has assumed a much broader role in the overall economy than so far envisaged, as an integral part of expanding and diversified agriculture. Vast livestock population of Jharkhand offers tremendous potential for meeting domestic demand for milk, egg, meat etc. Problem of Jharkhand in livestock is its growth pace. Growth in India is very fast whereas in Jharkhand it is very slow. In India there were 293 million livestock during 1951, which grew to 512 million in 2012. On the other hand, Jharkhand has recorded 28602 thousand livestock in 1982 and it has grown to only 29592 thousand during 2019. Slow growth rate of livestock is the main constraint of Jharkhand. Prospect for livestock is its production. Table 7.22 shows that milk production in Jharkhand is just half of the average in India. India produces milk per head per day around 307 grams while Jharkhand produces only 146 grams per head per day. Meat production is 45.5 thousand tons in Jharkhand. Problem is that the production is constant from 2009-10 and decreasing from all sources except goat and poultry.

Fishing industry in Jharkhand has been moving gradually towards the modern phase. Infrastructure, cooperatives and number of workers are gradually increasing in the state. More than 50 percent of the population in the state consume fish products. With an estimated annual demand of 1.40 lakh tons of fish, the state is producing 106430 tons of fish annually with rest of the demand being catered through fish from Andhra Pradesh and West Bengal. Thus, fisheries become an important economic activity in the state for additional employment and income generation. Employment generation activities through fisheries include harvesting, net and boat making, seed production, transportation and marketing of fish in wholesale and retail market.

## Conclusion

We have seen in the foregoing chapters that Geology, Terrain, Climate, Soil, Drainage, Population, Transportation etc., are, to a great extent, related to the factors of physical and cultural Geography of the area and all these are closely associated to the Agriculture of the area also. Jharkhand is a land of great inequalities. Great differences in elevation and relief bring about contrasts in climate, soil, drainage. All these in turn, influence the land use pattern, cropping pattern, land holding pattern, distribution of population (farmers) and various aspects of Agriculture. Extensive areas above 600 meters and greater part between 600 to 300 meters present a variety of morphological features which are clearly expressed in the cultural landscape. River valleys, most of them long, deep and terraced, a few of them wide with gentle slopes, alternate with low, elongated ridges. Quick runoff renders the stream ephemeral and ponds empty hollows in summer. The impervious character of gneissic rocks further aggravates the problem of underground water supply.

Physiographically Jharkhand has been divided into six major units - Central Plateau, Lower Plateau, South-Koel Sankh basin, Dharwarian country, Damodar basin, Rajmahal Upland and climatically which is now a days known as agroclimatic subzones, Jharkhand falls under Zone VIII i.e., Eastern Plateau and Hilly region, which has been further divided into three subzones namely, IV. Central and North Eastern Plateau, V. Western Plateau, and VI South-Eastern Plateau. The state receives annual rainfall of 1200-1600 mm and the climate range from dry semi humid to humid semi-arid types. Out of the average annual precipitation of 10 billion cubic meters (BCM), that the state receives, about 15% is lost in the atmosphere and 50 % flows out as surface run off. Thus, 35% of the total precipitants recharge the ground water.

Soil of Jharkhand are low to very low in available phosphorus and Sulphur, medium in available nitrogen and potassium, deficient in available boron in general. About 1.6 million hectares are acidic. The region has a major problem of slight (51% area) to moderate (36% area) soil erosion since about 43% of area is located on a very gentle slope (1.3%) and 31% of area on gentle slope (3 to 8 percent).

The drainage system in the state or water bodies consist of rivers, reservoirs, ponds, check dams and low-lying paddy fields. The major river system in the state is Subarnarekha, Damodar, Shankh, South Koel, North Koel, Ganga, Mayurakshi, Barakar, Konar etc. These river systems spread over 1800 Km is length. The state is having a number of reservoirs, created by damming of rivers. The total area under reservoirs is 108001 hectares. Under different development programs, the irrigation department and volunteer organizations have made check dams on the small rivers and on creeks. The water area under check dams is spread in 20000 hectares. In addition to these water bodies, majority of villages in the state have ponds and low-lying paddy fields. The area under ponds is 29000 hectares and the area under low-lying paddy fields is 200000 hectares.

Population is closely associated with agriculture because population is worker, producer and consumer of agricultural products. In Jharkhand highest concentration of population is in Dhanbad district where density is more than 1000 (1316). Next comes areas having 1000-500 persons per square Km. These districts are Deoghar (602), Godda (580), Sahebganj (558), Bokaro (715), East Singhbhum (644), Ramgarh (708) and Ranchi districts. The third category is between 500 to 200 persons per sq km. These are Garhwa (323), Chatra (280), Koderma (282), Giridih (492), Pakur (497), Lohardaga (307), Palamu (442), Hazaribagh (488), Dumka (351), Jamtara (437), Khunti (210), West Singhbhum (208), Saraikela (401). Below 200 persons per sq km has been recorded in Latehar (169), Gumla (191) and Simdega (159) districts.

Agricultural density (Agri Population/Total cropped area) is highest in Dhanbad (14.81), Giridih (12.93), Koderma (12.60) and Ramgarh. Lowest agricultural density has been recorded by Khunti district (2.77).

Total population of Jharkhand is 32988134 persons during the census of 2011. Among this population 3985644 persons are SC and 8645042 persons are Scheduled Tribes. Total tribal population of Jharkhand has been divided in 32 tribal groups among them 8 are primitive tribal groups.

The population composition of Jharkhand shows 39.70 percent Total workers, 52.05 percent Main workers and 47.94 percent marginal workers. About 60.29 percent population of Jharkhand are non-workers. Main working population of Jharkhand has been divided into 29.35 % cultivators, 18.16% agricultural laborers, and 3.65% are house hold workers. 48.82 percent are in the category of other workers. The constrain here is the decreasing number of cultivators, agricultural laborers and household workers. Only the Other Worker group is increasing day by day.

Land utilization is an integrated part of agriculture. Land use proportion during 2018-19 was as under - Forest area - 28%, Land not available for cultivation - 16.1 %. Other uncultivated land is recorded 8.19 % under which three sub categories of land are available - permanent pasture land and grazing ground is 1.59 percent, Land under miscellaneous tree crops not included in net sown area are 1.83 percent and culturable waste land is 4.76 percent. Similarly, Fallow land has been recorded 31.5 percent which has two categories - other fallow land 13.6% and current fallow land 17.9 percent of the total area of the state.

Net sown area has 16.07 percent and area sown more than once is 6.82 percent. Both are together known as total cropped area. It is 23.77 percent in Jharkhand. Total agricultural land/cultivable land/culturable land/arable land which includes four categories of land (land under misc. tree crop, grove + culturable wasted land + Fallow land + Net sown area) has recorded 54.19 percent area in Jharkhand in 2018-19 census.

When we calculate the change in land use of Jharkhand, we find no change in forest area and geographical area. But a growth of 0.1 percent can be seen in Land not available for cultivation area. It was 16 percent is 2012-13 and it was recorded 16.1 percent in 2018-19. Similarly, other uncultivated land was 7 percent to the total geographical area of the state during 2012-13 which grows to 8.19 percent during 2018-19. It is a growth of 1.19 percent of the total geographical area. Fallow land of the state has changed only 0.5 percent. It was 31 percent in 2012-13 which increased to 31.5 percent in 2018-19. Net sown area is an important aspect of agriculture. During census of 2012-13 net sown area of Jharkhand was 17.6 percent and now it is 16.07 percent during 2018-19. It is a change of 1.53 percent in the state. Total cropped area has changed around 3 percent. It is an increase in the total cropped area. It is due to the increase of are of sown more than once. It was 3.1 percent and now it has grown to 6.8 percent in the corresponding years. Therefore, the cropping intensity of the state has grown from 117.9 to 142.4.

Chapter two is related to the Land tenure and size of the holding. Total number of marginal holding (0-1 hectare) in Jharkhand is 18.48 lakh that covers an area of about 764 thousand hectares and this is 24.13 percent of the total area of the holdings. Small holdings (1-2 hectares) are 4.29 lakhs that cover an area of 591 thousand hectares which is 18.67 percent of the total area. Semi medium (2-4 hectare) and medium (4-10 hectare) holdings have their numbers 2.83 lakh and 1.29 lakh which cover the area of 775 and 725 thousand hectares respectively. These are about 24.48 percent and 22.90 percent of the total area of the state. Large land holdings are only 20 thousand in number that cover 311 thousand hectares area which is 9.82 percent of the total area. Jharkhand has maximum number of holdings of small category (1-2 hectares). If further fragmentation takes place it would be marginal holdings at the top.

Chapter three deals with Farm technology in Jharkhand. We have discussed various aspects of technology right from meaning, types, Farm power, Farm workers, fertilizers and seeds (HYV) etc. in detail. The interface between technological institutions and the farmers is rather weak in Jharkhand at present, which needs to be strengthened; list of appropriate technology/varieties should be regularly updated at the state level. The low level of technology adoption and the farmers not being fully equipped for technology adoption are considered an outcome of this. The state government should develop and strengthen institutional mechanism for quick transfer of appropriate and new technologies to the farmers. New technology has been adopted in soil management, water resources, seed supply, fertilizer use efficiency, organic farming, Agroforestry, Integrated farming systems (IFS), climate resilient agriculture etc. But a great constrain is the reducing number of cultivators and agricultural laborers. Now it is a separate subject of research that how much technology is replacing the man power in agriculture.

Chapter four has been concentrated on cropping intensity of Jharkhand. No doubt, cropping intensity is very closely associated to irrigation intensity and both are related to the Net area sown. Here it is a very interesting situation that year after year net sown area has been decreasing, but the cropping intensity of Jharkhand is increasing. Cropping intensity was 117.9 percent during 2012-13 and now during 2018-19 it has become 142.4. Net area sown was 17.6 percent of the total area and now it is 16.07 percent. The reason behind the increase in net area sown is related to the rise of area sown more than once which is increasing year by year. It was only 3 percent during 2012-13 but has risen to 6.8 percent during 2018-19. Similarly, irrigation facilities have been enhanced in the state. Nevertheless, cropping intensity is still less than India. It was 194.40 in India during 2012-13 and in the same year Jharkhand had recorded only 117.9 percent which was 76.5 percent less than the national average. There is a direct correlation between irrigation intensity and cropping intensity. Whenever the irrigation intensity increases, the cropping intensity of the state also increases because irrigation is the most influential factor in agricultural development.

Chapter five is about the cropping pattern and diversification in Jharkhand. In this chapter we have studied about the type of cultivated land, harvesting calendar, areal pattern of food grain cultivation and their diversification. In Jharkhand Rice is the first crop. Maize is the second crop and Wheat is the third food crop.

Area of the food crops has increased gradually. It was 1753 thousand hectares during 1985-86. It had grown to 1957 thousand hectares in 1999- 2000 and then reduced a little bit to 1933 thousand hectares during 2015-16. Area under pulses has grown significantly. It was only 77 thousand hectares during 1985-86 which has grown to 103 thousand hectares during 2000-2001 and 594.99 thousand hectares during 2015-16. Oil seeds production area has also grown like pulses. It was only 50.82 thousand hectares during 1988-89 but grown to 31.98 thousand hectares in 2001-02 and 259.68 thousand hectares during 2015-16.

Production of food crops was 1738 thousand tons during 1985-86 which raised by 2641.29 thousand tons during 2001 and 3128.77 thousand tons during 2015-16. Pulse production was only 45 thousand tons in 1985-86 which has grown to 53 thousand tons in 2000-01 and 547 thousand tons during 2015-16. Oil seeds production was 60.53 thousand tons during 1998-99 and decreased to 44.91 thousand tons during 2000-01. Now it has grown to 177 thousand tons during 2015-16. Overall growth in area and production has been recorded in all food crops but due to drought and uncertain monsoon, area and production both have recorded fluctuations in production and cropping area.

Jharkhand is a big producer of vegetables and it supplies regularly to neighboring states like West Bengal and Orissa. Vegetable growing area was 149.80 thousand hectares in 2000-01 and now the area is 229.63 thousand hectares in 2009-10. Similarly, vegetable production has recorded a significant rise. It was 20109.50 thousand tons in 2000-01 and now it has risen to 3827.01 thousand tons during 2009-10.

Chapter six has demarcated the agricultural regions

utilizing the methods like crop concentration, crop combination and cropping pattern. Crop concentration has demarcated three regions for each crop - high concentration, moderate concentration and low concentration. Crop combination has demarcated four regions - one crop, two crop, three crop and four crops combination regions. One crop combination region which is only rice region covers the 80 percent of the state. Cropping pattern has demarcated four regions i.e., High yield high spread area, High yield low spread, low yield high spread and low yield low spread area. All methodologies are very useful for other research students as well.

Chapter seven has discussed the animal husbandry of Jharkhand which is complimentary of agriculture. Jharkhand state has a total livestock and poultry population of 29.59 and 6.63 million, respectively; cattle (18.17 million) and goat (8.39 million) contribute maximum to the livestock population according to the 2019 census.

The productivity of existing livestock and poultry in Jharkhand is very poor and there is also a wide gap in production and requirement of livestock produce like milk, meat and eggs. Annual per capita milk availability in India (2013-14) is 307 grams/day whereas Jharkhand has only 146 gram/day. Egg production in India is 61 per capita per annum but in state it is only 14 per capita per annum. It is just about one-fourth of the country.

We have performed a very interesting and almost first-time work in Jharkhand i.e., Livestock combination just like crop combination and demarcated the livestock regions in Jharkhand. During the course of research, we have gone through a vast reference and previous literature related to agriculture but didn't find any such calculations. We have taken inspiration from Jasbir Singh who has mentioned this in his book - An Agricultural Geography of Haryana (1976. P.376). Jharkhand has been divided into four livestock combination regions i.e., one, two, three and four livestock regions. One livestock region is Pakur district which is a Poultry region. Two livestock regions have two sub-regions i.e., Cattle- Goat and Poultry-Goat. Cattle-goat region is spread over Chatra, Deoghar, Dumka, Garhwa, Palamu and Poultry-Goat region is only in Ramgarh district.

Three livestock regions have been divided into five sub-regions i.e.,

(i) Cattle-Goat-Poultry which spreads over Bokaro, Giridih, Hazaribagh, Jamtara, Koderma and Latehar.

(ii) Poultry-Cattle-Goat area spread over Dhanbad, East-Singhbhum and Khunti district.

(iii) Goat-Cattle-Poultry area which is Gumla, Lohardaga, Simdega

(iv) Poultry-Goat-Cattle in Ranchi and West Singhbhum district and

(v) Goat-Poultry-Cattle are which is Sahebganj and Saraikela districts.

Four livestock region which is known as Cattle-Goat-Poultry-Pig region is only in Godda district.

At the end of the chapter, we have also mentioned the various livestock development programs run by the government.

Fishing industry is growing in Jharkhand rapidly. It produces 2.23 lakh tons (2020) of fish every year. It is about 1.2 percent of the production of India. However, the production is still less than the demand.

Chapter eight is the problems and prospects of agriculture in Jharkhand. We have discussed several issues which are also constraints and also the prospects of Jharkhand. Here we highlight some of the challenges for the state in the concluding chapter.

- Jharkhand is by and large a mono cropping state and therefore the cropping intensity is very low.
- Agriculture is predominantly rainfed. 75 percent of the net sown area is rainfed in the state.
- Irrigated area is very low. According to 2015-16 data, only 235 thousand hectares area is irrigated by all sources. This is only 2.9 percent of the total geographical area and 16.19 percent of net sown area.

- The per capita availability of food grain is currently at 250 grams/day which needs to be increased to 400 grams/ day per head. Similarly, per capita availability of milk, fish, meat, poultry is also very low as compared to the national average.
- Farmers of Jharkhand are cultivating their fields traditionally and a few have partially adopted new technology.
- Jharkhand is continuously facing the problem of deforestation, overgrazing, loss of wetland due to siltation and over exploitation of forest.
- Heavy mining activities led to emergence of degraded land, which needs to be restored through planting.

# Bibliography


A. Prasad (1973); Chotanagpur: Geography of Rural Settlement, Ranchi University, Ranchi.

Ahmad, E. (1951); Geomorphic outline of Chotanagpur, Geographical Outlook, Vol-2, Department of Geography, Ranchi University, Ranchi.

Ahmad, Enayat (1965); Bihar: Physical, Economic, Regional Geography, Ranchi University, Ranchi.

Akhouri, Rajshree & Achyut, Keshav (2017); Chotanagpur Tenancy Act, 1908. Malhotra Book Agency, Ranchi.

Arther L. Bloom (1978); Geomorphology: A Systematic analysis of late Cenozoic landforms. Prentice Hall of India Pvt. Ltd. New Delhi-1.

Ayyappan S, & Kochhar, S. (2010); Indian Agriculture in the new millennium: Innovative strategies for productivity enhancement prospects and challenges. Indian Seed and Planning material, 3(1), P. 7002.

Bhatia, S.S. (1965); Pattern of Crop Concentration and Diversifications in India, Economic Geography, 41.

Buchanan, F. (1922); Quoted by Macpherson, H. Survey and settlement report of Santhal Pargana, Patna.

Camil, Bulke (1963); Yaksha Puja, Ramkatha, Allahabad University, Allahabad.

Chand, R (2007); The State of Indian agriculture and prospects for the future. In K. Chopra & C.H. Hanumantha Rao (Eds), Growth, Equity, Environment and Population: Economic and Sociological Perspectives, Sage Publication, New Delhi.

Chatterjee, S.C. (1949); The Physiographic Evolution of Chotanagpur, Calcutta University, Dept. Of Geography.

Chotanagpur Tenancy Act (1931 & 1946); Patna.

Choudhary, Rambalak & Marandi, Rosline Kusum (2017); Assessment of the status of dairying in the Eastern states and potential to improve socio- economic status of milk producer and convergence of all central and state schemes at district level in Jharkhand- Agro-economic Research Center for Bihar & Jharkhand, T.M Bhagalpur University Bhagalpur (Report No. 45/2017).

Cobley, LS (1962); An Introduction to Botany of Tropical crops, Longmans, London.

Dantawala, M.L. & W. B Donde (1965); The Economic Cultivator, The Journal of Agricultural Economics, Vol-15.

Dayal, P (2013); Agriculture Geography of Bihar, Rajesh Publications, New Delhi.

Dayal, P. (1953); Bihar in Maps, Kushum Prakashan, Patna.

Department of Agriculture Brochure (Govt. of Jharkhand); Dairying in Jharkhand, Jharkhand State Cooperative Milk Producer.

Devi, PB, Vijaybharti, R. Satyabhama, S. Malleshi, NG and Priyadarshini, VB (2014); Health Benefits of Finger Millets, Polyphenols and dietary fiber: A review. Journal of food science and technology, 51 (6) 1021-1040.

Dietary Manual (2011); Dietary Guidelines for Indians National Institute of Nutrition. Indian Council of Medical Research, Hyderabad.

District Gazetteers (1938); D.G. Santhal Pargana, Govt Press.

Dunn, J.A. (1941); Economic Geology and Mineral Resources of Bihar Memoirs, G.S.I. Volume-78, Calcutta.

Eddowes, Maurice (1969); Crop Technology, London, Hutchinson Educational.

Ghosh, B.N. (1987); Fundamentals of Population Geography, Sterling Publishers Pvt. Ltd. New Delhi/Bangalore.

Ghosh, G.N. (2013); Sustainable agriculture: Strategies and approaches. In P.K. Shetty, S. Ayyappan and M.S. Swamynathan (Eds) climate change and sustainable food security. Bangalore; National Institute of Advanced Studies and New Delhi: Indian Council of Agricultural Research

Gupta, PK (2017-18); Value Chain Study of Tomato of Ranchi, Jharkhand, National Horticultural Research & Development Foundation (NHRDF)

Hallet, M.G & Stewart Mcpherson (1917); Bihar and Orissa district Gazetteer: Ranchi, Province Govt Publication, Superintendent, Govt. Prints, Patna.

Hussain, Mazid (1996); Systematic Agriculture Geography, Rawat Publications, Jaipur/New Delhi.

Indian Council of Agricultural Research, New Delhi (1970); Pulse crops of India, PP. 10-11

Ivern, S.J.F (1969); Chotanagpur Survey, Indian Social Institute, New Delhi.

Jaiprakash & Singh, R.P. (2016); Growth performance of fish production in Jharkhand. Agro Economist-An International Journal, Renu Publications.

Johnson, B.L.C. (1979); India: Resource and Development. Anrold-Heinemann, London, First Indian Edition, 1980.

Johnson, R.J. (1986); The dictionary of Human Geography, National Publishing House, New Delhi.

Kautilyas; Arthshastra, Jaico Publishing House, Mumbai/Bhopal/ Kolkata.

Kendrew, W.G. (1961); The Climates of the continents, Clarendon Press, Oxford.

Klages, K.H.W. (1958); Ecological Crop Geography, Macmillan Co. New York.

Lekhi, R.K. & Singh, Joginder (2018); Agricultural Economics, An Indian Perspective, Kalyani Publishers, New Delhi.

Mandal, Banani & Others (2018); Chromosomal Study of Black Bengal Goats of Jharkhand, International Journal of Current Microbiology, Applied Science, Vol- 7 (Special Issue), P. 501-509.

Ministry of Agriculture (2015); Proceedings of the workshop. In union Agriculture Minister (Chair), Agriculture scenario in Eastern India: Review and future prospects. Workshop held at NASC Complex, DPS Marg, Pusa Complex; New Delhi. Retrieved from bgrei- rkvy.nic.in/minutes.

Ministry of Agriculture (2018); Final Report on Monitoring, Concurrent evaluation and impact assessment of Sub-Mission on Agricultural Mechanization.

Mishra, R.K & Choudary, J.C (2011-12); Agricultural Resource database of Jharkhand. SAMETI, Department of Agri. And

Cane development, Govt. of Jharkhand, Ranchi.

Mishra, R.K. (2006) & V. Jairam; Agricultural development in Jharkhand: At a Glance through figures:2006. Agricultural Data Bank, A/13 Krishi Vihar, Post Boreya, Ranchi.

Naylon, J (1959);Land Consolidation in Spain, Annals of the Association of American Geographers, Vol-49.

Negi, Balbir Singh (1971); Economic Geography of India, Kitab Mahal, Allahabad.

Nishar Akhtar, Anwesha Dey, Gufran Ahmad, Firoz Ahmad, Md. Mahtab Rashid (2020); Potential and Economics of Ragi cultivation under changing weather condition of Ranchi, Jharkhand, International Journal of Current Microbiology, Applied Science, 9, 2002.

NITI Ayog (2015); Comment on the concept note on agriculture: Challenges and way forward Jharkhand. Retrieved from http://nti.gov.in.

O'Malley, LSS (1938); Bihar District Gazetteers, Santhal Pargana (Revised by S.C. Mukherjee), Patna.

Pattnaik, B. B (2013); Quality Seeds: Contribution of national seeds corporation. In A.K. Singh (Ed.) Golden Jubilee Souvenir 2012-13, New Delhi National Seeds Corporation Limited.

Pearson, F.A. and Harper, F.A. (1945); The world's Hunger. New York, Cornel University Press, Ithaca.

Pokshishevskiy, V.V (1968); Foreword in Economic regionalization if India, Commissioner of India.

Prasad, Ayodhya (1973); Chaotanagpur: Geography of Rural Settlement, Ranchi University, Ranchi.

Randhawa, M.S. (1958); Agriculture and Animal Husbandry in India, ICAR, New Delhi.

Sengupta, P & Sdasyuk, Galina (1968); Principles and methods of Economic Regionalization, Census of India, 1961, Monograph Series, Vol-I, No.-8.

Simon Kuznets (1973); Population, Capital and Growth (Selected Essay), W.W. Norton & company, New York.

Singh, H & Chand, R (2011); The Seeds Bill, 2011: Some Reflections. Economic and Political Weekly, XLVI (51).

Singh, Jasbir & SS Dhillon (1984); Agriculture Geography, Tata Mc Graw Hill Publishing Co. Ltd, New Delhi.

Singh, Jasbir (1974); An Agricultural Atlas of India: A Geographical Analysis. Kurukshetra, Vishal Publications.

Singh, Jasbir (1974); The Green Revolution in India- How green it is, Kurukshetra, Vishal Publications.

Singh, Jasbir (1976); An Agricultural Geography of Haryana, Vishal Publications, Kurukshetra University Campus, Haryana.

Singh, K.N. & R.A. Siddiqui (2012); Economic Geography, Prayag Pustak Bhawan, Allahabad.

Singh, R. P & Singh, S. (2016); Optimizing seed replacement rates in Jharkhand: Present scenario, challenges and opportunities. Jharkhand Journal of Development and management studies, XISS, Ranchi, Vol. 14, No.-2 (June)

Singh, R.K. (2003); Socioeconomic Dynamics of Rice Production system in Jharkhand State. (ICAR) Central Rainfed Upland Rice Research Station CRRI), Hazaribagh, Jharkhand.

Singh, R.L. (1971) (Eds); India: A Regional Geography, National Geographical Society of India, Varanasi.

Singh, R.P. (1970); Monograph of Bihar, Bharti Bhawan, Patna.

Singh, R.P. (2012); New Frontiers in seed sector in Jharkhand - Role of SAU. Seed Times, 5 (3).

Singh, T.N.B & Purty, S.S. (2016); Ground water year book: Jharkhand, Central Ground Water Board, Patna/Ranchi.

Singh, V.P., Shastri, R.K, ASRAS, Baghel, SS and Choudhry, JL (1999); Rice growing environment in East India: An agroclimatic analysis. International Rice Research Institute, Philippines and Indira Gandhi Agricultural University, Raipur, Chattisgarh.

Singh, V.R. (1979); A method for analyzing agricultural productivity, Transactions, Vol-1, Pune.

Sinha, V.N.P Singh, L.K.P & Manoj Kumar (2018); Jharkhand: Land and People, Rajesh Publications, New Delhi.

Sinha, V.N.P. (2003); Jharkhand: Land and People, Rajesh Publications, New Delhi.

Slampa, O (1968); Agricultural Regions of India, Sbornik Ceskoslovenske Spolecsosti Zemepusne.

Thomas, D (1963); Agriculture in Wales during the Napoleonic War: a study in Geographical interpretation of historical sources, Cardiff, University of Wales Press.

Tiwari, R.C. (2010); Bharat ka Bhugol, Prayag Pustak Bhawan, Allahabad.

Tiwari, R.C. (2016); Krishi Bhugol, Prawalika Publications, Allahabad.

Tiwari, Ram Kumar (2015); Jansankhya Bhoogol, Prawalika Publications, Allahabad.

Tiwari, Ram Kumar (2019); Jharkhand Ka Bhoogol, Bihar Hindi Granth Academy, Patna.

Tiwari, Ram Kumar (2019); Jharkhand Ki Ruprekha, Bihar Hindi Granth Academy, Patna.

Tripathi, R.R (1970); Changing Pattern of Agricultural Land Use in Upper Ganga-Gomati Doab, Unpublished PhD Thesis, Agra University.

Wadia, D.N. (1953); Geology of India, Macmillan, London.

Wadia, D.N. and Others (1935); Records, G.S.I. Vol. LCVIII, Calcutta.

Weaver, J.C. (1954); Crop Combination Region in the Middle West. Geographical Review, Vol-44, No.-2 (Apr. 1954) P 175-200, Published by Taylor & Francis Ltd, United States.

Wolfe, TK and Kipps, MS (1959); Production of crops, Mc Graw Hill Book Company, INC, New York.

# Appendix-I

Correlation between Irrigation Intensity and cropping Intensity of 2004-05

| **Irrigation intensity** | **X-Ax=5** | | **Cropping Intensity** | | **y-Ay=105** | |
|---|---|---|---|---|---|---|
| **X** | $\delta x$ | $\delta x^2$ | y | $\delta y$ | $\delta y^2$ | $\delta x\ \delta y$ |
| 6 | 1 | 1 | 167 | 62 | 3844 | 62 |
| 5 | -1 | 1 | 113 | 108 | 11664 | -108 |
| 1 | -4 | 16 | 106 | 1 | 1 | -4 |
| 5 | 0 | 0 | 116 | 11 | 121 | 0 |
| 14 | 9 | 81 | 105 | 0 | 0 | 0 |
| 1 | -4 | 16 | 176 | 71 | 5041 | -284 |
| 21 | 16 | 256 | 138 | 33 | 1089 | 528 |
| 1 | -4 | 16 | 109 | 4 | 16 | -16 |
| 10 | 5 | 25 | 108 | 3 | 9 | 15 |
| 5 | 0 | 0 | 105 | 0 | 0 | 0 |
| 6 | 1 | 1 | 111 | 6 | 36 | 6 |
| 0 | -5 | 25 | 105 | 0 | 0 | 0 |
| 0 | -5 | 25 | 0 | -105 | 11025 | 525 |
| 1 | -4 | 16 | 126 | 21 | 441 | -84 |
| 1 | -4 | 16 | 121 | 16 | 256 | -64 |
| 5 | 0 | 0 | 115 | 10 | 100 | 0 |
| 2 | -3 | 9 | 109 | 4 | 16 | 0 |
| 6 | 1 | 1 | 121 | 16 | 256 | -12 |
| 7 | 2 | 4 | 107 | 2 | 4 | 16 |
| 2 | -3 | 9 | 108 | 3 | 9 | 4 |
| 8 | 3 | 9 | 106 | 1 | 1 | -9 |
| 3 | -2 | 4 | 110 | 5 | 25 | 3 |
| **N=24** | **Σ =+3** | **Σδx2= 547** | **N=24** | **Σδy =272** | **Σδy² =33904** | **Σδx δy =596** |

$$r = \frac{\sum \partial x \partial y - \frac{\sum \partial x \times \sum \partial y}{N}}{\sqrt{\left[\sum \partial x^2 - \frac{\sum \partial x^2}{N}\right]\left|\sum \partial x^2 - \frac{\sum \partial x^2}{N}\right|}}$$

$$r = \frac{596 - \frac{3 \times 272}{24}}{\sqrt{\left[547 - \frac{3^2}{24}\right]\left|33904 - \frac{272^2}{24}\right|}}$$

$$r = \frac{596 - 34}{\sqrt{[547 - 0.395]\,|33904 - 3082.6|}}$$

$$r = \frac{562}{\sqrt{[546.6] \times 30821.4}}$$

$$r = \frac{562}{\sqrt{16846977.24}}$$

$$r = \frac{562}{4104.50}$$

$$r = +0.136$$

Both have positive correlation but very low. It means irrigation intensity can change the cropping intensity can change the cropping intensity in positive direction but their correlation is very low.

# Appendix-II

Correlation between Irrigation Intensity and cropping Intensity of 2016-17.

| **Irrigation intensity** | **X-Ax=20** | | **Cropping Intensity** | **y-Ay=140** | | |
|---|---|---|---|---|---|---|
| **X** | $\delta x$ | $\delta x^2$ | y | $\delta y$ | $\delta y^2$ | $\delta x \ \delta y$ |
| 11 | -9 | 81 | 140 | 00 | 00 | 00 |
| 22 | +2 | 4 | 122 | -18 | 324 | -36 |
| 19 | -1 | 1 | 101 | -39 | 1521 | +39 |
| 9 | -11 | 121 | 139 | -1 | 1 | +11 |
| 12 | -8 | 64 | 152 | +12 | 144 | -96 |
| 5 | -15 | 225 | 102 | -38 | 1444 | +570 |
| 18 | -2 | 4 | 138 | -2 | 4 | 4 |
| 81 | +61 | 3721 | 350 | +210 | 44100 | +12810 |
| 20 | 00 | 00 | 119 | -21 | 441 | 00 |
| 6 | -14 | 196 | 220 | +80 | 6400 | -1120 |
| 22 | +2 | 4 | 120 | -20 | 400 | -40 |
| 7 | -13 | 169 | 120 | -20 | 400 | +250 |
| 32 | +12 | 144 | 125 | -15 | 225 | -180 |
| 22 | +2 | 4 | 102 | -38 | 1444 | -76 |
| 39 | +19 | 361 | 220 | +80 | 6400 | +1520 |
| 32 | +19 | 144 | 138 | -2 | 4 | -24 |
| 12 | +12 | 64 | 122 | -18 | 324 | +144 |
| 25 | -8 | 25 | 127 | -13 | 169 | -65 |
| 90 | +5 | 4900 | 101 | +506 | 256036 | +35420 |
| 7 | +70 | 169 | 107 | -13 | 169 | +169 |
| 7 | -13 | 169 | 104 | -39 | 152 | +507 |
| 11 | -13 | 81 | 137 | -33 | 1089 | +297 |
| 4 | -9 | 256 | 120 | -36 | 1296 | +576 |
| 2 | -16 | 324 | 138 | -3 | 9 | +54 |

| N=24 | Σδx =+52 | Σδx2= 11231 | N=24 | Σδy =+559 | Σδy² =323065 | Σδx δy =50744 |
|---|---|---|---|---|---|---|

$$r = \frac{\sum \partial x \partial y - \frac{\sum \partial x \times \sum \partial y}{N}}{\sqrt{\left[\sum \partial x^2 - \frac{\sum \partial x^2}{N}\right]\left|\sum \partial x^2 - \frac{\sum \partial x^2}{N}\right|}}$$

$$r = \frac{50744 - \frac{52 \times 559}{24}}{\sqrt{\left[11231 - \frac{52^2}{24}\right]\left|323065 - \frac{559^2}{24}\right|}}$$

$$r = \frac{50744 - 1211.16}{\sqrt{[11231 - 112.66]\,|323065 - 13020.04|}}$$

$$r = \frac{49532.84}{\sqrt{11118.34} \times 31044.96}$$

$$r = \frac{49532.84}{\sqrt{3447185281}}$$

$$r = \frac{49532.84}{58712.7}$$

$$r = +0.843$$

Correlation result shows a positive and high correlation in between irrigation intensity and cropping intensity. This shows that increase in irrigation will increase the crop production.

# Appendix-III

**Sample Survey for Research Work (Ph.D)**

Researcher: Raj Kamal Tiwari, Scholar, Department of Geography, Ranchi University

Supervisor: Lal G.S.N. Shahdeo, Asso. Prof. University Department of Geography, Ranchi University

| Name of Farmer | Village | Which Fertilizer is used in Paddy Field | How much use of Fertilizer for Paddy (Kg /ha) | Which Fertilizer in Maize Farm | How much fertilizer use in Maize Farm (Kg/ha) | What new equipment are you using | Name of High Yield Breed seeds used in Farm |
|---|---|---|---|---|---|---|---|
| | | | | | | | |
| | | | | | | | |
| | | | | | | | |
| | | | | | | | |
| | | | | | | | |
| | | | | | | | |
| | | | | | | | |
| | | | | | | | |

# Appendix-IV

## Crop Varieties for Jharkhand

### Rice

#### Birsa Dhan-101

Released in the year 1985 by SVRC for Bihar plateau (now Jharkhand) and notified vide no. 15(CE) dt. 17.08.1992. It has been developed through pedigree method of breeding from the cross Fine Gora x IET-2832. It is sown with the onset of monsoon at 20 cm row to row distance. Its plant height is around 72 cms and it matures in 80-85 days. It is tolerant to blast, BLB, stem borer and drought. It is very much suitable for rainfed upland ecosystems and yield potential of B D-101 is 20-25 q/ha.

#### Birsa Gora-102

Released in the year 1993 by SVRC and notified vide no. 15(CE) dt. 17.08.1992. It has been selected from Gora rice through pure line selection and it is suitable for Jharkhand state. Its plant height is 90-100 cm and it has maturity of about 90-95 days. It is sown with the onset of monsoon at 20 cm row to row distance. It is susceptible to diseases & pests. Its kernel color is red and yield potential of BG-102 is 18-20 q/ha.

#### Birsa Dhan-103

Released in the year 1993 by SVRC and notified vide no. 15(CE) dt 17.08.1992. It has been developed through pedigree method of breeding from cross Fine Gora x IET- 2832. Its plant height is around 85 cm and it matures in 95-110 days. Its planting time is 2nd fortnight of June (after onset of monsoon) at a distance of 20 cm row to row. The seed rate is 80 kg/ha. It is tolerant to blast, BLB, stem borer, Helminthosporium and Gallmidge. It is suitable for upland ecosystems. The yield potential of BD-103 is 20-35 q/ha.

**Birsa Dhan-104**

Released in the year 1993 by SVRC and notified vide no. 15(CE) dt. 17.08.1992. It has been selected from the cross Fine Gora x IET-2832 and it is suitable for Jharkhand state. Its plant height is semi dwarf. The grain is long and bold in size having maturity time of about 85-90 days. It is tolerant to blast, BLB and BPH, stem borer, Helminthosporium and Gallmidge and yield potential of BD-104 is 20-25 q/ha.

**Birsa Dhan-105**

Released in the year 1995 from SVRC. Notified vide no. 1(E) dt. 01.01.1996. It has been developed through pedigree method of breeding from the cross Fine Gora x IET-2832. Its planting time is 15-30 June and matures in 85-90 days. It is tolerant to blast, BLB, stem borer, Helminthosporium and Gallmidge. It is suitable for rainfed upland ecosystems and yield potential of BD-105 is 32-35 q/ha.

**Birsa Dhan-106**

Released in the year 1995 from SVRC. Notified vide no. 1 (E) dt. 1.1.1996. It is selection from cross Bala x Black Gora x OS-36 x CH-1039. Its plant height is 75-85 cm and planting time is 15-30 June. It matures in about 90-95 days. It is tolerant to blast, BLB, stem borer and Helminthosporium. It is suitable for rainfed upland ecosystems and yield potential of BD-106 is 35-40 q/ha.

**Birsa Dhan-107**

Released in the year 1995 by SVRC. Notified vide no. 1 (E) dt. 01.01.1996. It has been developed through pedigree method of breeding. Its planting time is 15-30 June and it matures in 90-95 days. It is resistant to stem borer, BLB, Helminthosporium and Gallmidge. It is suitable for rainfed upland ecosystems. Its yield potential is 25-30 q/ha.

**Birsa Dhan-108**

Released in the year 2003 by SVRC. Notified vide no. 1572 (E) dt. 20.9.2006. It has been developed through pedigree method of breeding. Its planting time is 15-30 June and it matures in 70 days (extra early) and plant height is 90-95 cm. It is sown at 20 cm row to row distance. It is tolerant to blast, BLB, stem borer, leaf spot, Helminthosporium and Gallmidge. It is suitable for rainfed upland ecosystems. Its yield potential is 25-30 q/ha.

**Birsa Vikash Dhan-109**

Released in the year 2003 from SVRC. It has been developed through pedigree method of breeding after using Participating Plant Breeding (PPB) and has been recommended for Jharkhand state. Its plant height is 95 cm. It is planted in the 2nd fortnight of June and matures in about 85 days. It is tolerant to major pests, disease and drought and it is suitable for rainfed upland ecosystems. It has superfine long slender grain and its yield potential is 25-29 q/ha.

**Birsa Vikash Dhan-110**

Released in the year 2003 from SVRC and it has been found suitable for Jharkhand state. It has also been developed through Participatory Plant Breeding (PPB). It has plant height of 95 cm and matures in 90-95 days. Its seed rate is 80 kg/ha. It is tolerant to blast, BLB and Helminthosporium leaf spot. It is suitable for rainfed upland ecosystems. It is a superfine and long slender variety and yield potential is 25-28 q/ha.

**Birsa Vikash Dhan-111**

Released in the year 2013 by SVRC. It has been developed i through Marker Assisted Back Crossing of Kalinga III I Azucena // KalingaIII / Azucena // K III. It is suitable for direct seeding of upland in Jharkhand. Its Flag leaf long, erect and broad. It is deep rooted having medium tall plant (Plant height 84-95 cm). It has weed smothering capabilities and its grain is long slender (length-9.30mm, breath 2.40mm, L: B ratio 3.87. The yield potential of BVD-111 is 23-25q/ha.

**Rajendra Dhan-202**

Released in the year 1979 by SVRC. It has been developed through pedigree method of breeding. Its plant height is 95 cm and matures in 125-130 days. It has good seedling vigor, long and heavy weight drooping panicle, straw colored hull, long bold grain with white kernel. It is resistant to blast, brown spot and Gallmidge. Its yield potential is 40- 45 q/ha.

**Birsa Dhan-201**

Released in the year 1985 by SVRC. Notified vide no. 1 (E) dt. 01.01.1996. It is suitable for medium land of Jharkhand state. It is semi-dwarf and having compact and long panicle, straw colored hull and long bold grain with white kernel. It matures in115-120 days. It is resistant to stem borer, blast, BLB and Gallmidge. Its yield potential is 35-40 q/ha.

**Birsa Dhan-202**

Released in the year 1985 by SVRC. Notified vide no. 5233/ Agril. dt. 30.04.1985. It has inter-mediate plant height, good seedling vigor, long and heavy weight drooping panicle, straw colored hull, long bold grain with white kernel. It is very much suitable for medium land. It matures in120-125 days. It is resistant to blast, brown spot and Gundhi bug. Its yield potential is 40-45 q/ha.

**Birsamati**

Released in the year 2003 by SVRC. It has been developed through pedigree method of breeding by crossing IR-36 x BR-9. Its plant height is approx 95 cm and it is suitable for rainfed medium low land. It is resistant to all major pests and diseases. It is aromatic and having long slender grain with medium maturity (130-135 days). It has yield potential of 40-45 q/ha.

**Birsa Vikas Sugandha-1**

Released in the year 2013 by SVRC. It has been developed through selection of gamma irradiated population of Pusa Basmati-1. It is suitable for medium land of Jharkhand state. Its leaves are broader, longer and droopy than most of the popular rice varieties. It has medium maturity and having long slender grain (white grains with aroma). It has excellent drought tolerance in farmers' fields. It is moderately resistant to blast, bacterial leaf blight, Helminthosporium spots, stem borer and Gundhi bug. It has yield potential of 40-45 q/ ha.

**Lalat**

Released in the year 2013 by SVRC. It has been developed

through bulk population breeding followed by pure line selection and previously released from OUA & T Bhubaneshwar in the year 1989. It is suitable for early and late sowing condition in mid and low land of Jharkhand. It has long flag leaf, erect and broad, semi-tall plant, non-lodging, long slender grain; whitish husk with white kernel. It is moderately resistant to blast, bacterial leaf blight and false smut and also moderately resistant to Stem borer, Gundhi bug and resistant to Gall midge. It has yield potential of 45-50 q/ha.

**Birsa Vikash Dhan -203**

Released in the year 2013 by SVRC. It has been developed by cross between. IR-64 x Kalinga-lll through bulk population breeding followed by pure line selection. It is suitable for transplanting in drought prone medium land of Jharkhand under rainfed condition. It has flag leaf erect, long and broad, showy, semi-dwarf plant, non-lodging, long slender grain; white husk with white kernel. It is moderately resistant to blast, bacterial leaf blight and Helminthosporium spots, Stem borer and Gundhi bug under field conditions. It is sown both by direct seeding and transplanting. The yield potential of BVD-203 is 40-45 q/ha.

## Wheat

**Birsa Gehun-3**

Released by SVRC in the year 2013, It has been developed by using modified Bulk method. The material was selected from CIMMYT nursery (IBWSN) having parentage of Attila/ 3/HUI/CARC/CHEN) CH0/4/ATTILA. It has the plant height of 90-95 cm with dark green and erect leaf. The flowers have non pubescent white glumes. It matures in 110-115 days. It is tolerant to drought and perform well under limited irrigated condition (3 irrigations). It is resistant to rust and leaf blight. The grain has the protein percentage of 12.30% with sedimentation value 39.5 m. It has the yield potential of 40.00 q/ha.

## Maize

**Birsa Makai-1**

Released in the year 1995 by SVRC. It has been developed

through population improvement of D-741 (Pantnagar) by half sib recurrent selection and it is recommended for Jharkhand state. Its seed is deep yellow in color and plant height is 160-180 cm. It matures in 80-85 days. The seed rate is 18 kg/ha. Row spacing of BM-1 is 65 cm, planting time 15 June in kharif and 15 Oct. to 15 Nov. in rabi. It is moderately resistant to maydis and highly resistant to leaf blight and tolerant to drought after germination. It is suitable for timely sowing in upland condition. Its yield potential is 35-40 q/ha.

**Birsa Vikash Makka-2 (BVM-2)**

Released in the year 2003 by SVRC. It has been developed by controlled hand pollination between three yellow-flint (Suwan, BM-1, Chandan-3) and three white flint (GDRM-187, RPL, SHWETA) composite varieties by using population improvement through PPB and it has bright yellow seed and plant height is 180-185 cm. It matures in 75-80 days during kharif and it is moderately resistant to leaf blight, highly resistant to sheath blight and tolerant to drought after germination. The yield potential of BVM- 2 is 40-45 q/ha.

**Suwan Composite-1**

Released in 1986 by SVRC. It has been developed by population improvement of Suwan 1. Its sowing time is 20 June-10 July in Kharif and 15 Oct.-15 Nov. in Rabi in well drained upland soil. Its plant height is 200 cm and seed are orange yellow. The seed rate is 18-20 kg/ha. It matures in 95-100 days. It is moderately resistant to stem borer, Banded Leaf and Sheath Blight and Leaf Blight. Its yield potential is 45-50 q/ha.

**HQPM 1**

It is Single Cross Hybrid (HKI 193-1xHKI 163-1). It is sown from 20th June-10th July in kharif and 15th Oct.-15th Nov. in Rabi. It is I recommended for well drained upland soil of Jharkhand state. Its I plant height is 190 cm and grain color is yellow. The seed rate is 12-15 kg/ha. It is moderately resistant to stem borer, resistant to leaf blight. It has high-tryptophan and lysine content in grain. It matures in 100-110 days. The yield potential of HQPM-1 is 55-60 q/ha.

**Amber pop corn**

Released from DMR, Hyderabad (composite). Its sowing time is 20th June-10th July in kharif and 15 Oct.-15 Nov. in rabi. The seed rate is 15-18 kg/ha and plants are tall. It has vigorously growing bearing broad dark green leaves. It has partial resistant to downy mildew, leaf blight and rust. Its grains are flintish and yellow-orange. It is suitable in well drained upland soil of Jharkhand state. It has good popping quality. It matures in 95-105 days and yield potential is SO-35 q/ha,

**Priya sweet corn**

Released from DMR, Hyderabad in 2005 and recommended for Jharkhand state. The time of sowing is 20th June-10th July in kharif and 15th Oct.-15th Nov. in Rabi season. It is suitable in well drained upland soil; its plant height is 180 cm and seed color are dull yellow. The seed rate is 2-15 kg/ha. It matures in 90-95 days. It has high sugar up to 30-36 %. It is suitable for table purpose after boiling. The yield potential of Priya sweet corn is 25 q/ha.

**Baby corn**

BVM-2, VL baby corn-1, HM-4 are some of promising varieties and recommended for Jharkhand state. The time of sowing is 20th June-10th July in kharif, 15th Oct.-15th Nov. in rabi season and 15th Jan.-30th Jan. in summer season. Its seed rate is 25-30 kg/ha. It is suitable in well drained upland soil, Good baby corn yield and adequate fodder yield.

## Finger Millet (Marua)

### A 404

Released by SVRC in the year 1979 and notified vide no. 17-2/93 SD IV dated 17.08.1993. It has been developed through selection of introduced germplasm from Anakapalle (AP). It has plant height of 100-115 cm with incurred panicle. It matures in 115-120 days under transplanted condition and 105-110 days under direct sown condition. It is moderately resistant to blast and fairly tolerant to drought and suitable for rainfed upland soils. Its yield potential is 36.00 q/ha.

### Birsa Marua-2

Released by SVRC in the year 1995 and notified vide no. 17-6/95 SD IV dated 01.01.1996. It has been developed by selection from the local germplasm. It has plant height of 100-105 cm and brownish grey in color. The variety is tolerant to finger and neck blast disease. It matures in 110-115 days under upland situation. It has yield potential of 36.00 q/ha.

## Little Millet (Gundli)

### Birsa Gundli-1

Released in the year 1993 by SVRC for Bihar plateau (now Jharkhand) and notified vide no. is 17-2/93 SDIV dated 17.08.1993. It has been developed through selection from local germplasm. It is sown with the onset of monsoon. Its plant height is around 100- 120 cm and it matures in 50-60 days. It is moderately resistant to stem borer. It is very much suitable for rainfed upland ecosystems and yield potential of Birsa Gundli-1 is 800 Kg/ha.

## Pigeonpea (Arhar)

### Birsa Arhar-1

Released in year1992 by selection from F 5 bulk population from the cross ICP x 78120- WB. Its seed color is light brown and plant height ranges from 200-220 cm. It matures in 180-190 days. The row to row and plant to plant spacing should be 70x20 cm. Its right time of sowing is from 15th June to 15th July. The seed rate is 20 kg/ha. It is suitable for upland. It is resistant to wilt disease under field condition and moderately resistant to pod borer. It is also tolerant to water stress and its yield potential is15- 20q/ha.

## Urad

### Birsa Urid 1

Released in the year 1995 by SVRC and notified vide no. 9256 dated 51" February 1996. It has been developed through Pedigree method of breeding and parentage are IC-11008 x PS-1. Its plant height is approx. 28 cm and it is suitable in upland soils of Jharkhand. It is resistant to YMV and powdery mildew disease and it matures in 81 days. Its yield potential is -15-18 q/ha.

## Soybean

### Birsa Soybean-1

Released in the year 1983. Notified vide no. 499 (E) dated 08.07.1983. It has been developed through mutation breeding of 'Sepaya black'. Birsa Soybean-1 is a black seeded variety which matures in 112-115 days. Its plant height is 49-55 cm. It is recommended for timely sown in rainfed upland condition. It is moderately resistant to bacterial postule and bacterial blight, resistant to mosaic and Cercospora leaf spot. Its protein content is 40% and oil content is 20%. Its yield potential is 28 q/ha.

### Birsa Safed Soybean-2

Released in the year 2013 by SVRC. It has been developed by pedigree method of breeding of cross JS 335 x MACS-58. It is a white seeded variety which matures in 105- 107 days. Its plant height is 60-63 cm. It is resistant to lodging, shattering and responsive to fertilizer application. It is suitable for early sown conditions. It is tolerant to Bacterial postule and frog eye leaf spot, moderately resistant to Cercospora leaf spot and immune to target leaf spot diseases and also found tolerant to Blue beetle and defoliator (Bihar hairy caterpillar) pest. Its protein content is 40% and oil content is 17%. Its yield potential is 30 q/ha.

## Chickpea

### Birsa Chana-3

Released in the year 2013 by SVRC. It has been developed through pedigree method of breeding. It has parentage of ICCCV-2 x Bhawanipatnam local. Its plant height is 41 cm. It is suitable in irrigated and rainfed condition. It is resistant to wilt disease and tolerant to gram pod borer. It has medium maturity (115-118). It has protein content of 20-21 % and yield potential of BG-3 is 18-20 q/ha.

## Niger (Sarguja)

### Birsa Niger-1

Released in 1995 by SVRC and notified vide number is 17-6/95 S.D IV dated 01.01.1996, It is a composite variety (CH-

26 +N-122+5-1+78-25+ RCR-2).lts plant height is 70-90 cm, stem color is light pink at maturity, ray florets are canary yellow. It matures in 95-100 days. It is suitable for sowing in 2nd week of August to 2nd week of September in upland. Its oil content is 41.70%. Its seed yield is 600-700 Kg ha.

**Birsa Niger-2**

Released in the year 2003 by CVRC and notified vide No 122 (E) dt 2.2.2005. it has been recommended for Jharkhand, MP, Odisha, Chhattisgarh, Maharashtra, AP and parts of Bihar. Its plants height is 85 - 110 cm and matures in 90-95 days. It is tolerant to major diseases and pest and tolerant to drought. It has oil of 39.1 % with yield potential of 7-8 q/ha.

**Birsa Niger-3**

Released in 2010 by SVRC. It has been developed through selection from local germplasm. Its plant height is 80-90 cm, stem color is light pink at maturity ray florets are canary yellow. It matures in 97-98 days. Its plant bears 65-80 capsules per plants. This variety is an early stage and most suitable for Niger growing states in India. It is also tolerant to drought like condition. Its oil content is 33.33%. Its yield potential is 600-700 kg/ha.

**Pooja-1**

A variety of Niger released by BAU in 2008 and notified vide no. 272 dated February 2009. It was developed through selection from local germplasm. This is early maturing variety and most suitable for Niger growing states in India. It has big and bold size capsules. Its plant bears 100-265 capsules per plant, seed yield 7-8 q/ha. It is responsive to chemical fertilizers and suitable for sowing in middle August to middle September in upland soil. Oil content 42-43 %.

## Groundnut

**Birsa Groundnut - 1**

Released in 1981 by SVRC and notified vide No. 493(E) dated 8.7.1983. It has been developed through mutation breeding (Gamma rays) in variety 41 C. It matures in 113- 115 days. It is tolerant to tikka disease. It has 49 - 50% Oil and Its Yield potential is 18- 20 q/ha

**Birsa Groundnut - 2**

Released in 1981 by SVRC and notified vide no. 492 (E) dt. 3.1.1983. It has been developed through mutation breeding (Gamma rays) in variety 41 - C. It matures in 113-115 days, It is tolerant to tikka disease, resistant to spodeptera litura and helicoverpa armigera., It has 49 - 50% oil and Its yield potential is 20- 22 q/ha.

**Birsa Groundnut - 3**

Released in 1985 by SVRC and Notified vide no. 386 (E) dated 15.5.1990. It has been developed through hybridization of cross Ashiriya Mwiturde x Early Runner. It matures in 115-120 days. It is tolerant to tikka disease and resistant to occasional drought spell. It has 49 - 50% oil and Its yield potential are 20- 22 q/ha.

**Birsa Bold**

Released in 1992 by SVRC and in 1994 by CVRC and Notified vide No 636 (E) dt. 2.9.1994. It has been developed through hybridization of cross derivative BAU-6 X M-13. It matures in 125-130 days, it is moderately resistant to tikka disease, col I or rot, and seed narcosis. Its yield potential is 20- 22 q/ha and It has 47 - 49% oil with confectionary quality.

**Birsa Groundnut - 4**

Released in the year 2013 by SVRC. It has been developed through hybridization of cross ROBOT -33-1X M-13. It has upright branching and foliage remain green at the time of maturity and it may be used as fodder. It is tolerant to tikka disease. It matures in 115-120 days and shelling percentage is 70-75 and HKW is 64- 69 gm. It has 50 to 52% oil with Its yield potential of 20- 22 q/ha.

**Mustard**

**(Shivani)**

Released by SVRC for Jharkhand in this year 2003 and notified vide no. 887/ S.O - 1177(E), 25.08.2005. This variety was developed through local germplasm selection followed by pure line selection. Its plant height is 125-130cm and has early maturity (100-105 days), Sparse leaf hairs, dentate leaf margin, long leaf, high number of lobes per leaf and 1000 seed weight varies between3.4 to 4.0 gms. It shows tolerant reaction to White rust and Alternaria Blight whereas resistant to Phyllody disease. It is also resistant to saw fly and moderately resistant to aphids.

It has oil content which varies between 41-42%, protein content -27%. different fatty acids viz., Palmitic acid-2.6%, Stearic acid-0.9%, Oleic acid-6.4%, Linoleic acid-16.3%, Linolenic acid-16.5%, Eicosenoic acid-6.0% and Erucic acid-51.3%. The yield potential of Shivani is 8-10q/ha under rainfed condition whereas under timely sown irrigated condition the yield harvested to the tune of 12-14 q/ ha.

## Fababean

### Vikrant

Released in 1999 for Haryana. It's number and date of notification is 425 (E), dated. 8.6.1999. It is collection from foothill region of Haryana and selected after pure line selection. It is recommended for rabi and in normal sowing in upland condition of Jharkhand. Its plant height is 60-70 cm having light yellow bold seeded. Its plant type is annual herb. Leaflet shape is six large oval shape, compound leaf type and dark green foliage color. Flower color is white with a black dot on the keel. Pod size is long (5.5-6.5 cms.) and greenish black pod color.

Seed shape is round, creamy white seed color and smooth slightly depressed surface with black hilum seed test a texture. Days to 50% flowering and days to maturity (medium maturity group) are 59 days and 122 days, respectively. 100-seed weight is 29.3g. Seed rate is 100 kg/ha. Protein percent is 24.85. Its yield potential is 12.1 q/ha.

## Rice Bean

### RBL-1

Released in 1985 from Punjab, and notified vide no 165

(E), dated 06.03.1987. It has been collected from Nagaur district of Rajasthan and selected after following single plant selection. It is recommended for kharif season in normal sown and for rainfed upland of Jharkhand. Its plant height is 104 cm having light green/ yellow smooth bold seeded and medium duration. It is immune to yellow mosaic virus. It has lobed leaflets and trifolier leaf. The stem is climbing nature and flower color is yellow in cluster. The pod size is long curved, seed shape is oblong and seed color is light yellowish green. Seed size is bold. Seed test texture is smooth. Days to flowering is 60 days and days to maturity is 116 days as medium maturity group. 100-seed weight is 6 g. Seed rate is 20- 25 kg/ha. It is highly resistant to yellow mosaic virus. It has 12.36 % protein content. Its yield potential is 9 q/ha.

## Horse Gram

### Birsa Kulthi-1

Released in the year 1985 by SVRC for Bihar plateau (now Jharkhand). It has been developed through selection from local germplasm. It is sown with the onset of monsoon. Its plant height is around 60-64 cm and it matures in 92-97 days. It is moderately resistant to micro femina leaf blight. It is responsive in late condition and yield potential of Birsa Kulthi-1 is 10.00 q/ha.

## Sugarcane

### BO-147

Released in year 2013 by clonal selection. It is the selection of BOHO. It matures in twelve (12) months. The cane is straight whitish green having normal to slight swollen nodes, small to medium size round onion shaped eye and have erect leaves. The seed rate is 50q/ha. It is suitable for upland and medium land. Its harvesting starts from 15th Jan onwards irrespective of time of sowing. It is tolerant to all major diseases and insect pests especially the red rot of sugarcane. It also performs well in drought situation. It can tolerate the water-logged condition. The sucrose content in this variety is around 16% in juice and its yield potential is 80-100t/ha.

## Crop Varieties and their Package of Practices

**Cereal Crops**

| Name of the Crop | Release/Recommended Varieties | Average Yield (q/ha) | Days to Maturity | Seed Rate (Kg/ha) | Time of Sowing | Sowing Distance (cm) | Amount of fertilizer (Kg/ha) |
|---|---|---|---|---|---|---|---|
| **Rice** | **Upland** | | | | | | |
| | Birsa Dhan-108 | 22-25 | 70-75 | 80 | 1st June to 25th June | 20x10 | 40:30:20 N P K (Rainfed) |
| | Birsa Vikas Dhan-109 | 25-30 | 85-90 | | | | |
| | Birsa Vikas Dhan-110 | 25-30 | 90-95 | | | | |
| | Birsa Vikas Dhan-111 | 23-25 | 90-95 | | | | |
| | Vandana | 25-30 | 90-95 | | | | |
| | Birsa Goda-102 | 20-25 | 95-100 | | | | |
| | **Medium Land** | | | | | | |
| | IR-64 | 40-45 | 120-125 | 40 | 1st June to 30th June | 20x10 | 80:40:20 N P K |
| | Lalat | 40-45 | 120-125 | | | | |
| | Birsa Dhan-202 | 35-40 | 115-120 | | | | |
| | Birsa Vikas Dhan-203 | 40-45 | 120-125 | | | | |
| | Birsamati | 35-40 | 130-135 | | | | |
| | Birsa Vikas Sugandha-1 | 40-45 | 120-125 | | | | |

| | | | | | | | |
|---|---|---|---|---|---|---|---|
| | **Low Land** | | | | | | |
| | MTU-7029 (Swarna) | 50-55 | 145-150 | 40 | 15th June | 20x10 | 100:60:40 |
| | BPT-5204 (Sambha | 50-55 | 140-145 | | to 30th | | N P K |
| | Mahsuri) | | | | June | | |
| | Rajshree | 45-50 | 140-145 | | | | |
| | **Hybrid Rice** | | | | | | |
| | Proagro-6444 | 60-70 | 125-130 | 12-15 | 15th May | 20x5 | 150:70:90 |
| | PHB-71 | | | | to 30th | | N P K |
| | Pusa RH-10 | | | | June | | |
| Wheat | **Timely Sowing (Irrigated)** | | | | | | |
| | K-9107 | 40 | 125 | 125 | 10th to | 22x7 | 120:60:40 |
| | Birsa Genun-3 | 38-40 | 110-115 | | 20th Nov | | N P K |
| | **Late Sowing (Irrigated)** | | | | | | |
| | NW-1014 | 40 | 115 | 125-150 | 20th Nov | 18-20x7 | 80:40:30 |
| | Raj-3765 | 40 | 110 | | to 15th | | N P K |
| | HW-2045 | 40 | 115 | | Dec | | |
| | **Timely Sowing (Rainfed)** | | | | | | |
| | HDR-77 | 25 | 105 | 125 | 1st Oct | 22x7 | 40:20:20 |
| | K-8027 | 25 | 115 | | to 10th | | N P K |
| | C-306 | 20-25 | 115 | | Nov | | |
| | Birsa Gehun-3 | 23 | 122-125 | | | | |

| | | | | | | | |
|---|---|---|---|---|---|---|---|
| Barley | **Irrigated** | | | | | | |
| | Jyoti | 30-35 | 108-115 | 100 | Nov to Dec | 22x7 | 100:30:50 N P K |
| | Ratna | | | | | | |
| | DL-36 | | | | | | |
| | BR-32 | | | | | | |
| | **Rainfed** | | | | | | |
| | BR-31 | 12-18 | 108-115 | 100 | November | 22x7 | 40:20:20 N P K |
| | K-125 | 15 | | | | | |
| | Ratna | 16 | | | | | |
| Maize | | K/R | K/R | | | | |
| | Birsa Vikas Makka-2 | 40/45 | 75/105 | 18-20 | 15th June to 30th June (Kharif) 15th Oct to 15th Nov (Rabi) | 70x20 (Kharif) | 120:60:40 N P K |
| | Birsa Makai-1 | 35/40 | 80/100 | | | | |
| | Suwan Composite-1 | 45/50 | 100/130 | | | | |
| | HQPM-1 | 55/60 | 105/135 | | | | |
| | HM-4 | 50/55 | 105/135 | | | | |
| | Sweet Corn (Priya) | 20/25 | 90/120 | | | | |
| | Popcorn (Amber) | 25/30 | 90/120 | | | | |
| | Babycorn (BVM-2) | - | - | 25-30 | | 60x10 | 150:60:40 N P K |
| | Babycorn (HM-4) | - | - | | | | |
| | VL Babycorn | - | - | | | | |

| | | | | | | | |
|---|---|---|---|---|---|---|---|
| Marua | A-404 | 30-32 | 115-120 | 8-10 | 15th June-<br>30th June | 20x15 | 60:40:20<br>N P K |
| | Birsa Marua-2 | 24-36 | 105-110 | | | | |
| | HR-374 | 24-28 | 100-105 | | | | |
| | Birsa Marua-1 | 20-25 | 80-90 | | | | |
| Gundli | Birsa Gundli-1 | 7-8 | 60 | 8-10 | 23rd May-<br>23rd June | 20 row<br>to row | 40:20:20<br>N P K |
| Jowar | CSV-17 | 30 | 95-100 | 10-12 | 15th June<br>to 15th<br>July | 45x15 | 80:40:20<br>N P K |
| | SPV-1616 | 38 | 115 | | | | |
| | CSH-16 | 42 | 110-115 | | | | |

# Appendix-V

## Transect of Agricultural and Forest Land

| | Forest (Non-Agricultural) | Upland (Tanr II, III) | Bari land (Upland) (Tanr I) | Medium (DON III,DON II) | LOW (DON I) |
|---|---|---|---|---|---|
| **Soil** | Stoney | Sandy loam | Loam | Clay loam | Clay loam |
| **Crops** | | Rice, Tomato, Zinger, Tarmeric Pigeonpea, Maize, Black gram, Millets, Niger, | Maize, Tomato, Cauliflower, Potato, Beans, Brinjal, Radish, Peas, Onion, Fenugreek, Spinach, Pumpkin, Spinach, Chillies, Ridge gourd, Carrot, Garlic, Coriander, Cucumber, Bitter gourd, Bottle gourd, Ladies finger, Rrice seedlings | Rice – Wheat<br>Rice – Summer tomato | Rice |
| **Water Resour-ces** | Rainfed | Rainfed | Irrigated with wells | Rainfed and partial irrigated with well/ ponds | Rainfed |
| **Trees** | Forest trees | Mango, Mahua, Black berry | Mango, Papaya, Bamboo, Drumstick, Guava, Jackfruit, Banana, Citrus, | | |
| **Livestock** | Cattle grazing | Cow, Cattle, Buffaloes and Bull grazing. | Cow, Buffaloes, Fowl, Bullock, Goats | Cattle grazing | |
| **Problems** | Deforestation | Lack of irrigation, Fragmented, Scattered land holding, Poor fertility, Run off & soil erosion, Cattle grazing, Paucity of quality seeds of HYV's | Quality vegetable seeds, Termites | Lack of irrigation, Paucity of quality seeds of HYV, Poor fertility, Scattered land and monocropping. | Paucity of HYV seeds,Monocropped due to no irrigation, Water -logging |
| **Opportu-nities** | 1. Aforestation<br>2. Multi-tier agro forestry development | 3. Short duration improved rices<br>4. Intercropping<br>5. Sequence cropping (linseed)<br>6. Land consolidation<br>7. Composting & green manuring<br>8. Timely supply of seeds<br>9. Ridge & Hedge methods/Intercepts for water conservation<br>10. Fodder cultivation<br>11. Rice based multi tier agro forestry system development<br>12. Orchards | Improved pure seed supply | 1. Drought resistant improved rice variety<br>2. Green manuring<br>3. Composting<br>4. Sequence/relay cropping(cereals, oilseeds, pulses)<br>5. Land consolidation<br>6. Open dug well<br>7. Small farm ponds for irrigation | 1. Improved variety<br>2. Green manuring<br>3. Composting<br>4. Land consolidation<br>5. Sequence cropping (cereals/vegetables)<br>6. Pond construction<br>7. Paddy cum fish culture<br>8. Proper drainage<br>9. Summer rice |

# Appendix-VI

## Methods of Rice Establishment

In Jharkhand, the farming community owned agricultural land under three ecological conditions. They are

- Forest ecology
- Open ecology
- Mixed ecology (Forest + open)

Under forest ecology, the majority of settlers are either tribal people or the migrated community from different parts of the eastern India or from within the Jharkhand state. The migration took place hundreds of years ago from outside of state. The people within the state migrated from one district to other during forties, either due to population pressure or due to job opportunity. Their good service and hard work pleased the Zamindars (Landlord) and they had allotted the land within the forest area for cultivation with the conditions that the settlers will use the forest wood only for house consumption (fuel, house building construction, agricultural tools, making fence, animal shade etc.) but not for commercial use. In open ecology, the majority of settlers belong to the farming community of different caste and migrated from out of Jharkhand hundreds of years ago.

They acquired majority of agricultural land. In the mixed ecology the agricultural land was acquired by cutting the forest trees, terracing and bunding the field. All three ecologies are having both the land situation namely uplands and lowlands and every farm family owns both types of land. The majority of these lands are commonly used for rice cultivation during wet seasons (mid-June to November). The farming communities have developed different type of rice establishment methods as per ecosystem. In upland and lowlands, the farmers commonly used the following rice establishment methods.

## Upland

In upland situation the farmers sow the rice following

- direct seeding followed by tiwai
- seeding behind plow (line seeding)
- lewa/kado seeding with water-soaked seed
- lewa/kado seeding with sprouted seed

## Direct Seeding Followed by Tiwai

Under this method, the fanners prepare the field by two to three plowing after onset of monsoon. After last plowing, the farmer broadcast the rice seed and mixed the seed in the soil by plowing. Third day after seeding rice, the farmers again plow the seeded field. Before this plowing, the farmers apply DAP (Diammonium phosphate) @ 31.75 kg per hectare. This operation is called the "Tewai" operation. They believe that the tewai operation helps in reducing, the weed population and providing the good tilth for rice growth. The farmers used seed rates ranging from 120 to 150 kg ha. The advantage of this method is less labor requirement and low weed population. The disadvantages are, difficulty in weeding and harvesting if the crop lodges and seed damage by bird.

## Seeding Behind Plow

The farmers seed the rice by putting the rice in the furrow opened by the bullock drawn wooden plow. This method involves a pair of bullocks with plow man for opening the furrow and a female farmer for dropping the rice seed. This method is also very common in the farming community. The advantages of this method are, easy to weed the field, easy to apply the hand tools for removing the weeds, easy to harvest, no bird seed damage and less seed requirement per hectare. The disadvantages are, more manpower requires for seeding and high weed population requires more labor for weeding.

## Seeding by Lewa/Kado

This method is not very common but it is used under

contingency planning. When there is continues rain and the farmers are not able to seed the upland rice either under direct seeding or behind the plow, the farmers go for lewa seeding. Under this process, the fields are plowed in rainy condition. After proper plowing, the field are puddle and the leveling off the field was done by laddering. Laddering is an operation of leveling the puddled field using a flat heavy wooden plank. In leveled field the water soaked or sprouted rice seeds are broadcasted. The water soaked or sprouted seeds used in lewa seeding, assure the uniform seed contact with soil, otherwise rice seed may float in rain water and lost with flowing water or accumulate in depressed or lower portion of the field. Seed waking water also help in quick germination and fast establishment of rice After broadcasting the rice seeds, the care is taken to drain the rain from the field, otherwise the seeded rice will rot if standing water allowed in the field. Lewa seeding rarely needs weeding. The merits of lewa seeding are

- Less labor requirement
- Less weed infestation
- Good crop growth
- High grain and straw yield

## Lowland

Similar to upland, the farmers practice different type of rice establishment methods in lowlands because the lowlands are characterized with the quantum of standing water during crop growth period. They are drought prone shallow lowland (Chater), shallow lowland (Tarkha) and lowland (Garha). In revenue class these lands are known as Don-1 (lowland), Don-2 (shallow lowland) and Don-3 (drought prone shallow lowland). The farm families in Jharkhand commonly use the following methods of rice crop establishment in lowlands. They are

- Lewa/Kado
- Transplanting
- Broadcasting followed by beushening

**Lewa/Kado**

This method is already discussed in earlier paragraph. The method is very common in drought prone shallow lowland (Don - 3) with short duration traditional varieties. To avoid the effect of drought the farmers generally use short duration rice varieties namely Shathi and Karahani. The newly harvested rice of these varieties are available for eating during food shortage and for festival (Chhath puja). Under lewa seeding, the weed problem is very minimum.

**Transplanting**

In shallow low land (Don-2), locally called the tarkha land, the majority of rice is cultivated by transplanting. In local dialect tarkha means the land with standing / flowing water. The drought is rare phenomenon in these

The farmers plant 25 to 35 days old seedlings with best possible management, because in farmers perception, this land will produce harvest of rice. Farmers apply higher dose of organic (FYM) and inorganic fertilizers in transplanted rice. Few farmers also practice transplanting as well as beushening methods of rice establishment in these lands. They believe that alternate use of two methods helps in sustaining the rice productivity. Otherwise, the continuous use of single method reduces the rice yield. Transplanting of rice has its merits and demerits.

- Less seed requirement
- Less or no weeding requires
- Easy to harvest
- More rice production
- More labor demand for a single operation, leads to hiring of labor on payment
- High input
- Transplanting coincide with upland rice weeding, upland rice suffers and yields loss
- Traditional tall varieties lodge and yield loss

**Beushening Practice**

Beushening is also locally known as 'baug' or 'bidauni' and involves two broad series of field operations: direct seeding of mostly traditional tall rice cultivars using higher seed rates than for direct seeding without beushening or for transplanting, and wet plowing and laddering of the field, generally 25-35 days after germination of the rice, when about 15 cm of rain water is accumulated in the crop field. In some cases, seedling redistribution is also done after this operation.

The fields to be beushened are plowed immediately after the winter or summer rain and two or three times thereafter, during summer, to control weeds, insects and rodents. Plowing is done by animal-drawn wooden plow, which makes a 5-7 cm deep furrow without inverting the soil. If weeds are seen after the first monsoon rain, additional plowings are done to suppress them. Rice seeds are then broadcasted in the dry soil and mixed by harrowing in the last week of May ('Rohini Nakshatra': local calendar followed by the farmers for cultivation practices). Farmers had found that sowing in May resulted in higher grain and straw yields because the crop matured before the occurrence of late season drought in early to mid-October ('Hathia Nakshatra'). and in early November and thereafter, farmers used seed rates ranging from 130 to 240 Kg hr.

Wet plowing was done 25-35 days after germination, but only when there was 15-20 cm of standing water in the rice field. Fanners let their cattle graze rice seedlings 1 week before plowing to detop seedlings to improve tillering and to maintain optimum plant population, and using the crop foliage as fodder which otherwise would be lost during plowing and laddering. Laddering is an operation of leveling the land after plowing using a flat heavy wooden plank. Apparently, grazing was the farmers' strategy to minimize the vegetative growth constraint inflicted by the damage (crushing) of the upper part of the stem from plowing, laddering and animal trampling. Grazing is probably an economic way of detopping the tall cultivars, thereby preventing plant mortality from mud sealing when the top is bent down.

Rice fields were plowed after grazing followed by one or two laddering with a 15-20 kg load (the usual plow load) on the ladder, only to break the 'soil slice' and loosen it without damaging the rice plants. In non-beushened field laddering, the plowman generally puts a heavy load on the ladder, often by standing on it (to pulverize soil finely), but in the case of beushening, the plow alone was fixed on the ladder removing its iron share.

In some cases, seedlings were redistributed to fill in the gaps caused by laddering. However, seedling redistribution was not practiced where farmers believed a higher seed rate helped good distribution of seedlings. Seedling redistribution also coincided with the labor-scare period in the area.

Farmers observed the weed condition in the fields for 2-3 days after laddering. If the weeds were not well incorporated into the soil, they repeated laddering after 2-3 days without plowing but this time with a heavy load on the ladder (plowman standing on it). Two to three ladderings were generally sufficient to damage and incorporate the weeds, especially Echinochloa colona ('jhimpa' or 'jharua'), which was difficult to distinguish from rice plants. Seedling redistribution was done to fill up blank patches and to thin out the dense patches caused by repeated laddering. Family members hand-weeded the still remaining weeds 8-10 days after laddering.

Rice Cultivars mostly used in beushening are of local origin, with maturity duration of 150-170 days and medium tillering capacity. Improved varieties are rarely used for beushening.

**Merits of Beushening**

- Less labor requirement
- Labor demands spread over time
- Low fertilizer requirement
- No need of nursery
- Puddling is not required
- Less pest problems
- Less cash inputs required
- Appropriate alternative in the drought-and submergence-prone areas
- Stable yield
- Easy and timely establishment of non-rice crop

# Appendix-VII

## Rice Cultivars Grown in Jharkhand (Traditional and Improved)

| Traditional | Improved |
|---|---|
| Ratgoli | CR 314-5-10 |
| Kalamdani | IR 36 |
| Haskalma | Sita |
| Rata | Sarju 52 |
| Jhingasol | Mansuri |
| Laldhan | Hybrid 6201 |
| Kapusay | Sujata |
| Sadama | IR 64 |
| Dudhkandar | Ratna |
| Katki | Hybrid 801 |
| White Gora | Rajshri |
| Amgoudh | Basmati |
| Mainathor | CR 306 |
| Ramsindha | Kanak |
| Bherakabar | Pant 4 |
| Koharaphool | Vandana |
| Parmal | Hybrid 6102 |
| Kherkakochi | Super Sita |
| Kanika | Jaya |
| Bandphool | IR 36 |
| Sufala | Mahsuri |
| Nardha | |
| Jonga | |
| Black Gora | |
| Jabakusum | |
| Bhadai | |
| Khirasar | |
| Charka | |
| Dehradun | |
| Haldiphool | |